Application of Microfluidic Technology in Bioengineering

Application of Microfluidic Technology in Bioengineering

Guest Editors

Shuli Wang
Yigang Shen

Basel • Beijing • Wuhan • Barcelona • Belgrade • Novi Sad • Cluj • Manchester

Guest Editors
Shuli Wang
School of Electronic Science and Engineering
Xiamen University
Xiamen
China

Yigang Shen
College of Chemistry and Chemical Engineering
Xiamen University
Xiamen
China

Editorial Office
MDPI AG
Grosspeteranlage 5
4052 Basel, Switzerland

This is a reprint of the Special Issue, published open access by the journal *Micromachines* (ISSN 2072-666X), freely accessible at: https://www.mdpi.com/journal/micromachines/special_issues/065N55S5Y6.

For citation purposes, cite each article independently as indicated on the article page online and as indicated below:

Lastname, A.A.; Lastname, B.B. Article Title. *Journal Name* **Year**, *Volume Number*, Page Range.

ISBN 978-3-7258-6203-0 (Hbk)
ISBN 978-3-7258-6204-7 (PDF)
https://doi.org/10.3390/books978-3-7258-6204-7

Contents

About the Editors

Shuli Wang

Shuli Wang is an Assistant Professor at Xiamen University. His research endeavors primarily revolve around the interfacial design of microfluidic chips for biomedical applications, and the design and patterning technologies of quantum dots for micro-LED displays. He leads several research projects, including Youth Project of the National Natural Science Foundation of China, the Youth Project of the Fujian Provincial Natural Science Foundation of China, the Xiamen Natural Science Foundation Youth Project, and the Fundamental Research Funds for the Central Universities. To date, he has contributed over 60 research papers in journals such as *Nature, Advanced Materials, Lab on a Chip, Nano Letters, Small, CCS Chemistry, Laser & Photonics Reviews,* and *ACS Photonics.* In addition, he holds 10 authorized patents on microfluidics and micro-LED displays. He serves as the Associate Editor of *Engineering Reports,* the Academic Editor of *Exploration,* and is a member of the Youth Editorial Board of *Nano-Micro Letters, Nano Materials Science, Chinese Chemical Letters, Carbon Neutralization,* and *The Innovation Materials.* Moreover, he has acted as a reviewer for more than 40 academic journals, such as *Nano-Micro Letters, Journal of Colloid and Interface Science, Biomaterials,* and *Nanoscale.* He is regarded as a high-level talent in both Fujian province (Class C) and Xiamen (Class C).

Yigang Shen

Yigang Shen is currently an Associate Professor at Zhejiang Normal University (his previous affiliation is Xiamen University). His research focuses on sensing and microfluidic technologies for single-cell analysis, thermal manipulation, and integrated microsystems. He leads the National Natural Science Foundation of China Youth project and has contributed to multiple NSFC and provincial grants. His experience includes research stays at Cardiff University, a postdoctoral appointment at Xiamen University, and a PhD from Osaka University. He has published over 20 papers in journals such as *Matter, Biosensors and Bioelectronics, Sensors and Actuators B, Lab on a Chip,* and *Analytical Chemistry.* He has filed five patents and serves as a reviewer for *Chinese Chemical Letters, IEEE Sensors Journal, Smart Materials and Structures, Journal of Micromechanics and Microengineering,* and *Journal of Applied Physics.*

Preface

In the ever-evolving landscape of scientific research, microfluidic technology has emerged as a revolutionary force. This Special Issue Reprint titled "Application of Microfluidic Technology in Bioengineering" is a collection of cutting-edge research papers that explore the vast potential of this technology in the bioengineering realm.

The subject of this Special Issue is firmly rooted in the application of microfluidic technology within bioengineering. Microfluidics, at its core, focuses on the manipulation and precise control of tiny volumes of fluids in micrometer-sized channels. The scope of this issue is broad, encompassing a wide range of topics such as design of microfluidic chips, microscale flow control, single-cell manipulation, liquid biopsy, molecular diagnosis, and simulation modeling. The aim of this Special Issue is to provide a comprehensive overview of the latest advancements in microfluidics as they pertain to bioengineering. By gathering research from multiple disciplines, we hope to foster cross-disciplinary collaboration and inspire new ideas. The purpose is to disseminate knowledge about the practical applications of microfluidic technology, enabling researchers, engineers, and students to stay abreast of the most recent developments and trends in this field.

The motivation behind this Special Issue stems from the remarkable potential of microfluidic technology. Its unique characteristics, including miniaturization, integration, high precision, automation, and versatility, make it an invaluable tool for fluid manipulation and sample analysis. In bioengineering, these features translate into numerous benefits, such as more efficient diagnostic methods, better understanding of cellular behavior at the single-cell level, and the development of novel therapeutic strategies.

This Special Issue is addressed to a wide audience, including researchers in bioengineering, biomedical engineering, chemistry, materials science, and environmental sciences. It will also be of great interest to engineers and scientists working in the field of microfluidics, as well as students who are interested in learning about the latest developments in this exciting area. Additionally, it can serve as a valuable resource for industry professionals looking to explore the potential applications of microfluidic technology in product development and innovation.

In conclusion, we believe that this Special Issue on the Application of Microfluidic Technology in Bioengineering contribute significantly to the field. We hope that it will inspire further research and development, leading to new breakthroughs and applications that will ultimately benefit society.

We would like to express our sincere gratitude to all the authors who have contributed their valuable research to this Special Issue, as well as the reviewers for their insightful comments and suggestions. Their efforts have been instrumental in ensuring the high quality of this collection of papers.

Shuli Wang and Yigang Shen
Guest Editors

micromachines

MDPI

Editorial

Editorial for the Special Issue on the Application of Microfluidic Technology in Bioengineering

Shuli Wang [1,*] and Yigang Shen [2,*]

[1] Fujian Engineering Research Center for Solid-State Lighting, Department of Electronic Science, School of Electronic Science and Engineering, Xiamen University, Xiamen 361005, China

[2] The Institute of Precision Machinery and Smart Structure, College of Engineering, Zhejiang Normal University, Jinhua 321004, China

* Correspondence: slwang@xmu.edu.cn (S.W.); yigangshen@zjnu.edu.cn (Y.S.)

1. Introduction

Microfluidics, also called lab-on-a-chip, is a cutting-edge technology in contemporary interdisciplinary science. Its development stems from the rapid progress of microelectronics manufacturing technology, as well as from the integration of multiple disciplines such as biomedicine, materials science, fluid mechanics, and interface science [1–6]. Its core technology involves constructing a chip system with micron-scale channels and structures to manipulate nanoliter or even picoliter fluids to complete a series of biochemical laboratory functions, including sample transmission, mixing, reaction, sorting, and detection [7–14]. At present, microfluidics has progressed from theoretical exploration to the forefront of clinical transformation and industrial application, playing an increasingly vital role in modern bioengineering.

The unique advantages of microfluidics technology show great potential in many research fields of bioengineering [15–23]. First, its highly integrated micro-systems effectively reduce sample usage and reagent consumption, significantly lowering experimental costs while enabling automated detection [24,25]. Secondly, its excellent control precision and rapid response speed facilitate real-time dynamic detection, making it particularly valuable for high-throughput single-cell analysis, rapid nucleic acid detection, and point-of-care testing (POCT) systems [26–29]. In addition, microfluidic technology can be flexibly designed into multi-channel and multi-modal information outputs, supporting comprehensive multi-parameter monitoring that provides crucial capabilities for precision medicine [30–33], early disease screening [34–39], and environmental toxicology detection [23,40–43].

In recent years, with the emergence of new materials and improved multi-physical-field coupling mechanisms, microfluidic technology has advanced toward intelligence, systematization, biomimicry, and clinical applications [44–54]. In microstructure manufacturing, traditional polydimethylsiloxane (PDMS) materials are being replaced by polymer materials characterized by biocompatibility, photoresponsiveness, and electrical responsiveness, creating more stable and reliable chip systems. For driving mechanisms, the integration of acoustic, electric, magnetic, and light fields enhances the flexibility of fluid manipulation while enabling precise control of biological cells and micro-nanoparticles [55–58]. The use of artificial intelligence algorithms for flow field modeling and image recognition has significantly improved system automation and data processing efficiency [59–64].

Against this background, this Special Issue of *Micromachines*, "Application of Microfluidic Technology in Bioengineering", presents ten selected papers (seven interesting original articles and three reviews) from various research institutions, covering the latest progress of microfluidics in single-cell manipulation and detection, liquid biopsy, molecular diagnosis, interface design, and simulation modeling. These papers not only reflect the current multi-faceted research landscape of microfluidics integration but also present the development trends in microfluidics at both the basic innovation and clinical application levels.

2. Overview of Published Articles

The eleven papers featured in this Special Issue cover five aspects of microfluidic technology, including single-cell manipulation and detection, liquid biopsy, molecular diagnosis, interface design, and simulation modeling.

2.1. Clinical Cell Analysis: Liquid Biopsy, Prenatal Testing, and Hematology

In the field of cell manipulation and liquid biopsy, Chu et al. proposed a microfluidic chip system based on optically induced dielectrophoresis (ODEP) which achieved efficient, label-free enrichment and size-selective separation of circulating tumor cells (CTCs) [65]. This system relies on micromirror arrays to regulate spatial electric field distribution, enabling automatic screening of different-sized tumor cell subgroups in a very short time, with a separation purity reaching 93.5%. It has significant potential applications in the fields of liquid biopsy and early cancer screening. Yang et al. developed a self-assembled cell array chip (SACA) that combines multi-marker immunorecognition with microfluidic array design, achieving automated identification and extraction of fetal nucleated red blood cells (fnRBCs) [66]. The chip utilizes a joint discrimination strategy with markers such as Hoechst, CD71, and HbF to rapidly complete the extraction and paternity analysis of the extremely small number of fnRBCs in maternal blood, with a total operation time of less than 2 h and a sorting efficiency of 97.85%, providing a feasible new technical approach for non-invasive prenatal diagnosis. Hassan et al. conducted systematic modeling and experimental analysis on the impact of carrier fluid elasticity on cellular mechanical characterization in deformability cytometry [67]. They discovered significant errors in cell stiffness calculations when fluid viscoelastic effects are ignored. Simulation results based on CFD models indicate that fluid environment parameters significantly influence deformation test outcomes, suggesting that material rheological properties should be incorporated into future microfluidic chip structure optimization processes. Jiang et al. developed a microfluidic platform to evaluate red blood cell (RBC) deformability in healthy and thalassemia samples [68]. The chip, featuring capillary-mimicking microchannels, combined high-speed imaging with deep learning algorithms for automated cell tracking and contour analysis. The results confirmed significantly reduced deformability and slower recovery in thalassemic RBCs, especially in narrow channels, establishing a standardized and high-throughput method with strong diagnostic potential for hematological disorders.

2.2. Molecular Diagnostics and Extracellular-Vesicle Analysis

For molecular detection and nucleic acid analysis, Xian et al. proposed a MEMS digital PCR system based on superhydrophilic microarray structures, which can complete rapid, multi-target nucleic acid detection in reaction chambers as small as 120 pL [69]. The system features high integration and easy operation and enables multi-channel amplification without droplet encapsulation, demonstrating excellent detection sensitivity and specificity for clinically relevant markers such as hepatitis B virus and EGFR mutations, with significant practical value for point-of-care testing (POCT). Chen et al. authored a review systematically examining recent major strategies and technological platforms for exosome separation and analysis in microfluidics [70]. The article evaluates various approaches, including micro/nanopore arrays, acoustic chips, dielectric enrichment, and microbead capture, while pointing out the current contradiction between sample throughput and enrichment purity.

It suggests that future systems could integrate electroacoustic synergistic strategies with AI-assisted data analysis to enhance overall system performance.

2.3. Interface and Materials Engineering for Robust Microfluidic

Regarding microfluidic chip interface design, Wu et al. employed a single-cycle brush-grafted polymer modification strategy to construct liquid bionic interfaces on PDMS surfaces, effectively enhancing the anti-fouling performance of microfluidic chips in complex blood environments [71]. Through droplet contact angle and fluorescent labeling experiments, this interface significantly inhibited protein adsorption and cell adhesion, improving system reusability stability and making it suitable for multiple sampling and analysis scenarios. From a fluid dynamics modeling perspective, Ayeni et al. analyzed the effects of microfiber bundles with different arrangements and densities on oil–water interface morphology and flow behavior [72]. The research revealed that controlling fiber spacing within the range of 10–50 μm can maximize interface area, helping to improve interface reaction efficiency and providing engineering optimization ideas for microfluidic extraction, emulsification, and phase separation process design.

2.4. System Modeling and Thermal Engineering

In the field of microfluidic chip system construction, Thiem et al. developed a thermal conduction model for a microfluidic cooling system based on a micro-hotplate substrate, addressing the needs of cryo-electron microscopy sample preparation [73]. The simulation results showed that when using a silicon-based substrate with a thickness of less than 5 μm, instantaneous cooling rates exceeding 10^6 K/s can be achieved, potentially meeting the extreme freezing requirements for preserving biological macromolecular structures, thus providing theoretical support for organ-on-chip freezing processes and cell preservation technologies.

2.5. Methodological Reviews: Microfluidics-Enabled Omics and Impedance Sensing

The other two reviews in the Special Issue focus on the introductory biological applications of microfluidics. Sun et al. conducted a comprehensive review of the application of microfluidic technology in single-cell RNA sequencing and spatial omics research [74]. The article reviewed the microfluidic chip structure and application scenarios of technical platforms such as Drop-seq, 10× Genomics, and MERFISH, pointing out that microfluidics has shown extremely high value in revealing cell heterogeneity, reconstructing developmental lineages, and studying the tumor microenvironment. The authors look forward to the trend of spatial resolution improvement and multimodal collaborative development. Finally, Shen et al. focus on the basic principles, system construction, and biological applications of microfluidic impedance detection technology [75]. The review systematically introduces the cell equivalent circuit modeling, impedance spectrum feature extraction, and multi-parameter detection methods; sorts out the typical application cases of this technology in multiple dimensions, such as tumor cell identification, blood component analysis, and microbial monitoring; and proposes the future integration of flexible electronics and AI algorithms as a development direction.

3. Conclusions

The ten papers included in this Special Issue cover multiple areas, from modeling microscopic mechanisms to clinical transformations, reflecting the multi-dimensional capabilities of pathfinding and system construction in microfluidic technology for bioengineering applications. Whether in experimental design, material innovation, or analysis strategies and data interpretation, each research team has demonstrated solid technical skills and forward-looking application perspectives, highlighting the significant value of microfluidic technology in promoting medical intelligence, systematization, and individualization. Finally, we would like to highlight that this Special Issue collects cross-disciplinary and complementary contributions from China, the USA, and Germany, representing a shared and enhanced cross-technological research effort regarding microfluidic technology.

Author Contributions: The authors equally contributed to the Editorial. All authors have read and agreed to the published version of the manuscript.

Data Availability Statement: Data underlying the results presented in this paper are not publicly available at this time but may be obtained from the authors upon reasonable request.

Acknowledgments: We wish to thank all authors for their valuable contributions to this Special Issue, all reviewers for their valuable support and efforts in improving the quality of the submitted papers, and our Assistant Editor for their support in the production of this Special Issue.

Conflicts of Interest: The authors declare no conflicts of interest.

References

1. Gopinathan, K.A.; Mishra, A.; Mutlu, B.R.; Edd, J.F.; Toner, M. A Microfluidic Transistor for Automatic Control of Liquids. *Nature* **2023**, *622*, 735–741. [CrossRef] [PubMed]
2. Yafia, M.; Ymbern, O.; Olanrewaju, A.O.; Parandakh, A.; Sohrabi Kashani, A.; Renault, J.; Jin, Z.; Kim, G.; Ng, A.; Juncker, D. Microfluidic Chain Reaction of Structurally Programmed Capillary Flow Events. *Nature* **2022**, *605*, 464–469. [CrossRef] [PubMed]
3. Lin, H.; Yu, W.; Sabet, K.A.; Bogumil, M.; Zhao, Y.; Hambalek, J.; Lin, S.; Chandrasekaran, S.; Garner, O.; Di Carlo, D.; et al. Ferrobotic Swarms Enable Accessible and Adaptable Automated Viral Testing. *Nature* **2022**, *611*, 570–577. [CrossRef] [PubMed]
4. Whitesides, G.M. The Origins and the Future of Microfluidics. *Nature* **2006**, *442*, 368–373. [CrossRef]
5. Yu, W.; Lin, H.; Wang, Y.; He, X.; Chen, N.; Sun, K.; Lo, D.; Cheng, B.; Yeung, C.; Tan, J.; et al. A Ferrobotic System for Automated Microfluidic Logistics. *Sci. Robot.* **2020**, *5*, eaba4411. [CrossRef]
6. Zhu, P.; Wang, L. Microfluidics-Enabled Soft Manufacture of Materials with Tailorable Wettability. *Chem. Rev.* **2022**, *122*, 7010–7060. [CrossRef]
7. Wan, J.; Zhou, S.; Mea, H.J.; Guo, Y.; Ku, H.; Urbina, B.M. Emerging Roles of Microfluidics in Brain Research: From Cerebral Fluids Manipulation to Brain-on-a-Chip and Neuroelectronic Devices Engineering. *Chem. Rev.* **2022**, *122*, 7142–7181. [CrossRef]
8. Sznitman, J. Revisiting Airflow and Aerosol Transport Phenomena in the Deep Lungs with Microfluidics. *Chem. Rev.* **2022**, *122*, 7182–7204. [CrossRef]
9. Akif Sahin, M.; Werner, H.; Udani, S.; Carlo, D.D.; Destgeer, G. Flow Lithography for Structured Microparticles: Fundamentals, Methods and Applications. *Lab. A Chip* **2022**, *22*, 4007–4042. [CrossRef]
10. Zhang, Q.; Feng, S.; Lin, L.; Mao, S.; Lin, J.-M. Emerging Open Microfluidics for Cell Manipulation. *Chem. Soc. Rev.* **2021**, *50*, 5333–5348. [CrossRef]
11. Grisoni, F.; Huisman, B.J.H.; Button, A.L.; Moret, M.; Atz, K.; Merk, D.; Schneider, G. Combining Generative Artificial Intelligence and On-Chip Synthesis for de Novo Drug Design. *Sci. Adv.* **2021**, *7*, eabg3338. [CrossRef]
12. Mercado-Perez, A.; Beyder, A. Gut Feelings: Mechanosensing in the Gastrointestinal Tract. *Phys. Rev. Fluids* **2022**, *19*, 283–296. [CrossRef]
13. Özkan, A.; LoGrande, N.T.; Feitor, J.F.; Goyal, G.; Ingber, D.E. Intestinal Organ Chips for Disease Modelling and Personalized Medicine. *Nat. Rev. Gastroenterol. Hepatol.* **2024**, *21*, 751–773. [CrossRef] [PubMed]
14. Shen, Y.; Yuan, Y.; Liu, X.; Tang, T.; Yalikun, Y.; Tanaka, Y. Multiscale Integrated Temperature/Flow Velocity Sensor Patches for Microfluidic Applications. *IEEE Sens. J.* **2023**, *23*, 1795–1802. [CrossRef]
15. Convery, N.; Gadegaard, N. 30 Years of Microfluidics. *Micro Nano Eng.* **2019**, *2*, 76–91. [CrossRef]
16. Zhang, K.S.; Nadkarni, A.V.; Paul, R.; Martin, A.M.; Tang, S.K.Y. Microfluidic Surgery in Single Cells and Multicellular Systems. *Chem. Rev.* **2022**, *122*, 7097–7141. [CrossRef]
17. Dong, H.; Lin, J.; Tao, Y.; Jia, Y.; Sun, L.; Li, W.J.; Sun, H. AI-Enhanced Biomedical Micro/Nanorobots in Microfluidics. *Lab. Chip* **2024**, *24*, 1419–1440. [CrossRef]
18. Zhang, J.; Zhan, K.; Zhang, S.; Shen, Y.; Hou, Y.; Liu, J.; Fan, Y.; Zhang, Y.; Wang, S.; Xie, Y.; et al. Discontinuous Streaming Potential via Liquid Gate. *eScience* **2022**, *2*, 615–622. [CrossRef]
19. Wang, S.; Zhang, Y.; Han, Y.; Hou, Y.; Fan, Y.; Hou, X. Design of Porous Membranes by Liquid Gating Technology. *Acc. Mater. Res.* **2021**, *2*, 407–419. [CrossRef]
20. Wang, S.; Yang, X.; Wu, F.; Min, L.; Chen, X.; Hou, X. Inner Surface Design of Functional Microchannels for Microscale Flow Control. *Small* **2020**, *16*, 1905318. [CrossRef]

21. Wu, F.; Chen, X.; Wang, S.; Zhou, R.; Wang, C.; Yu, L.; Zheng, J.; Yang, C.; Hou, X. Green Synthesized Liquid-like Dynamic Polymer Chains with Decreased Nonspecific Adhesivity for High-Purity Capture of Circulating Tumor Cells. *CCS Chem.* **2024**, *6*, 507–517. [CrossRef]
22. Wu, F.; Kong, X.; Liu, Y.; Wang, S.; Chen, Z.; Hou, X. Microfluidic-Based Isolation of Circulating Tumor Cells with High-Efficiency and High-Purity. *Chin. Chem. Lett.* **2024**, *35*, 109754. [CrossRef]
23. Ota, N.; Tanaka, N.; Sato, A.; Shen, Y.; Yalikun, Y.; Tanaka, Y. Microenvironmental Analysis and Control for Local Cells under Confluent Conditions via a Capillary-Based Microfluidic Device. *Anal. Chem.* **2022**, *94*, 16299–16307. [CrossRef] [PubMed]
24. Shen, Y.; Tanaka, N.; Yamazoe, H.; Furutani, S.; Nagai, H.; Kawai, T.; Tanaka, Y. Flow Analysis on Microcasting with Degassed Polydimethylsiloxane Micro-Channels for Cell Patterning with Cross-Linked Albumin. *PLoS ONE* **2020**, *15*, e0232518. [CrossRef]
25. Filippi, M.; Yasa, O.; Kamm, R.D.; Raman, R.; Katzschmann, R.K. Will Microfluidics Enable Functionally Integrated Biohybrid Robots? *Proc. Natl. Acad. Sci. USA* **2022**, *119*, e2200741119. [CrossRef]
26. Yuan, H.; Miao, Z.; Wan, C.; Wang, J.; Liu, J.; Li, Y.; Xiao, Y.; Chen, P.; Liu, B.-F. Recent Advances in Centrifugal Microfluidics for Point-of-Care Testing. *Lab. Chip* **2025**, *25*, 1015–1046. [CrossRef]
27. Ren, S.; Liu, D.; Xu, S. Development and Challenges of Pathogen Molecular Point-Of-Care Testing Systems Based on Microfluidic Technology. *iLABMED* **2025**, *3*, 21–28. [CrossRef]
28. Kulkarni, M.B.; Goel, S. Recent Advancements in Integrated Microthermofluidic Systems for Biochemical and Biomedical Applications–A Review. *Sens. Actuators A Phys.* **2022**, *341*, 113590. [CrossRef]
29. Hu, J.; Bai, H.; Wang, L.; Li, J.; Shen, Y.; Zhang, L.; Tang, J.; Wang, M.; Liu, Q.; Zhou, J.; et al. Hermetic Microfluidic Device for Point-of-Care Viral Nucleic Acid Testing. *Sens. Actuators B Chem.* **2024**, *411*, 135740. [CrossRef]
30. Lan, Z.; Chen, R.; Zou, D.; Zhao, C. Microfluidic Nanoparticle Separation for Precision Medicine. *Adv. Sci.* **2025**, *12*, 2411278. [CrossRef]
31. Khoo, B.L.; Grenci, G.; Lim, Y.B.; Lee, S.C.; Han, J.; Lim, C.T. Expansion of Patient-Derived Circulating Tumor Cells from Liquid Biopsies Using a CTC Microfluidic Culture Device. *Nat. Protoc.* **2018**, *13*, 34–58. [CrossRef] [PubMed]
32. Wang, Y.I.; Abaci, H.E.; Shuler, M.L. Microfluidic Blood–Brain Barrier Model Provides in Vivo-like Barrier Properties for Drug Permeability Screening. *Biotech. Bioeng.* **2017**, *114*, 184–194. [CrossRef] [PubMed]
33. Cai, Z.; Fan, S.; Sun, X.; Mo, X.; Yang, G. Novel Microfluidic Device for Measurable Residual Disease Detection in Acute Leukemia. *Innov.* **2023**, *4*, 100408. [CrossRef]
34. Zhai, J.; Liu, Y.; Ji, W.; Huang, X.; Wang, P.; Li, Y.; Li, H.; Wong, A.H.-H.; Zhou, X.; Chen, P.; et al. Drug Screening on Digital Microfluidics for Cancer Precision Medicine. *Nat. Commun.* **2024**, *15*, 4363. [CrossRef]
35. Offord, C.; Cohen, J.; Enserink, M.; Barrett, R.J.; Shirure, V.S.; Bi, Y.; Curtis, M.B.; Lezia, A.; Goedegebuure, M.M.; Goedegebuure, S.P.; et al. Establishment of Colorectal Cancer Organoids in Microfluidic-Based System. *Nat. Commun.* **2025**, *12*, 497. [CrossRef]
36. Xiang, T.; Wang, J.; Li, H. Current Applications of Intestinal Organoids: A Review. *Stem Cell Res. Ther.* **2024**, *15*, 155. [CrossRef]
37. Zhao, Y.; Wang, E.Y.; Lai, F.B.L.; Cheung, K.; Radisic, M. Organs-on-a-Chip: A Union of Tissue Engineering and Microfabrication. *Trends Biotechnol.* **2023**, *41*, 410–424. [CrossRef]
38. Haque, M.R.; Wessel, C.R.; Leary, D.D.; Wang, C.; Bhushan, A.; Bishehsari, F. Patient-Derived Pancreatic Cancer-on-a-Chip Recapitulates the Tumor Microenvironment. *Microsyst. Nanoeng.* **2022**, *8*, 36. [CrossRef]
39. de Haan, P.; Santbergen, M.J.C.; van der Zande, M.; Bouwmeester, H.; Nielen, M.W.F.; Verpoorte, E. A Versatile, Compartmentalised Gut-on-a-Chip System for Pharmacological and Toxicological Analyses. *Sci. Rep.* **2021**, *11*, 4920. [CrossRef]
40. Aryal, P.; Hefner, C.; Martinez, B.; Henry, C.S. Microfluidics in Environmental Analysis: Advancements, Challenges, and Future Prospects for Rapid and Efficient Monitoring. *Lab. Chip* **2024**, *24*, 1175–1206. [CrossRef]
41. He, Q.; Wang, B.; Liang, J.; Liu, J.; Liang, B.; Li, G.; Long, Y.; Zhang, G.; Liu, H. Research on the Construction of Portable Electrochemical Sensors for Environmental Compounds Quality Monitoring. *Mater. Today Adv.* **2023**, *17*, 100340. [CrossRef]
42. Gao, Y.; Wang, Y.; Wang, Y.; Magaud, P.; Liu, Y.; Zeng, F.; Yang, J.; Baldas, L.; Song, Y. Nanocatalysis Meets Microfluidics: A Powerful Platform for Sensitive Bioanalysis. *TrAC Trends Anal. Chem.* **2023**, *158*, 116887. [CrossRef]
43. Zhang, J.; Zhang, X.; Zhang, Y.; Yang, X.; Guo, L.; Man, C.; Jiang, Y.; Zhang, W.; Zhang, X. Emerging Biosensors Integrated with Microfluidic Devices: A Promising Analytical Tool for on-Site Detection of Mycotoxins. *npj Sci. Food* **2025**, *9*, 84. [CrossRef] [PubMed]
44. Li, S.; Zhang, R.; Zhang, G.; Shuai, L.; Chang, W.; Hu, X.; Zou, M.; Zhou, X.; An, B.; Qian, D.; et al. Microfluidic Manipulation by Spiral Hollow-Fibre Actuators. *Nat. Commun.* **2022**, *13*, 1331. [CrossRef] [PubMed]
45. Ibrahim, O.A.; Navarro-Segarra, M.; Sadeghi, P.; Sabaté, N.; Esquivel, J.P.; Kjeang, E. Microfluidics for Electrochemical Energy Conversion. *Chem. Rev.* **2022**, *122*, 7236–7266. [CrossRef]
46. Zhou, J.; Mukherjee, P.; Gao, H.; Luan, Q.; Papautsky, I. Label-Free Microfluidic Sorting of Microparticles. *APL Bioeng.* **2019**, *3*, 041504. [CrossRef]
47. Yazicigil, R.T.; Bali, A.; Caygara, D.; Densmore, D. Improving Engineered Biological Systems with Electronics and Microfluidics. *Nat. Biotechnol.* **2025**, *43*, 1067–1083. [CrossRef]

48. Shen, Y.; Yuan, Y.; Tang, T.; Ota, N.; Tanaka, N.; Hosokawa, Y.; Yalikun, Y.; Tanaka, Y. Continuous 3D Particles Manipulation Based on Cooling Thermal Convection. *Sens. Actuators B Chem.* **2022**, *358*, 131511. [CrossRef]
49. Ashammakhi, N.; Nasiri, R.; de Barros, N.R.; Tebon, P.; Thakor, J.; Goudie, M.; Shamloo, A.; Martin, M.G.; Khademhosseini, A. Gut-on-a-Chip: Current Progress and Future Opportunities. *Biomaterials* **2020**, *255*, 120196. [CrossRef]
50. Kaden, T.; Alonso-Román, R.; Stallhofer, J.; Gresnigt, M.S.; Hube, B.; Mosig, A.S. Leveraging Organ-on-Chip Models to Investigate Host–Microbiota Dynamics and Targeted Therapies for Inflammatory Bowel Disease. *Adv. Healthc. Mater.* **2025**, *14*, 2402756. [CrossRef]
51. Jomezadeh Kheibary, N.; Abolfazli Esfahani, J.; Mousavi Shaegh, S.A. Analysis of Oxygen Transport in Microfluidic Bioreactors for Cell Culture and Organ-on-chip Applications. *Eng. Rep.* **2020**, *2*, e12062. [CrossRef]
52. Chen, Z.; Lv, Z.; Zhang, Z.; Weitz, D.A.; Zhang, H.; Zhang, Y.; Cui, W. Advanced Microfluidic Devices for Fabricating Multi-structural Hydrogel Microsphere. *Exploration* **2021**, *1*, 20210036. [CrossRef]
53. Li, J.; Jamieson, W.D.; Dimitriou, P.; Xu, W.; Rohde, P.; Martinac, B.; Baker, M.; Drinkwater, B.W.; Castell, O.K.; Barrow, D.A. Building Programmable Multicompartment Artificial Cells Incorporating Remotely Activated Protein Channels Using Microfluidics and Acoustic Levitation. *Nat. Commun.* **2022**, *13*, 4125. [CrossRef]
54. Li, J.; Baxani, D.K.; Jamieson, W.D.; Xu, W.; Rocha, V.G.; Barrow, D.A.; Castell, O.K. Formation of Polarized, Functional Artificial Cells from Compartmentalized Droplet Networks and Nanomaterials, Using One-Step, Dual-Material 3D-Printed Microfluidics. *Adv. Sci.* **2020**, *7*, 1901719. [CrossRef] [PubMed]
55. Shen, Y.; Yalikun, Y.; Tanaka, Y. Recent Advances in Microfluidic Cell Sorting Systems. *Sens. Actuators B Chem.* **2019**, *282*, 268–281. [CrossRef]
56. Cai, H.; Ao, Z.; Wu, Z.; Song, S.; Mackie, K.; Guo, F. Intelligent Acoustofluidics Enabled Mini-Bioreactors for Human Brain Organoids. *Lab. Chip* **2021**, *21*, 2194–2205. [CrossRef] [PubMed]
57. Tian, Z.; Yang, S.; Huang, P.-H.; Wang, Z.; Zhang, P.; Gu, Y.; Bachman, H.; Chen, C.; Wu, M.; Xie, Y.; et al. Wave Number–Spiral Acoustic Tweezers for Dynamic and Reconfigurable Manipulation of Particles and Cells. *Sci. Adv.* **2019**, *5*, 5. [CrossRef]
58. Hu, X.; Gao, X.; Chen, S.; Guo, J.; Zhang, Y. DropLab: An Automated Magnetic Digital Microfluidic Platform for Sample-to-Answer Point-of-Care Testing—Development and Application to Quantitative Immunodiagnostics. *Microsyst. Nanoeng.* **2023**, *9*, 10. [CrossRef]
59. Zhou, J.; Dong, J.; Hou, H.; Huang, L.; Li, J. High-Throughput Microfluidic Systems Accelerated by Artificial Intelligence for Biomedical Applications. *Lab. Chip* **2024**, *24*, 1307–1326. [CrossRef]
60. Nelson, M.D.; Goenner, B.L.; Gale, B.K. Utilizing ChatGPT to Assist CAD Design for Microfluidic Devices. *Lab. Chip* **2023**, *23*, 3778–3784. [CrossRef]
61. Liu, H.; Nan, L.; Chen, F.; Zhao, Y.; Zhao, Y. Functions and Applications of Artificial Intelligence in Droplet Microfluidics. *Lab. Chip* **2023**, *23*, 2497–2513. [CrossRef]
62. Siu, D.M.D.; Lee, K.C.M.; Chung, B.M.F.; Wong, J.S.J.; Zheng, G.; Tsia, K.K. Optofluidic Imaging Meets Deep Learning: From Merging to Emerging. *Lab. Chip* **2023**, *23*, 1011–1033. [CrossRef] [PubMed]
63. Jeon, H.; Han, J. Microfluidics with Machine Learning for Biophysical Characterization of Cells. *Annu. Rev. Anal. Chem.* **2025**, *18*, 447–472. [CrossRef]
64. Liu, L.; Bi, M.; Wang, Y.; Liu, J.; Jiang, X.; Xu, Z.; Zhang, X. Artificial Intelligence-Powered Microfluidics for Nanomedicine and Materials Synthesis. *Nanoscale* **2021**, *13*, 19352–19366. [CrossRef] [PubMed]
65. Chu, P.-Y.; Nguyen, T.N.A.; Wu, A.-Y.; Huang, P.-S.; Huang, K.-L.; Liao, C.-J.; Hsieh, C.-H.; Wu, M.-H. The Utilization of Optically Induced Dielectrophoresis (ODEP)-Based Cell Manipulation in a Microfluidic System for the Purification and Sorting of Circulating Tumor Cells (CTCs) with Different Sizes. *Micromachines* **2023**, *14*, 2170. [CrossRef] [PubMed]
66. Yang, H.-Y.; Lin, C.-H.; Hu, Y.-W.; Chien, C.-H.; Huang, M.-C.; Lai, C.-H.; Wu, J.-K.; Tseng, F.-G. Automatic Single-Cell Harvesting for Fetal Nucleated Red Blood Cell Isolation on a Self-Assemble Cell Array (SACA) Chip. *Micromachines* **2024**, *15*, 1515. [CrossRef]
67. Pouraria, H.; Houston, J.P. Elasticity of Carrier Fluid: A Key Factor Affecting Mechanical Phenotyping in Deformability Cytometry. *Micromachines* **2024**, *15*, 822. [CrossRef]
68. Jiang, H.; Li, X.; Liu, Z.; Luo, S.; Huang, J.; Chen, C.; Chen, R.; Li, F. A Microfluidic Approach for Assessing the Rheological Properties of Healthy Versus Thalassemic Red Blood Cells. *Micromachines* **2025**, *16*, 957. [CrossRef]
69. Xian, Q.; Zhang, J.; Wong, Y.C.; Gao, Y.; Song, Q.; Xu, N.; Wen, W. Pico-Scale Digital PCR on a Super-Hydrophilic Microarray Chip for Multi-Target Detection. *Micromachines* **2025**, *16*, 407. [CrossRef]
70. Chen, J.; Zheng, M.; Xiao, Q.; Wang, H.; Chi, C.; Lin, T.; Wang, Y.; Yi, X.; Zhu, L. Recent Advances in Microfluidic-Based Extracellular Vesicle Analysis. *Micromachines* **2024**, *15*, 630. [CrossRef]
71. Wu, F.; Xu, J.; Liu, Y.; Sun, H.; Zhang, L.; Liu, Y.; Wang, W.; Chong, F.; Zou, D.; Wang, S. Rapid Construction of Liquid-like Surfaces via Single-Cycle Polymer Brush Grafting for Enhanced Antifouling in Microfluidic Systems. *Micromachines* **2024**, *15*, 1241. [CrossRef]

72. Ayeni, O.O.; Stretz, H.A.; Vasel-Be-Hagh, A. Effects of Fiber Arrangement on Flow Characteristics Along a Four-Fiber Element of Fiber Extractors. *Micromachines* **2025**, *16*, 425. [CrossRef]
73. Thiem, D.B.; Szabo, G.; Burg, T.P. Model-Based Optimization of Solid-Supported Micro-Hotplates for Microfluidic Cryofixation. *Micromachines* **2024**, *15*, 1069. [CrossRef]
74. Sun, Y.; Yu, N.; Zhang, J.; Yang, B. Advances in Microfluidic Single-Cell RNA Sequencing and Spatial Transcriptomics. *Micromachines* **2025**, *16*, 426. [CrossRef]
75. Shen, Y.; Wang, Z.; Ren, T.; Wen, J.; Li, J.; Tang, T. Recent Advances in Microfluidic Impedance Detection: Principle, Design and Applications. *Micromachines* **2006**, *16*, 683. [CrossRef]

micromachines

MDPI

Article

The Utilization of Optically Induced Dielectrophoresis (ODEP)-Based Cell Manipulation in a Microfluidic System for the Purification and Sorting of Circulating Tumor Cells (CTCs) with Different Sizes

Po-Yu Chu [1,†], Thi Ngoc Anh Nguyen [1,†], Ai-Yun Wu [1,†], Po-Shuan Huang [1], Kai-Lin Huang [1], Chia-Jung Liao [2], Chia-Hsun Hsieh [3,4] and Min-Hsien Wu [1,3,4,*]

[1] Graduate Institute of Biomedical Engineering, Chang Gung University, Taoyuan City 33302, Taiwan; d000018394@cgu.edu.tw (P.-Y.C.); d1131003@cgu.edu.tw (T.N.A.N.); m1031001@cgu.edu.tw (A.-Y.W.); leo_6813@cgu.edu.tw (P.-S.H.); m1031009@cgu.edu.tw (K.-L.H.)

[2] Department of Biomedical Sciences, College of Medicine, Chang-Gung University, Taoyuan 333, Taiwan; l329735@cgu.edu.tw

[3] Division of Hematology-Oncology, Department of Internal Medicine, New Taipei City Municipal Tucheng Hospital, New Taipei City 23652, Taiwan; wisdom5000@cgmh.org.tw

[4] Division of Hematology-Oncology, Department of Internal Medicine, Chang Gung Memorial Hospital at Linkou, Taoyuan City 33302, Taiwan

[*] Correspondence: mhwu@mail.cgu.edu.tw; Tel.: +886-3211-8800 (ext. 3599)

[†] These authors contributed equally to this work.

Abstract: The analysis of circulating tumor cells (CTCs) at the molecular level holds great promise for several clinical applications. For this goal, the harvest of high-purity, size-sorted CTCs with different subtypes from a blood sample are important. For this purpose, a two-step CTC isolation protocol was proposed, by which the immunomagnetic beads-based cell separation was first utilized to remove the majority of blood cells. After that, an optically induced dielectrophoresis (ODEP) microfluidic system was developed to (1) purify the CTCs from the remaining magnetic microbeads-bound blood cells and to (2) sort and separate the CTCs with different sizes. In this study, the ODEP microfluidic system was designed and fabricated. Moreover, its optimum operation conditions and performance were explored. The results exhibited that the presented technique was able to purify and sort the cancer cells with two different sizes from a tested cell suspension in a high-purity (93.5% and 90.1% for the OECM 1 and HA22T cancer cells, respectively) manner. Overall, this study presented a technique for the purification and sorting of cancer cells with different sizes. Apart from this application, the technique is also useful for other applications in which the high-purity and label-free purification and sorting of cells with different sizes is required.

Keywords: optically induced dielectrophoresis (ODEP); microfluidic technology; circulating tumor cells (CTCs); cell isolation; cell sorting

1. Introduction

Cancer has been a primary threat to global health in recent decades. It is well known that the majority of cancer deaths are due to cancer metastasis [1,2]. Circulating tumor cells (CTCs) are the rare cancer cells existing in the peripheral blood of cancer patients, which were first described in 1869 [3]. Several studies have well demonstrated that the presence of CTCs in blood circulation is relevant to cancer metastasis [4]. Therefore, the analysis of CTCs holds great promise for several clinical utilizations (e.g., cancer detection [4,5], prognosis evaluation of cancer [6,7], therapeutic response evaluation of cancer [8–10], or monitoring of cancer metastasis [11,12]). Compared with the clinical applications of CTCs simply based on the quantification of their cell number [6,11], the analysis of CTCs at molecular level (e.g., gene expression analysis) could pave a more promising route to comprehensively understand CTC heterogeneity [13,14], cancer progression [15,16], or mechanisms of therapeutic resistance [17,18]. To achieve this goal, the harvest of high-purity and all possible CTCs with different subtypes from the blood samples of cancer patients is crucially important.

However, CTCs are naturally rare in a blood sample (e.g., approximately 1 CTC per 10^5–10^7 leukocytes [19]), making them technically challenging to isolate and purify. In terms of the CTC isolation and purification schemes, they can be generally categorized into cell biophysical- and immunoaffinity-based mechanisms [20]. For the former, the physical (e.g., size or density) differences between CTCs and the background blood cells [21] are mainly used to separate and then purify CTCs from a blood sample. This strategy is generally regarded to be simple and low cost to operate. However, its purity and specificity for target cell isolation are commonly inferior to the immunoaffinity-based strategy [22]. For the immunoaffinity-based methods, their working principles are based on the use of specific antibodies to selectively bind to the surface antigens of cells to distinguish the target CTCs from the other blood cells. CTCs normally express specific cellular surface antigens (e.g., EpCAM and CKs) that set them apart from the other blood cells (i.e., positive immunoselection). Alternatively, the CTC isolation and purification can be achieved by specifically removing the majority of blood cells, leaving the cell population that do not express any blood cell-related antigens for the subsequent collection (i.e., negative immunoselection) [23].

Due to the heterogeneity issue of CTCs regarding their surface antigens, the utilization of the positive immunoselection-based strategy might only be able to harvest some possible CTCs with different subtypes [24]. Conversely, the negative immunoselection-based CTC isolation and purification techniques are believed to obtain all possible and label-free CTCs from a blood sample [25]. Due to this technical advantage, the subsequent CTC analysis could produce more comprehensive information, revealing more real situations of cancer status. In practice, briefly, only blood cells (e.g., leukocytes) are targeted for removal via a standard immunomagnetic microbeads-based cell separation and isolation technique [25]. Nevertheless, the CTC isolation and purification based on this strategy generally comes across the technical problem of low CTC purity [25]. In this situation, the contamination of leukocytes in the obtained cell sample could, therefore, complicate the subsequent CTC analysis. Apart from the abovementioned issues, it was also reported that the CTCs in blood samples have a diverse array of sizes (e.g., diameter: 12~30 μm) [26,27], reflecting that cancer cells maintain their high phenotypic plasticity in the bloodstream. Moreover, by analyzing the morphological features of CTCs in different cancer types, the results reveal that the CTCs isolated from hepatocellular and breast cancer patients exhibit larger cell size compared with those obtained from the other types of cancers [28–30]. This finding indicates that CTCs may display unique morphological features (e.g., size) depending on the specific origin of the tumor. As a result, the initial sorting and separation of CTCs based on their size differences during the CTC isolation and purification process might facilitate the following CTC analytical work.

Moreover, with the advances in microfluidic technology, the microfluidic systems integrated with various techniques (e.g., fluidic control- [31], acoustophoresis- [32], dielectrophoresis (DEP)- [33], or optically induced dielectrophoresis (ODEP)- [34] based mechanism) for cell manipulation have been presented for the isolation and purification of biological cells. Due to the feature of miniaturization, overall, these microfluidic systems have been proved to have superior cell isolation and purification performance over conventional cell isolation and purification schemes [31]. Among the cell manipulation mechanisms, ODEP-based techniques are particularly suitable for biologists to use owing to their ease of fabrication and operation. The microparticle manipulation using ODEP was first presented in 2005 [35]. Its working principle is well discussed elsewhere and briefly described herein. An electric voltage is applied within the solution layer of an ODEP system to generate a uniform electric field. This in turn causes the dielectric microparticles (e.g., cells) within the solution to be electrically polarized. When the photoconductive substrate of the ODEP system is projected with light, the illuminated light can cause the electrical impedance at the light-projected area to decrease. This phenomenon in turn causes the applied electric voltage to decline across the solution layer within the light-illuminated region. Overall, the abovementioned phenomenon creates a locally non-uniform electric field within an ODEP system. Based on these facts, the interaction between the non-uniform electric field created and an electrically polarized microparticle is used to manipulate the microparticles. In terms of operation, a scientist can simply control the dynamic optical images projected on an ODEP system to manipulate the microparticles in a manageable manner.

To harvest high-purity and all possible CTCs with different sizes from a blood sample, a two-step CTC isolation and purification protocol was proposed in this study. For the first step, the negative selection-based CTC isolation scheme based on the immunomagnetic beads-based cell separation technique was adopted to remove 99% of blood cells [36]. In the subsequent step, the combination of ODEP-based cell manipulation and the laminar flow phenomenon in the presented microfluidic system was utilized to (1) purify the label-free CTCs from the remaining magnetic microbeads-bound blood cells, and to (2) further sort and separate the CTCs with different sizes. In this study, research was mainly focused on the second step. For the ODEP microfluidic system, its working principles are mainly based on the phenomena described in the following. First, the ODEP force acting on the cells with varied sizes is different [37]. Second, previous studies have demonstrated that the ODEP force generated on a cell is influenced by the intensity (or color brightness ratio) of illuminated light [38–40]. Based on these facts, two dynamic light image arrays were designed in the microchannel (i.e., the defined zone for cell separation) to first gather all cells to one side of the microchannel and then to sort and separate the magnetic microbeads-bound blood cells and the cancer cells with different sizes, respectively. Moreover, due to the laminar flow pattern in a microchannel, the separated cells were then transported in a partitioned manner for the subsequent cell collections via two static light bar arrays.

In this work, the ODEP microfluidic system was designed and fabricated. Moreover, the ODEP operating conditions (i.e., the moving velocities and color brightness of the light bars used in the two dynamic light image arrays, and the color brightness of the light bars used in the two static light bar arrays) for effective purification and sorting of cancer cells with two different sizes were explored experimentally. For performance evaluation purposes, two cancer cell lines (OECM1 and HA22T cells) and a Jurkat cell line (used as model leukocytes) were used as the model cells representing the CTCs with different sizes and the remaining blood cells in the processed sample. Finally, the performance of the proposed microfluidic system for the purification and sorting of cancer cells with two different sizes was assessed. The results demonstrated that the ODEP-based cell manipulation scheme was able to effectively purify and sort the cancer cells with two different sizes from a tested cell suspension model in a high-purity (achieved purity for the OECM1 and HA22T cancer cells: 93.5% and 90.1%, respectively) manner. Overall, this study presented an ODEP microfluidic system for the purification and sorting of cancer cells with different sizes. In addition to the application for CTC purification and sorting, the proposed approach is also useful in other applications in which the high-purity and label-free purification and sorting of cells with different sizes are required.

2. Materials and Methods

2.1. Design, Fabrication, and Experimental Setup of ODEP-Based Microfluidic System for the Purification and Sorting of CTCs with Different Sizes

In this work, an ODEP-based microfluidic chip was designed to (1) purify the CTCs from the magnetic microbeads-bound blood cells and to (2) further sort and isolate the CTCs with different sizes in a continuous manner. The design of the ODEP microfluidic chip is schematically shown in Figure 1a. Briefly, a microchannel (L: 25 mm, W: 1 mm, H: 50 μm) was designed to transport cell suspension samples. In this work, the ODEP-based cell manipulation was conducted within the defined cell separation zone of the microchannel, as indicated in Figure 1a. In the design (Figure 1a), two through-holes (D: 1 mm) for tubing connection were used for the fresh cell suspension sample input and the waste sample output, respectively. The layer-by-layer structure of the ODEP microfluidic chip is illustrated in Figure 1b. Briefly, the microfluidic chip consisted of a top processed polydimethylsiloxane (PDMS) substrate (Layer A), an indium-tin-oxide (ITO) glass (Layer B), a processed double-sided adhesive tape (thickness: 50 μm) with a hollow microchannel (Layer C), and a bottom ITO glass coated with a layer of photoconductive material (Layer D).

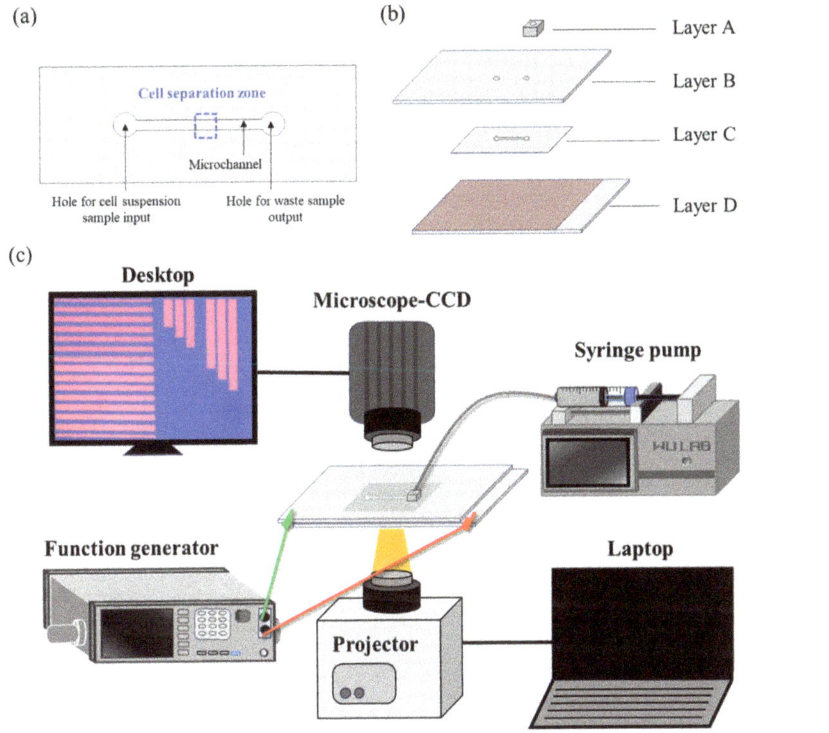

Figure 1. Schematic illustration of the (**a**) layout and (**b**) layer-by-layer structure [Layer A: a processed PDMS substrate; Layer B: an indium-tin-oxide (ITO) glass; Layer C: a processed double-sided adhesive tape with a hollow main microchannel; Layer D: a bottom ITO glass coated with a layer of photoconductive material] of the ODEP microfluidic chip, and (**c**) the overall experimental setup.

The fabrication and experimental setup have been well described previously [36,38]. Briefly, the PDMS tube connector (Layer A; Figure 1b) was made by the CNC (EGX-400, Roland Inc., Hamamatsu, Japan) machining for positive polymethylmethacrylate (PMMA) mold fabrication and the subsequent PDMS (Sylgard® 184, Dow Corning, Midland, MI, USA) replica molding for the tube connector making. For the ITO glass (InnoLux Corporation, ZMI, Taiwan) with two through-holes (Layer B; Figure 1b), the holes were fabricated via mechanical drilling. For Layer C (Figure 1b), the hollow structure of the microchannel was fabricated in the double-sided adhesive tape (L298, Sun-yieh, TYN, Taiwan) simply via a manual punching process using a custom-made metal mold. For Layer D (Figure 1b), an ITO glass was deposited with a photoconductive material (i.e., a 20 nm thick n+ hydrogenated amorphous silicon layer (n^+ a-Si:H) and a 1 μm thick intrinsic hydrogenated amorphous silicon (a-Si:H) layer) via plasma-enhanced chemical vapor deposition (PECVD).

For the following assembly process, the Layer A was bonded with the Layer B via O_2 plasma surface treatment. This was followed by the assembly with the Layer D via the Layer C (i.e., a double-sided adhesive tape). In terms of operation, a syringe pump (KDS LEGATO 180, KD Scientific, Holliston, MA, USA) was used to transport the cell suspension sample in the microchannel. For cell manipulation using ODEP, a function generator (AFG-2125, Good Will Instrument Co., Ltd., New Taipei City, NTPC, Taiwan) was used to generate an alternating current (AC) voltage between the two ITO glass substrates (Figure 1b). In the following step, a computer-controlled projector (EB-X05, Epson, Suwa, Japan) was used to illuminate dynamic light images onto the Layer D. In this work, the observation and recording of the cell manipulation process was achieved by using a CCD-equipped microscope (Zoom 160, OPTEM, Medina, OH, USA). The overall experimental setup is schematically illustrated in Figure 1c.

2.2. The Working Principle of the ODEP-Based Cell Manipulation Scheme for the Purification and Sorting of CTCs with Different Sizes

The mechanism of the cell manipulation using ODEP is well discussed elsewhere [36,41] and has also been briefly described in the introduction section. The ODEP force generated on a microparticle can be expressed by the Equation (1), also used to describe the DEP force [36,41]:

$$F_{DEP} = 2\pi r^3 \varepsilon_0 \varepsilon_m \mathrm{Re}[f_{CM}] \nabla |E|^2 \quad (1)$$

where r, ε_0, ε_m, $\nabla |E|^2$, and Re[f_{CM}] denote the microparticle's radius, vacuum permittivity, relative permittivity of the surrounding solution, gradient of the exerted electrical voltage squared, and real part of the Clausius–Mossotti factor (f_{CM}), respectively. It can be found from Equation (1) that the ODEP force generated on a microparticle is proportional to its radius cubic under the given operation conditions (e.g., the applied electrical voltage or the property of solution) [36,41]. Therefore, one can use ODEP-based cell manipulation to sort and separate the cells with different sizes based on their different ODEP manipulation force [36,41]. Apart from the factors described in Equation (1), the ODEP force generated on a microparticle is also increased with the intensity increase in the illuminated light, which was reported to be proportional to its color brightness ratio according to our previous study [38,39].

Based on the phenomena abovementioned, this study proposed a two-step CTC isolation and purification protocol for the isolation, purification, and sorting of CTCs with different sizes from a blood sample. For the first step, a blood sample is treated with the negative selection-based CTC isolation process utilizing a standard immunomagnetic microbeads-based cell isolation approach, which is well described elsewhere [36], to remove the majority of the blood cells in a blood sample. In the second step, the combination of ODEP cell manipulation and laminar flow phenomenon in the presented ODEP microfluidic system was used to (1) purify all possible CTCs from the remaining magnetic microbead-bound blood cells, and to (2) further sort and separate the CTCs with different sizes. In this study, research mainly focused on the second step. For the development work, two cancer cell lines (OECM1 and HA22T cancer cells; diameter: 20.9 ± 2.6 and 26.6 ± 3.3 μm, respectively) and a Jurkat cell line (used as model leukocytes [36]) were used as the model cells representing the CTCs with different sizes and the remaining blood cells in the processed sample. In the ODEP microfluidic system, briefly, two dynamic light image arrays were designed in the defined cell separation zone, first to gather all cells to one side of the microchannel and then to sort and separate the magnetic microbeads-bound cells and the cancer cells with different sizes, respectively. Furthermore, due to the laminar flow pattern in a microchannel, the separated cells as abovementioned were then transported in a partitioned manner for the subsequent cell collections via two static light bar arrays. The entire operation procedures are illustrated in Figure 2.

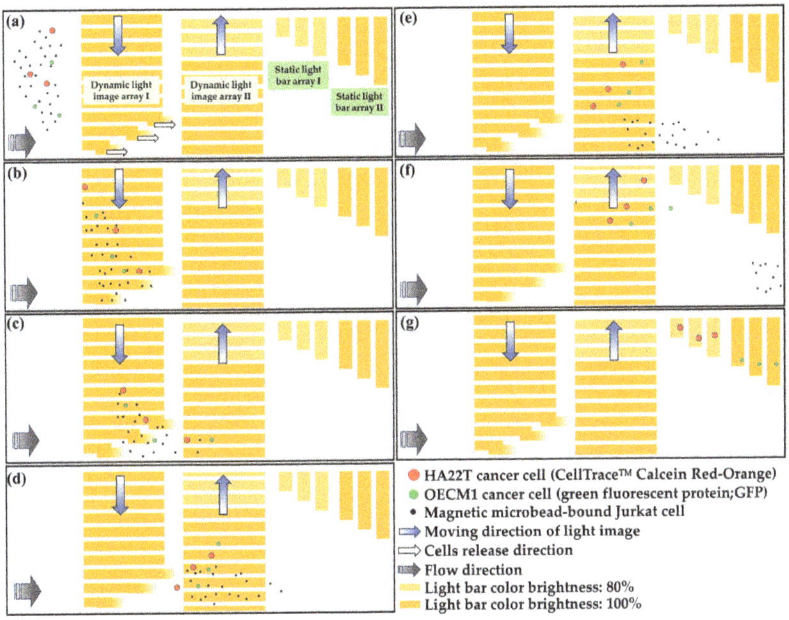

Figure 2. Schematic illustration of the entire ODEP manipulation process for the purification and sorting of cancer cells with different size: (**a**) the designed light image arrays including two dynamic light image arrays and two static light bar arrays (as indicated) in the cell separation zone of the microchannel (the color brightness is as indicated); (**b**,**c**): the dynamic light image array I was used to dynamically pool all cells, continuously transported to the cell separation zone, to one side of the microchannel; (**c**,**d**): the cells manipulated to the side of the microchannel were then released by the three light bars near the side of the microchannel in the flow direction; (**d**,**e**): the cells flowed to the dynamic light image array II were then sorted and separated by the ODEP manipulation, by which the magnetic microbeads-bound Jurkat cells could not be manipulated by ODEP and were thus released by the light bars near the side of the microchannel in the flow direction; (**f**,**g**): the cancer cells (i.e., the HA22T and OECM1 cancer cells) were manipulated by ODEP and thus were attracted and pulled to the another side of the microchannel. During this process, they were sorted and separated based on their size difference by the designed ODEP mechanism and then were collected by the downstream static light bar array I and II, respectively.

2.3. The ODEP Operation Conditions for the Purification and Sorting of Cancer Cells with Different Sizes

In the ODEP microfluidic system, ODEP cell manipulation was utilized to continuously purify and sort the cancer cells from the background magnetic microbeads-bound Jurkat cells, as illustrated in Figure 2. In this work, the electric voltage and frequency were set at 10 Vpp and 3 MHz, respectively, which was previously demonstrated to be a cell-friendly condition for cells [41]. Under the set electric condition, the ODEP manipulation force (i.e., the net force between the ODEP force and friction force producing on a cell) of the cells manipulated (i.e., the Jurkat, magnetic microbeads-bound Jurkat, OECM1, and HA22T cells) was then experimentally evaluated based on the method described previously [36,41]. In a steady state, the ODEP manipulation force acting on a cell is balanced by the viscous drag of fluid. As a result, the hydrodynamic drag force of a moving cell was used to evaluate the net ODEP manipulation force of a cell [36,41]. Stokes' law (Equation (2)) describes the hydrodynamic drag force (F) acting on a spherical particle under continuous flow conditions [36,41]:

$$F = 6\pi r \eta v \quad (2)$$

where r, η, and v represent the cell radius, fluid viscosity, and maximum velocity of the cell, respectively. According to Stokes' law, thus, the ODEP manipulation force acting on the cell investigated can be experimentally assessed through measurements of the maximum velocity of a moving light image that can manipulate the cell [36,41]. Moreover, as described earlier, the ODEP force generated on a cell is influenced by the intensity (or color brightness ratio) of illuminated light, according to our previous study [38,39]. The previous evaluation of the ODEP manipulation force of the manipulated magnetic microbeads-bound Jurkat, OECM1, and HA22T cells was based on the use of the light image with 100% color brightness. In this study, similar evaluations were also carried out using the light images with varied percentages (i.e., 60, 70, 80, 90, and 100%) of color brightness.

2.4. Design of the Light Image Arrays in ODEP Microfluidic System for the Purification and Sorting of Cancer Cells with Two Different Sizes

In this study, the combination of ODEP cell manipulation and laminar flow phenomenon in the ODEP microfluidic system was designed to (1) purify the cancer cells (i.e., OECM1 and HA22T cells) from the magnetic microbead-bound Jurkat cells, and to (2) further sort and separate the cancer cells with different sizes. As described in Figure 2, two dynamic light image arrays (i.e., the dynamic light image array I (light bar number: 14; color brightness: 100%; size of light bar: 50 μm (W) and 1316 μm (L); 50 μm (W) and 1459, 1005, and 580 μm (L) for the three bars near the side of the microchannel) and the dynamic light image array II (light bar number: 14; size of light bar: 50 μm (W) and 870 μm (L); color brightness: 100% and 80% as indicated in Figure 2)) were designed in the defined cell separation zone of the microchannel to first gather all cells to one side of the microchannel, and then to sort and separate the magnetic microbeads-bound Jurkat cells and the cancer cells with different sizes, respectively. Moreover, two static light bar arrays (i.e., the static light bar array I (light bar number: 3; color brightness: 80%; size of light bar: 75 μm (W) and 210, 295, and 380 μm (L) for the three bars, respectively) and the static light bar array II (light bar number: 3; color brightness: 100%; size of light bar: 75 μm (W) and 567, 650, and 738 μm (L) for the three bars, respectively)) were designed at the downstream part of the microchannel for the collection of the sorted and separated cancer cells with different sizes, respectively. To achieve this goal, the operation conditions of the two dynamic light image arrays (i.e., the percentage of color brightness and the moving velocity) and the two static light bar arrays (i.e., the percentage of color brightness) were determined based on the previous experimental evaluations, as described in the Section 2.3. The determined operation conditions for the four light image arrays were described and discussed in the Section 3.3.

2.5. Performance Evaluation of the ODEP Microfluidic System for the Purification and Sorting of Cancer Cells with Two Different Sizes

After the dynamic and static light image arrays were designed, their function for the manipulation of the three types of cells as aforementioned was then tested. The function of the dynamic light image array I to manipulate the microbead-bound Jurkat cells to one side of the microchannel (Figure 2b,c) was first experimentally evaluated. In this study, the magnetic microbead-bound Jurkat cell suspension with different concentrations ($2.5 \sim 10 \times 10^4$ cells 100 μL^{-1}) was loaded in the microfluidic chip. The magnetic microbead-bound Jurkat cells transported to the area of the dynamic light image array I was then manipulated using the designed array. The video of the operation process was recorded for quantifying the cells transported to the dynamic light image array I as well as the cells released by the three light bars most near the side of the microchannel (Figure 2c). Based on the data, the removal rate of magnetic microbead-bound Jurkat cells [(the number of cells released by the three light bars most near the side of the microchannel)/(the number of cells entered to the area of dynamic light image array I) $\times$ 100%] under different concentrations of cell suspension was then calculated. After understanding the working capacity of the dynamic light image array I to manipulate the magnetic microbead-bound Jurkat cells to one side of the microchannel, the function of the dynamic light image array

II to sort and separate the cancer cells with different sizes was experimentally evaluated. Briefly, the magnetic microbead-bound Jurkat, HA22T, and OECM1 cells (concentration: 5×10^4, 5×10^3, and 5×10^3 cells 100 μL^{-1}, respectively) were individually loaded into the ODEP microfluidic chip. The loaded cells were then treated with the proposed ODEP manipulation scheme as illustrated in Figure 2. For evaluating the distribution of the cells after ODEP operation, the number of the loaded cells and the number of the cells finally collected by the static light bar arrays I and II were quantified with the aid of video recording.

After the basic function tests as described above, the performance of the proposed ODEP microfluidic system for the purification and sorting of cancer cells with two different sizes was then evaluated. As described earlier, the ODEP microfluidic system was mainly designed to process the blood sample treated with the immunomagnetic beads-based cell separation for removing 99% of blood cells. Therefore, the main purpose of the ODEP microfluidic system was to (1) purify the cancer cells from the magnetic microbead-bound cells, and (2) further sort and separate the cancer cells with different sizes. For evaluating its working performance, the cell suspension tested, containing the magnetic microbead-bound Jurkat, HA22T, and OECM1 cells (ratio: 1:1:5) mimicking a blood sample processed with the immunomagnetic beads-based cell separation process, was prepared and then loaded into the ODEP microfluidic system. After the ODEP manipulation as described in Figure 2, the numbers of the three types of cells collected in the two static light bar arrays were quantified microscopically. For identifying the cell species, HA22T and OECM1 cancer cells were pre-stained with CellTrace™ Calcein Red-Orange (C34851, Invitrogen, Carlsbad, CA, USA) or expressed with GFP (green fluorescent protein), respectively. After that, the purity of the two cancer cells collected in the two static light image arrays was then evaluated.

2.6. Statistical Analysis

In this study, the data were presented as the mean ± standard deviation (3 separate experiments). One-way ANOVA was used to evaluate the effect of the experimental factors tested on the outcomes. Tukey's honestly significant difference (HSD) post hoc test was used to compare differences between the two conditions tested when the null hypothesis of the ANOVA was rejected.

3. Results and Discussion

3.1. Technical Advantages of the ODEP Microfluidic System for the Purification and Sorting of CTCs with Different Sizes

In terms of application, an ODEP microfluidic system (Figure 1) was developed and used to process further the cell suspension sample obtained from the blood sample treated with the immunomagnetic beads-based cell separation process to specifically remove the majority of blood cells (i.e., the proposed two-step CTC isolation and purification protocol as described earlier). In this work, the proposed ODEP microfluidic system was to (1) purify the CTCs from the remaining magnetic microbeads-bound blood cells, and (2) further sort and separate the CTCs with different sizes (Figure 2). Based on the operations, high-purity, label-free, and all possible CTCs with different size from a blood sample can be harvested, which is both biologically and clinically meaningful for the subsequent medical applications (e.g., prognosis evaluation of cancer [6,7], therapeutic response evaluation of cancer [8–10], or monitoring of cancer metastasis [11,12]) or fundamental studies (e.g., mechanisms behind cancer metastasis [15] or anti-cancer drug resistance [17,18]). This technical advantage is beyond what is possible by using other CTC isolation and purification methods by which the CTCs harvested are limited to certain CTC subtypes (e.g., EpCAM-positive CTCs [42]), are labeled with magnetic microbeads [42], or are in a low-purity manner (e.g., biophysical- and negative immunoselection-based CTC enrichment techniques [22,36]).

Among the techniques (e.g., fluidic control- [31], acoustophoresis- [32], DEP- [33], or ODEP- [34] based techniques) available for cell manipulation in microfluidic systems,

DEP-based cell or bacteria manipulation is commonly utilized for various applications [33]. Nevertheless, it generally requires a time-consuming, technically demanding, and costly microfabrication to create a special metal electrode array. This technical requirement could therefore restrict its widespread applications. Conversely, the key technical advantage of using ODEP cell manipulation in a microfluidic system is that it can easily and quickly generate or modify an electrode array in a virtual manner via the control of the optical patterns, functioning as a virtual electrode, in an ODEP system. In practice, therefore, a scientist can use a commercial digital projector to display optical images on an ODEP system to manipulate cells in a flexible, manageable, and user-friendly manner through a computer-interfaced control [34]. Overall, this technical feature is particularly suitable for biologists to use owning to their ease of fabrication and operation.

3.2. The ODEP Operation Conditions for the Manipulation of Cells

In the ODEP microfluidic system, two dynamic light image arrays were designed in the defined cell separation zone of the microchannel to, first, gather all cells to one side of the microchannel and, then, to sort and separate the magnetic microbeads-bound Jurkat cells and the OECM1 and HA22T cancer cells, as illustrated in Figure 2. Moreover, two static light bar arrays were designed at the downstream part of the microchannel for the collection of the sorted and separated cancer cells with different size, respectively. To achieve this goal, the ODEP operation conditions for the manipulation of the cells (i.e., magnetic microbeads-bound Jurkat, OECM1, and HA22T cells) were first explored. In this study, the electric voltage and frequency for ODEP-based cell manipulation were set at 10 Vpp and 3 MHz, respectively, which was demonstrated to be cell friendly [41]. In this study, the sizes of the model cells used were first evaluated microscopically. The results (Figure 3a) showed that the sizes of Jurkat, OECM1, and HA22T cells were measured to be 13.8 $\pm$ 1.7, 20.9 $\pm$ 2.6, and 26.6 $\pm$ 3.3 μm in diameter, respectively, which were evaluated to have significant difference ($p < 0.05$). After that, the evaluation of their ODEP manipulation force using the indicator of the maximum velocity of a light image that can manipulate the cells [36,41] was carried out. The results (Figure 3b) exhibited that the maximum velocities of a light image that can manipulate the cells (i.e., Jurkat cells and the OECM1 and HA22T cancer cells) were measured to be 105.7 $\pm$ 21.2, 138.8 $\pm$ 16.5, and 184 $\pm$ 20.3 μm s^{-1}, respectively. Overall, the measured maximum velocities of a light image that can manipulate the cells increased with the increase in cellular sizes (Figure 3a,b), which were statistically evaluated to have significant differences ($p < 0.05$). This finding is in line with the fact that the ODEP force acting on a cell is proportional to its size, as described in the Equation (1) [36,41]. As discussed previously, a two-step CTC isolation and purification protocol was proposed in this study for harvesting high-purity and all possible CTCs with different sizes. After the first step process as described earlier, the Jurkat cells, the tested model cells representing the leucocytes in this study, should be labeled with magnetic beads. In this situation, the size or electric property of the Jurkat cells might be altered, which could in turn affect their ODEP manipulation force. Nevertheless, the maximum velocities of a light image that can manipulate the Jurkat cells with or without magnetic beads binding were evaluated to have no significant difference (118 $\pm$ 21.2 and 105.7 $\pm$ 21.2 μm s^{-1}, respectively) ($p > 0.05$) (Figure 3b). This result could be partially due to their similar size (i.e., 14.1 $\pm$ 1 and 13.8 $\pm$ 1.7 μm in diameter for the Jurkat cells with or without magnetic beads binding, respectively) ($p > 0.05$).

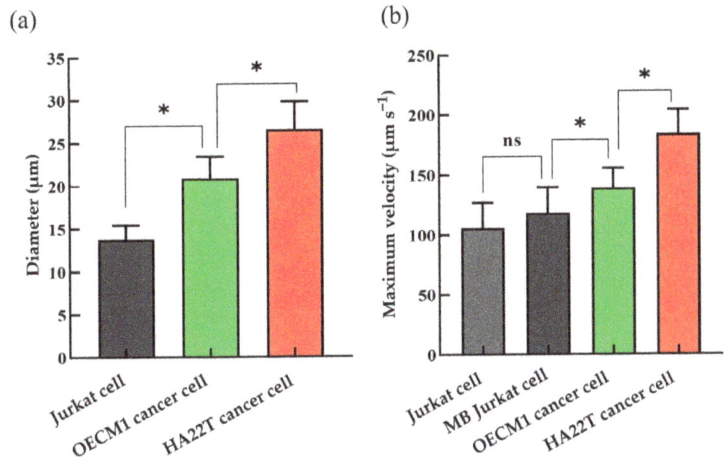

Figure 3. (**a**) The size (diameter) of the model cells (i.e, Jurkat cells and the OECM-1 and HA22T cancer cells) used in this study, and (**b**) evaluation of the maximum velocities of a light image that can manipulate these model cells (i.e., the Jurkat, magnetic microbead-bound Jurkat (MB Jukat), and HA22T, and OECM1 cells). *: significant difference ($p < 0.05$) and ns: not significant.

Apart from the fundamental investigation of the maximum velocities of the light images (100% color brightness) that can manipulate the cells, the similar evaluations were also carried out using the light images with varied percentages (i.e., 60, 70, 80, 90, and 100%) of color brightness. Figure 4a–c exhibited the relationship between the maximum velocities of light images that can manipulate the magnetic microbeads-bound Jurkat, OECM1, and HA22T cells, respectively, under varied percentages of color brightness. The results demonstrated that the ODEP force generated on a cell was influenced by the color brightness of light images (Figure 4a–c; $p < 0.05$). The overall trend showed that the ODEP force generated on a cell increased with the percentage increase in color brightness of light images, although some data points revealed that the increase was not statistically significant. As a whole, this finding is line with the result revealed in our previous study [38,39].

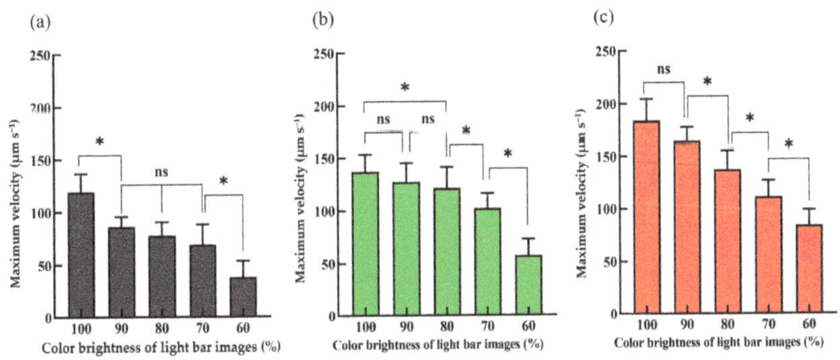

Figure 4. The relationship between the maximum velocities of light images that can manipulate the (**a**) magnetic microbeads-bound Jurkat (MB Jukat), (**b**) OECM1, and (**c**) HA22T cells under varied percentages (60, 70, 80, 90, 100%) of color brightness of light bar images. *: significant difference ($p < 0.05$) and ns: not significant.

3.3. Design of the Light Image Arrays in the ODEP Microfluidic System for the Purification and Sorting of Cancer Cells with Two Different Sizes

After fundamentally understanding the ODEP operation conditions of the manipulated cells under varied percentages of color brightness (Figures 3 and 4), the light image arrays including two dynamic light image arrays and two static light bar arrays were designed for the purification and sorting of cancer cells (i.e., OECM1 and HA22T cells) with different sizes and for the individual collection of the separated cancer cells, respectively, as illustrated in Figure 2. For the two dynamic light image arrays, the dynamic light image array I with multiple light bars (100% color brightness) was designed to dynamically pool the cells, continuously transported to the cell separation zone, to one side of the microchannel (Figure 2b,c). After that, the dynamic light image array II, encompassing multiple light bars with two different (i.e., 100 and 80%) color brightnesses, was designed to first sort and separate the cancer cells (i.e., OECM1 and HA22T cells) from the magnetic microbeads-bound Jurkat cells and was followed by sorting and separating the OECM1 and HA22T cancer cells by their sizes, respectively. To achieve this goal, the moving velocity of the dynamic light bars of the array I was set at 70 $\mu m\ s^{-1}$, which was much lower than the maximum velocities (i.e., 118 $\pm$ 21.2, 138.8 $\pm$ 16.5, and 184 $\pm$ 20.3 $\mu m\ s^{-1}$; Figure 3b) of the light bar images (100% color brightness) that can manipulate the magnetic microbeads-bound Jurkat cells and the OECM1 and HA22T cancer cells, respectively. This design ensured that the three types of cells tested can be effectively manipulated and then gathered to the side of the microchannel. The cells manipulated to the side of the microchannel were then released by the three light bars near the side of the microchannel in the flow direction owing to the gradual reduction in light intensities at the end of three light bar images (please also refer to Figure 2 for illustration). In this study, the cell suspension sample was delivered in the microchannel at the set flow rate of 0.4 $\mu L\ s^{-1}$.

After being released by the dynamic light image array I, the cells further flowed to the zone of dynamic light image array II, by which they were sorted and separated. In this work, the moving velocity of the dynamic light image array II was set at 130 $\mu m\ s^{-1}$, which was higher than the maximum velocities (i.e., 118 $\pm$ 21.2 $\mu m\ s^{-1}$; Figure 3b) of the light bar images that can manipulate the magnetic microbeads-bound Jurkat cells. Under this circumstance, the magnetic microbeads-bound Jurkat cells could not be attracted and manipulated by ODEP and were thus released by the light bars (100% color brightness) near the side of the microchannel in the flow direction, as schematically illustrated in Figure 2d–f. Conversely, the cancer cells (i.e., the OECM-1 and HA22T cancer cells) were manipulated by ODEP due to the fact that the maximum velocities (i.e., 138.8 $\pm$ 16.5, 184 $\pm$ 20.3 $\mu m\ s^{-1}$, respectively; Figure 3) of the light bar images that can manipulate them were higher than the set moving velocity of the dynamic light image array II (i.e., 130 $\mu m\ s^{-1}$). In this situation, the dynamic light bar images (100% color brightness) attracted and pulled the cancer cells to another side of the microchannel (Figure 2d,e). However, when these manipulated cancer cells entered the area where the dynamic light bar images had lower (i.e., 80%) color brightness (Figure 2f), the cancer cells with smaller size (i.e., the OECM1 cancer cells) were not effectively manipulated. This was mainly due to the maximum velocity (i.e., 120.9 $\pm$ 20.4 $\mu m\ s^{-1}$; Figure 4b) of the dynamic light bar images (80% color brightness) that can manipulate the OECM1 cancer cells being lower than the set moving velocity of the dynamic light images (i.e., 130 $\mu m\ s^{-1}$). Therefore, the OECM1 cancer cells were released from the dynamic light image array II and then captured by the static light bar array II (i.e., the three static light bars with 100% color brightness), as illustrated in Figure 2f,g. For the cancer cells with larger size (i.e., HA22T cancer cells), conversely, they were manipulated by the dynamic light bar images with 80% color brightness because the maximum velocity (i.e., 137.1 $\pm$ 17.8 $\mu m\ s^{-1}$; Figure 4c) of such light bar images that can manipulate the HA22T cancer cells was higher than the set moving velocity (i.e., 130 $\mu m\ s^{-1}$) of the dynamic light image array II. As a result, the HA22T cancer cells were manipulated to another side of the microchannel and finally released from the dynamic light bar images with 80% color brightness due to the effect of fluidic flow, as illustrated in Figure 2f. The released

HA22T cancer cells were soon captured and collected by the static light bar array I (i.e., the three static light bars with 80% color brightness) (Figure 2g). Based on the design as abovementioned, the presented approach was capable of first purifying the cancer cells from the magnetic microbeads-bound Jurkat cells, followed by sorting and separating the cancer cells with different sizes. The photograph of the designed light image arrays in the ODEP microfluidic system is shown in Figure 5a (a video clip was provided as the Video S1).

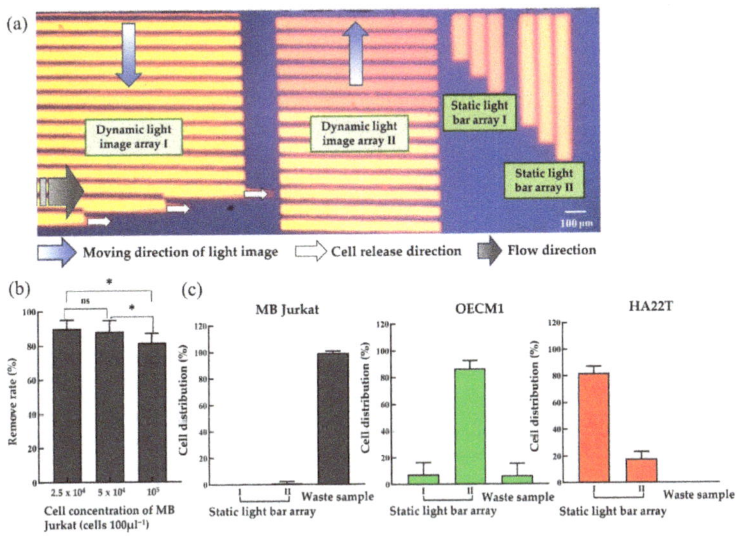

Figure 5. (**a**) The photograph of the designed light image arrays in the ODEP microfluidic system (a video clip was provided as the Video S1), (**b**) the performance (removal rate %) evaluation of the dynamic light image array I to manipulate the magnetic microbead-bound Jurkat cells to one side of the microchannel under different cell concentration conditions, (**c**) the distribution (%) of the magnetic microbeads-bound Jurkat (MB Jukat) (the left figure), OECM1 (the middle figure), and HA22T (the right figure) cells in the static light bar array I and II and in the waste sample after they were individually treated with the proposed ODEP manipulation scheme, as described in Figure 2. *: significant difference ($p < 0.05$) and ns: not significant.

3.4. *Performance Evaluation for the Purification and Sorting of Cancer Cells with Two Different Sizes*

After the dynamic and static light image arrays were designed, their function for the manipulation of the three types of cells was then tested. First, the dynamic light image array I was designed to pool all the cells (the majority of them were magnetic microbead-bound Jurkat cells) to one side of the microchannel. Due to the laminar flow pattern in a microchannel, the cells pooled to one side of the microchannel would flow along the side of the microchannel, making room in the microchannel for the following cell sorting and separation operation carried out by the dynamic light image array II, as illustrated in Figure 2. In order to test the abovementioned function and its working capacity, the magnetic microbead-bound Jurkat cell suspension with different concentrations ($2.5{\sim}10 \times 10^4$ cells 100 μL^{-1}) was loaded in the microfluidic chip. The magnetic microbead-bound Jurkat cells transported to the area of dynamic light image array I was then manipulated. The removal rate of magnetic microbead-bound Jurkat cells under different concentrations of cell suspension was then experimentally evaluated. Within the experimental conditions tested, the result (Figure 5b) showed that the dynamic light image array I was able to effectively manipulate the magnetic microbead-bound Jurkat cells to the side of the microchannel with a high removal rate (88.1~89.8%) under the cell concentration range of $2.5{\sim}5 \times 10^4$

cells 100 μL^{-1}. When the cell concentration was higher than 10^5 cells 100 μL^{-1}, however, the corresponding removal rate might be significantly affected (e.g., the resulting removal rate: 81.5%) ($p < 0.05$). Based on the evaluation, therefore, the cell concentration used in this work was set as 5×10^4 cells 100 μL^{-1}.

After investigating the function and working capacity of the dynamic light image array I, the performance of the dynamic light image array II to sort and separate the cells was then experimentally evaluated. In this work, the magnetic microbead-bound Jurkat, HA22T, and OECM1 cells (concentration: 5×10^4, 5×10^3, and 5×10^3 cells 100 μL^{-1}, respectively) were individually loaded into the ODEP microfluidic chip. The loaded cells were then treated with the proposed ODEP manipulation scheme (Figure 2). The distribution of the cells tested was then evaluated. The results (Figure 5c, the left figure) revealed that almost all the magnetic microbead-bound Jurkat cells were not found in the two static light bar arrays. Conversely, most of them (99.2%) were collected in the waste sample. Similar evaluations were performed for the OECM1 and HA22T cancer cells. The results (Figure 5c, the middle figure) exhibited that 86.4% of OECM1 cancer cells were collected by the static light bar array II (7.3% of them were collected by the static light bar array I). In this test, only 6.3% of OECM1 was found in the waste sample. The similar result (Figure 5c, the right figure) showed that 82% of HA22T cancer cells were collected by the static light bar array I (18% of them were collected by the static light bar array II). In this evaluation, none of HA22T cancer cells were found in the waste sample. These individual evaluations, overall, demonstrated that the proposed ODEP cell manipulation scheme (Figure 2) and the operation conditions adopted in this study are promising to effectively purify the cancer cells from the background cells (i.e., the magnetic microbead-bound Jurkat cells in this study) and to further sort and separate cancer cells with different sizes.

After the function tests, as described above, the performance of the ODEP microfluidic system for the purification and sorting of cancer cells with two different sizes was then evaluated. For identifying the model cells tested in the work, the HA22T and OECM1 cancer cells were pre-stained with CellTrace™ Calcein Red-Orange (red dots) or expressed with GFP (green dots), respectively. Moreover, a tested cell suspension sample containing the HA22T, OECM1 cells, and magnetic microbead-bound Jurkat cells (ratio: 1:1:5) mimicking a blood sample treated with the immunomagnetic beads-based cell separation process was prepared. After that, the prepared cell suspension sample was loaded into the ODEP microfluidic system and processed with the ODEP manipulation scheme (Figure 2) (a video clip was provided as the Video S2). After the operation, the cells collected in the two static light bar arrays were observed via fluorescent microscopic imaging to evaluate their cell purity. Figure 6a exhibited the fluorescent microscopic images of the cells collected in the two static light bar arrays. It can be found that most of the cells in the array I and II are HA22T and OECM1 cells, respectively. This result was in line with the goal to be achieved in this study (Figure 2). Moreover, the purities of the cells in the array I (i.e., the HA22T cancer cells) and II (i.e., the OECM1 cancer cells) were evaluated to be 90.1 ± 11.3 and 93.5 ± 10.7%, respectively. Furthermore, the cells in the two static light bar arrays can then be harvested individually via further ODEP-based cell manipulation. In practical operation, the static light arrays II can be removed to release the smaller cancer cells (e.g., OECM1 cancer cells in this study) by eliminating the light-illuminated ODEP force blocking. After the smaller cancer cells flow away and are collected at an exit hole, the removal of static light array I facilitates the release and collection of larger cancer cells (e.g., HA22T cancer cells in this study). Using the differences in cell release time would be beneficial for the subsequent collection of cancer cells with different sizes in a single exit channel. A similar concept to the abovementioned process has also been demonstrated in our previously published paper [38]. These purified and sorted cancer cells are found to be valuable for the subsequent applied or fundamental cancer research works. Overall, this study has demonstrated that the proposed ODEP microfluidic system was capable of (1) purifying the cancer cells from the magnetic microbead-bound cells, and (2) further separating and sorting the cancer cells with different sizes in an effective manner.

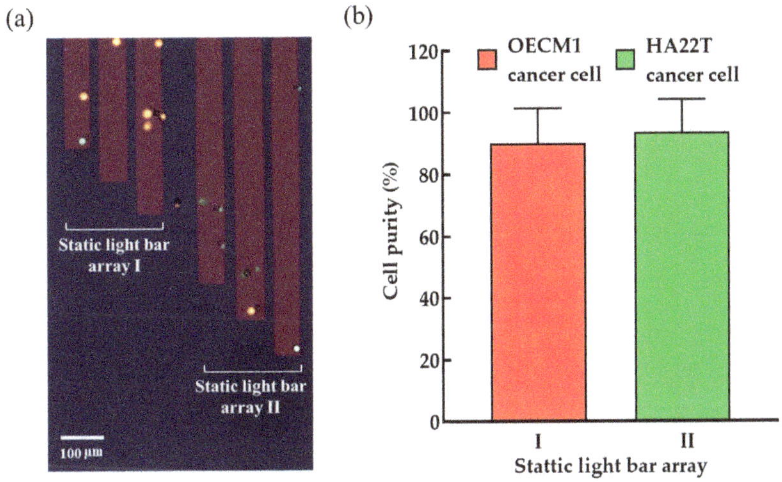

Figure 6. (**a**) The microscopic fluorescent images of the cells collected in the two static light bar arrays (HA22T cancer cells: the red dots; OECM1 cancer cells: the green dots), and (**b**) the evaluated cell purities of the HA22T and OECM1 cancer cells in the static light bar array I and II, respectively.

4. Conclusions

The analysis of CTCs could pave a promising route to comprehensively understand the CTC properties that might be relevant to cancer status. For this goal, the harvest of high-purity and all possible CTCs with different subtypes from the blood samples of cancer patients is crucially important. In addition, the size of CTCs is reported to be relevant to the specific origin of the tumor. Therefore, the initial sorting and separation of CTCs based on their size difference during the CTC isolation and purification process might facilitate the following CTC analytical work. To realize the goals abovementioned, a two-step CTC isolation and purification protocol was proposed. For the first step, the immunomagnetic beads-based cell separation technique was utilized to deplete the majority of blood cells. After that, an ODEP microfluidic system was developed to (1) purify the label-free CTCs from the remaining magnetic microbeads-bound blood cells, and to (2) further sort and separate the CTCs with different sizes. In this study, the ODEP microfluidic system was designed and fabricated. Moreover, the optimum ODEP operating conditions for effective purification and sorting of cancer cells with two different sizes were explored experimentally. Finally, its performance for the purification and sorting of cancer cells with two different sizes was experimentally assessed. The results demonstrated that the presented ODEP-based cell manipulation scheme was able to effectively purify and sort the cancer cells with two different sizes from a tested cell suspension model in a high-purity (achieved purity for the OECM-1 and HA22T cancer cells: 93.5% and 90.1%, respectively) manner. Overall, this study presented an ODEP microfluidic system for the purification and sorting of cancer cells with different sizes. In addition to the application for CTC purification and sorting, the presented method is also useful in other research areas in which the high-purity and label-free purification and sorting of cells with different sizes is required.

Supplementary Materials: The following supporting information can be downloaded at: https://www.mdpi.com/article/10.3390/mi14122170/s1, Video S1: The overall view of two dynamic light image arrays and two static light bar arrays used in this study; Video S2: The close-up view of the dynamic light image array II and two static light bar arrays for sorting and separating the cancer cells with different sizes.

Author Contributions: Conceptualization, P.-Y.C., T.N.A.N., A.-Y.W. and M.-H.W.; methodology, P.-Y.C., T.N.A.N., A.-Y.W., P.-S.H., K.-L.H. and C.-J.L.; formal analysis, investigation, data curation, and validation, P.-Y.C., T.N.A.N. and A.-Y.W.; visualization, P.-Y.C., T.N.A.N., A.-Y.W., P.-S.H. and K.-L.H.; writing—original draft preparation and original manuscript writing, P.-Y.C., T.N.A.N. and A.-Y.W.; writing—review and editing, C.-H.H. and M.-H.W.; supervision, M.-H.W. All authors have read and agreed to the published version of the manuscript.

Funding: This work was funded by the National Science and Technology Council, R.O.C. (110-2221-E-182-004-MY3, 110-2811-E-182-507, 111-2811-E-182-009) and the Chang Gung Memorial Hospital (CMRPD1M0561, CMRPD1M0562, CMRPD2M0291).

Data Availability Statement: Data are contained within the article.

Acknowledgments: We would like to thank Kwang-Huei Lin for providing the liver cancer cell lines (i.e., HA22T cancer cells) for our research work.

Conflicts of Interest: The authors declare no conflict of interest.

References

1. Fares, J.; Fares, M.Y.; Khachfe, H.H.; Salhab, H.A.; Fares, Y. Molecular principles of metastasis: A hallmark of cancer revisited. *Signal Transduct. Target. Ther.* **2020**, *5*, 28. [CrossRef]
2. Crosby, D.; Bhatia, S.; Brindle, K.M.; Coussens, L.M.; Dive, C.; Emberton, M.; Esener, S.; Fitzgerald, R.C.; Gambhir, S.S.; Kuhn, P.; et al. Early detection of cancer. *Science* **2022**, *375*, eaay9040. [CrossRef]
3. Ashworth, T.R. A case of cancer in which cells similar to those in the tumours were seen in the blood after death. *Aust. Med. J.* **1869**, *14*, 146.
4. De Wit, S.; van Dalum, G.; Terstappen, L.W. Detection of circulating tumor cells. *Scientifica* **2014**, *2014*, 819362. [CrossRef] [PubMed]
5. Duan, G.-C.; Zhang, X.-P.; Wang, H.-E.; Wang, Z.-K.; Zhang, H.; Yu, L.; Xue, W.-F.; Xin, Z.-F.; Hu, Z.-H.; Zhao, Q.-T. Circulating tumor cells as a screening and diagnostic marker for early-stage non-small cell lung cancer. *OncoTargets Ther.* **2020**, *13*, 1931–1939. [CrossRef]
6. Riethdorf, S.; O'Flaherty, L.; Hille, C.; Pantel, K. Clinical applications of the CellSearch platform in cancer patients. *Adv. Drug Deliv. Rev.* **2018**, *125*, 102–121. [CrossRef]
7. Li, Z.; Xu, K.; Tartarone, A.; Santarpia, M.; Zhu, Y.; Jiang, G. Circulating tumor cells can predict the prognosis of patients with non-small cell lung cancer after resection: A retrospective study. *Transl. Lung Cancer Res.* **2021**, *10*, 995–1006. [CrossRef] [PubMed]
8. Yu, M.; Bardia, A.; Wittner, B.S.; Stott, S.L.; Smas, M.E.; Ting, D.T.; Isakoff, S.J.; Ciciliano, J.C.; Wells, M.N.; Shah, A.M.; et al. Circulating breast tumor cells exhibit dynamic changes in epithelial and mesenchymal composition. *Science* **2013**, *339*, 580–584. [CrossRef]
9. Zhou, J.; Dong, F.; Cui, F.; Xu, R.; Tang, X. The role of circulating tumor cells in evaluation of prognosis and treatment response in advanced non-small-cell lung cancer. *Cancer Chemother. Pharmacol.* **2017**, *79*, 825–833. [CrossRef] [PubMed]
10. Deutsch, T.M.; Stefanovic, S.; Feisst, M.; Fischer, C.; Riedel, F.; Fremd, C.; Domschke, C.; Pantel, K.; Hartkopf, A.D.; Sutterlin, M.; et al. Cut-Off Analysis of CTC Change under Systemic Therapy for Defining Early Therapy Response in Metastatic Breast Cancer. *Cancers* **2020**, *12*, 1055. [CrossRef]
11. Chiu, S.Y.; Hsieh, C.H.; You, J.F.; Chu, P.Y.; Hung, H.Y.; Chu, P.H.; Wu, M.H. Enhancing Prediction Performance by Add-On Combining Circulating Tumor Cell Count, CD45^{neg} EpCAM^{neg} Cell Count on Colorectal Cancer, Advance, and Metastasis. *Cancers* **2021**, *13*, 2521. [CrossRef]
12. Hu, C.-L.; Zhang, Y.-J.; Zhang, X.-F.; Fei, X.; Zhang, H.; Li, C.-G.; Sun, B. 3D Culture of Circulating Tumor Cells for Evaluating Early Recurrence and Metastasis in Patients with Hepatocellular Carcinoma. *OncoTargets Ther.* **2021**, *14*, 2673–2688. [CrossRef]
13. Keller, L.; Pantel, K. Unravelling tumour heterogeneity by single-cell profiling of circulating tumour cells. *Nat. Rev. Cancer* **2019**, *19*, 553–567. [CrossRef]
14. Bulfoni, M.; Turetta, M.; Del Ben, F.; Di Loreto, C.; Beltrami, A.P.; Cesselli, D. Dissecting the heterogeneity of circulating tumor cells in metastatic breast cancer: Going far beyond the needle in the haystack. *Int. J. Mol. Sci.* **2016**, *17*, 1775. [CrossRef]
15. Hwang, W.L.; Pleskow, H.M.; Miyamoto, D.T. Molecular analysis of circulating tumors cells: Biomarkers beyond enumeration. *Adv. Drug Deliv. Rev.* **2018**, *125*, 122–131. [CrossRef]
16. Kolovskaya, O.S.; Zyuzyukina, A.V.; Dassie, J.P.; Zamay, G.S.; Zamay, T.N.; Boyakova, N.V.; Khorzhevskii, V.A.; Kirichenko, D.A.; Lapin, I.N.; Shchugoreva, I.A.; et al. Monitoring of breast cancer progression via aptamer-based detection of circulating tumor cells in clinical blood samples. *Front. Mol. Biosci.* **2023**, *10*, 1184285. [CrossRef] [PubMed]
17. Lisencu, L.A.; Bonci, E.A.; Irimie, A.; Balacescu, O.; Lisencu, C. The Role of Circulating Tumor Cells in Chemoresistant Metastatic Breast Cancer. *J. Clin. Med.* **2021**, *10*, 684. [CrossRef] [PubMed]
18. Lowes, L.E.; Allan, A.L. Recent advances in the molecular characterization of circulating tumor cells. *Cancers* **2014**, *6*, 595–624. [CrossRef] [PubMed]

19. Hong, B.; Zu, Y. Detecting circulating tumor cells: Current challenges and new trends. *Theranostics* **2013**, *3*, 377–394. [CrossRef] [PubMed]
20. Espejo-Cruz, M.L.; Gonzalez-Rubio, S.; Zamora-Olaya, J.; Amado-Torres, V.; Alejandre, R.; Sanchez-Frias, M.; Ciria, R.; De la Mata, M.; Rodriguez-Peralvarez, M.; Ferrin, G. Circulating Tumor Cells in Hepatocellular Carcinoma: A Comprehensive Review and Critical Appraisal. *Int. J. Mol. Sci.* **2021**, *22*, 13073. [CrossRef] [PubMed]
21. Low, W.S.; Wan Abas, W.A. Benchtop technologies for circulating tumor cells separation based on biophysical properties. *BioMed Res. Int.* **2015**, *2015*, 239362. [CrossRef] [PubMed]
22. Murlidhar, V.; Rivera-Baez, L.; Nagrath, S. Affinity Versus Label-Free Isolation of Circulating Tumor Cells: Who Wins? *Small* **2016**, *12*, 4450–4463. [CrossRef] [PubMed]
23. Banko, P.; Lee, S.Y.; Nagygyorgy, V.; Zrinyi, M.; Chae, C.H.; Cho, D.H.; Telekes, A. Technologies for circulating tumor cell separation from whole blood. *J. Hematol. Oncol.* **2019**, *12*, 48. [CrossRef]
24. Gabriel, M.T.; Calleja, L.R.; Chalopin, A.; Ory, B.; Heymann, D. Circulating Tumor Cells: A Review of Non-EpCAM-Based Approaches for Cell Enrichment and Isolation. *Clin. Chem.* **2016**, *62*, 571–581. [CrossRef] [PubMed]
25. Chen, L.C.; Bode, A.M.; Dong, Z.G. Circulating Tumor Cells: Moving Biological Insights into Detection. *Theranostics* **2017**, *7*, 2606–2619. [CrossRef] [PubMed]
26. Rostami, P.; Kashaninejad, N.; Moshksayan, K.; Saidi, M.S.; Firoozabadi, B.; Nguyen, N.T. Novel approaches in cancer management with circulating tumor cell clusters. *J. Sci. Adv. Mater. Dev.* **2019**, *4*, 1–18. [CrossRef]
27. Zhao, W.; Liu, Y.; Jenkins, B.D.; Cheng, R.; Harris, B.N.; Zhang, W.; Xie, J.; Murrow, J.R.; Hodgson, J.; Egan, M.; et al. Tumor antigen-independent and cell size variation-inclusive enrichment of viable circulating tumor cells. *Lab Chip* **2019**, *19*, 1860–1876. [CrossRef]
28. Dent, B.M.; Ogle, L.F.; O'Donnell, R.L.; Hayes, N.; Malik, U.; Curtin, N.J.; Boddy, A.V.; Plummer, E.R.; Edmondson, R.J.; Reeves, H.L.; et al. High-resolution imaging for the detection and characterisation of circulating tumour cells from patients with oesophageal, hepatocellular, thyroid and ovarian cancers. *Int. J. Cancer* **2016**, *138*, 206–216. [CrossRef]
29. Mendelaar, P.A.J.; Kraan, J.; Van, M.; Zeune, L.L.; Terstappen, L.; Oomen-de Hoop, E.; Martens, J.W.M.; Sleijfer, S. Defining the dimensions of circulating tumor cells in a large series of breast, prostate, colon, and bladder cancer patients. *Mol. Oncol.* **2021**, *15*, 116–125. [CrossRef]
30. Ligthart, S.T.; Coumans, F.A.; Bidard, F.C.; Simkens, L.H.; Punt, C.J.; de Groot, M.R.; Attard, G.; de Bono, J.S.; Pierga, J.Y.; Terstappen, L.W. Circulating Tumor Cells Count and Morphological Features in Breast, Colorectal and Prostate Cancer. *PLoS ONE* **2013**, *8*, e67148. [CrossRef]
31. Yousuff, C.M.; Ho, E.T.W.; Hussain, K.I.; Hamid, N.H.B. Microfluidic Platform for Cell Isolation and Manipulation Based on Cell Properties. *Micromachines* **2017**, *8*, 15. [CrossRef]
32. Wu, M.; Ozcelik, A.; Rufo, J.; Wang, Z.; Fang, R.; Jun Huang, T. Acoustofluidic separation of cells and particles. *Microsyst. Nanoeng.* **2019**, *5*, 32. [CrossRef] [PubMed]
33. Abd Rahman, N.; Ibrahim, F.; Yafouz, B. Dielectrophoresis for Biomedical Sciences Applications: A Review. *Sensors* **2017**, *17*, 449. [CrossRef] [PubMed]
34. Zhang, S.; Xu, B.; Elsayed, M.; Nan, F.; Liang, W.; Valley, J.K.; Liu, L.; Huang, Q.; Wu, M.C.; Wheeler, A.R. Optoelectronic tweezers: A versatile toolbox for nano-/micro-manipulation. *Chem. Soc. Rev.* **2022**, *51*, 9203–9242. [CrossRef] [PubMed]
35. Chiou, P.Y.; Ohta, A.T.; Wu, M.C. Massively parallel manipulation of single cells and microparticles using optical images. *Nature* **2005**, *436*, 370–372. [CrossRef] [PubMed]
36. Chu, P.Y.; Hsieh, C.H.; Wu, M.H. The Combination of Immunomagnetic Bead-Based Cell Isolation and Optically Induced Dielectrophoresis (ODEP)-Based Microfluidic Device for the Negative Selection-Based Isolation of Circulating Tumor Cells (CTCs). *Front. Bioeng. Biotechnol.* **2020**, *8*, 921. [CrossRef] [PubMed]
37. Chou, W.P.; Wang, H.M.; Chang, J.H.; Chiu, T.K.; Hsieh, C.H.; Liao, C.J.; Wu, M.H. The utilization of optically-induced-dielectrophoresis (ODEP)-based virtual cell filters in a microfluidic system for continuous isolation and purification of circulating tumour cells (CTCs) based on their size characteristics. *Sens. Actuators B Chem.* **2017**, *241*, 245–254. [CrossRef]
38. Yang, C.-M.; Wu, A.-Y.; Yu, J.-C.; Chu, P.-Y.; Hsieh, C.-H.; Wu, M.-H. Virtual Filter Membranes in a Microfluidic System for Sorting and Separating Size-Based Micro Polystyrene Beads by Illumination Intensity Design in Optically Induced Dielectrophoresis (ODEP). *Chemosensors* **2022**, *10*, 540. [CrossRef]
39. Yang, C.-M.; Chu, P.-Y.; Wu, A.-Y.; Hsieh, P.-H.; Hsieh, C.-H.; Wu, M.-H. Development of an optically induced dielectrophoresis (ODEP) microfluidic system with a virtual gel filtration chromatography (GFC)-inspired mechanism for the high-performance sorting and separation of microparticles based on their size differences. *Sens. Actuators B Chem.* **2023**, *395*, 134443. [CrossRef]
40. Filippi, J.; Di Giuseppe, D.; Casti, P.; Mencattini, A.; Antonelli, G.; D'Orazio, M.; Corsi, F.; Della-Morte Canosci, D.; Ghibelli, L.; Witte, C.; et al. Exploiting spectral information in Opto-Electronic Tweezers for cell classification and drug response evaluation. *Sens. Actuators B Chem.* **2022**, *368*, 132200. [CrossRef]

41. Chu, P.Y.; Hsieh, C.H.; Lin, C.R.; Wu, M.H. The Effect of Optically Induced Dielectrophoresis (ODEP)-Based Cell Manipulation in a Microfluidic System on the Properties of Biological Cells. *Biosensors* **2020**, *10*, 65. [CrossRef] [PubMed]
42. Rushton, A.J.; Nteliopoulos, G.; Shaw, J.A.; Coombes, R.C. A Review of Circulating Tumour Cell Enrichment Technologies. *Cancers* **2021**, *13*, 970. [CrossRef] [PubMed]

micromachines

Article

Automatic Single-Cell Harvesting for Fetal Nucleated Red Blood Cell Isolation on a Self-Assemble Cell Array (SACA) Chip

Hsin-Yu Yang [1,2], Che-Hsien Lin [1], Yi-Wen Hu [1], Chih-Hsuan Chien [1], Mu-Chi Huang [1], Chun-Hao Lai [1], Jen-Kuei Wu [3] and Fan-Gang Tseng [1,4,5,*]

1 Department of Engineering and System Science, National Tsing Hua University, Hsinchu 30013, Taiwan
2 Nano Science and Technology Program, Taiwan International Graduate Program, Academia Sinica and National Tsing Hua University, Taipei 11529, Taiwan
3 Biomedical Science and Engineering Center, National Tsing Hua University, Hsinchu 30013, Taiwan
4 Institute of Nano Engineering and Micro Systems (NEMS), National Tsing Hua University, Hsinchu 30013, Taiwan
5 Frontier Research Center on Fundamental and Applied Sciences of Matters, National Tsing Hua University, Hsinchu 30013, Taiwan
* Correspondence: fangang@ess.nthu.edu.tw

Abstract: (1) Background: Fetal chromosomal examination is a critical component of modern prenatal testing. Traditionally, maternal serum biomarkers such as free β-human chorionic gonadotropin (Free β-HCG) and pregnancy-associated plasma protein A (PAPPA) have been employed for screening, achieving a detection rate of approximately 90% for fetuses with Down syndrome, albeit with a false positive rate of 5%. While amniocentesis remains the gold standard for the prenatal diagnosis of chromosomal abnormalities, including Down syndrome and Edwards syndrome, its invasive nature carries a significant risk of complications, such as infection, preterm labor, or miscarriage, occurring at a rate of 7 per 1000 procedures. Beyond Down syndrome and Edwards syndrome, other chromosomal abnormalities, such as trisomy of chromosomes 9, 16, or Barr bodies, pose additional diagnostic challenges. Non-invasive prenatal testing (NIPT) has emerged as a powerful alternative for fetal genetic screening by leveraging maternal blood sampling. However, due to the extremely low abundance of fetal cells in maternal circulation, NIPT based on fetal cells faces substantial technical challenges. (2) Methods: Fetal nucleated red blood cells (FnRBCs) were first identified in maternal circulation in a landmark study published in *The Lancet* in 1959. Due to their fetal origin and presence in maternal peripheral blood, FnRBCs represent an ideal target for non-invasive prenatal testing (NIPT). In this study, we introduce a novel self-assembled cell array (SACA) chip system, a microfluidic-based platform designed to efficiently settle and align cells into a monolayer at the chip's base within five minutes using lateral flow dynamics and gravity. This system is integrated with a fully automated, multi-channel fluorescence scanning module, enabling the real-time imaging and molecular profiling of fetal cells through fluorescence-tagged antibodies. By employing a combination of Hoechst+/CD71+/HbF+/CD45− markers, the platform achieves the precise enrichment and isolation of FnRBCs at the single-cell level from maternal peripheral blood. (3) Results: The SACA chip system effectively reduces the displacement of non-target cells by 31.2%, achieving a single-cell capture accuracy of 97.85%. This isolation and enrichment system for single cells is well suited for subsequent genetic analysis. Furthermore, the platform achieves a high purity of isolated cells, overcoming the concentration detection limit of short tandem repeat (STR) analysis, demonstrating its capability for reliable non-invasive prenatal testing. (4) Conclusions: This study demonstrates that the SACA chip, combined with an automated image positioning system, can efficiently isolate single fetal nucleated red blood cells (FnRBCs) from 50 million PBMCs in 2 mL of maternal blood, completing STR analysis within 120 min. With higher purification efficiency compared to existing NIPT methods, this platform shows great promise for prenatal diagnostics and potential applications in other clinical fields.

Keywords: FnRBC; single-cell isolation; STR; NIPT

1. Introduction

The increasing maternal age in many parts of the world has significantly elevated the demand for prenatal examinations, which have become a cornerstone of maternal–fetal medicine. Over the past decade, technological advancements have revolutionized prenatal diagnostics, enabling earlier and more accurate evaluations of fetal health. Among these, serum-based biomarkers such as pregnancy-associated plasma protein-A (PAPP-A), first introduced in 1993, have served as predictors of pregnancy complications [1]. When combined with Free β-HCG, these markers achieved detection rates for trisomy 21 of approximately 90%, with false positive rates between 2% and 3%, depending on gestational age [2]. However, the scope of detectable conditions using these biomarkers remained narrow, paving the way for the development of non-invasive prenatal testing (NIPT) using cell-free DNA (cfDNA).

CfDNA-based NIPT involves the massively parallel sequencing of DNA fragments extracted from maternal plasma, allowing for the detection of fetal chromosomal abnormalities and small (>5 Mb) sub-chromosomal copy number changes. The abundance of cfDNA in maternal circulation and its relatively straightforward extraction process are notable advantages, making it one of the most widely used methods in clinical practice. However, cfDNA also has inherent limitations. Its short half-life (<2 h) and low fetal fraction (5–20%) can lead to false negative results. Additionally, cfDNA-based NIPT is considered a screening test, necessitating confirmation through invasive diagnostic methods [3]. While cfDNA demonstrates high sensitivity and specificity for detecting Down syndrome, it has slightly reduced sensitivity for Edwards and Patau syndromes. Moreover, the short fragment size of cfDNA restricts its utility in detecting a broader spectrum of genetic abnormalities [4].

An alternative to cfDNA for NIPT is found in fetal nucleated red blood cells (FnRBCs), which were first identified in maternal circulation in 1959 (*The Lancet*) and successfully isolated in 1990 (PNAS) [5]. FnRBCs have intact nuclei, offering a more comprehensive genetic blueprint compared to fragmented cfDNA. Early studies reported detection rates of 41.4% for fetal gender determination and 74.4% for chromosomal aneuploidy, with false positive rates ranging from 0.6% to 11.1% [6]. Despite its potential, FnRBC-based NIPT has faced challenges due to the rarity of these cells in maternal blood and technical difficulties in isolating them with sufficient purity for reliable analysis.

Cell sorting methods for FnRBCs can be categorized into fluorescence marker-based techniques [4,5], microbead capture [6–8], and marker-free physical sorting [9,10]. Fluorescence-based methods use immunofluorescent antibodies to label target cells, often in conjunction with flow cytometry, which enables sorting based on signal detection. Microbead capture methods employ antibody-coated beads, including magnetic beads, to selectively isolate FnRBCs. Marker-free physical sorting leverages intrinsic cell properties such as size, deformability, and density to distinguish and isolate specific cell populations [11]. Each of these approaches has strengths but also limitations, such as low throughput, time inefficiency, and high operational costs. Additionally, issues like contamination risks, difficulty in maintaining sterility, and large equipment size pose barriers to the widespread adoption of these methods in clinical settings.

Efforts to improve the precision, speed, and efficiency of cell sorting platforms have resulted in new technologies capable of addressing these challenges. Modern platforms aim to enhance throughput, reduce contamination risks, and streamline operations, while ensuring high purity and minimal cell loss. For example, advancements in automated imaging systems integrated with cell sorting mechanisms have significantly improved the performance of NIPT workflows. These systems not only enhance sorting purity but also minimize human error and reduce aerosol exposure, improving safety in laboratory settings. Despite these innovations, the low abundance of FnRBCs in maternal circulation remains a significant barrier to the broader application of cell-based NIPT.

During pregnancy, maternal erythrocytes may express CD71 (transferrin receptor) under specific physiological and pathological conditions, which can interfere with FnRBC isolation. For instance, increased erythropoiesis in pregnancy-related anemia elevates the proportion of reticulocytes that exhibit higher levels of CD71. Similarly, inflammatory conditions like preeclampsia disrupt erythropoiesis and lead to CD71 expression in maternal erythrocytes [12]. Additionally, fetal–maternal hemorrhage introduces fetal erythrocytes with naturally high CD71 expression into maternal circulation, complicating the differentiation of FnRBCs from maternal cells [13]. To address these challenges, combining specific markers such as HbF, CD71, and CD CD45− has been proposed to enhance the specificity of FnRBC isolation [14].

Moreover, selecting appropriate fluorescence-labeled antibodies is critical for improving the accuracy of FnRBC detection. CD147, for example, has been suggested as a potential marker for FnRBCs due to its expression in fetal cells [8,15]. However, its expression in other maternal blood components, including erythrocytes, leukocytes (e.g., lymphocytes and monocytes), extracellular vesicles, and endothelial cells, can lead to background signal interference, necessitating the careful optimization of antibody selection [16,17].

This study aims to advance the use of FnRBCs as a reliable target for NIPT by integrating a novel self-assembled cell array (SACA) chip with an automated imaging and sorting system. Our approach leverages a combination of HbF+, CD71+, and CD45− markers to achieve the high-purity isolation of FnRBCs, overcoming the limitations of traditional sorting methods. By utilizing this advanced microfluidic platform, we propose a robust and efficient solution for non-invasive prenatal diagnostics, offering the potential for more comprehensive genetic analysis and improved clinical outcomes.

2. Materials and Methods

2.1. Materials

Lymphoprep™ (PG-1114547-1) was obtained from Blossom Biotechnologies Inc., Taipei, Taiwan. ε-Polylysine was purchased from the JNC Corporation, Tokyo, Japan. PE Mouse Anti-Human Fetal Hemoglobin (560041) was supplied by BD Pharmingen through the UNIMED Corp., Taipei, Taiwan. Human TRA-1-85/CD147 Alexa Fluor® 350-conjugated Antibody (FAB3195U) and APC Anti-Human CD45 Antibody (AF18395) were acquired from the TAQKEY Corp., Taipei, Taiwan. FITC Anti-CD71 (ARG62923) was sourced from the Arigo Biolaboratories Corp., Taipei, Taiwan. Hoechst dye was purchased from ABP Biosciences, LLC., Taipei, Taiwan.

2.2. Blood Sample Pretreatment

The blood samples were processed according to a density gradient centrifugation protocol. First, a 2 mL blood sample was centrifuged horizontally at 800× *g* for 15 min, at 4 °C to separate the peripheral blood mononuclear cells (PBMC). The mononuclear cells containing FnRBCs were transferred into a 15 mL centrifuge tube and centrifuged horizontally at 300× *g* for 10 min. The supernatant was removed and PBS solution of 100 µL total volume was added. The process flow is shown in Figure S1.

2.3. FnRBC Immunofluorescent Staining

Immunofluorescent staining was performed to identify FnRBCs. After blood sample pretreatment, the cells were incubated in immunofluorescent staining solution containing 20 µL PE-conjugated antibody HbF (0.2% *w*/*v*), 20 µL APC-conjugated antibody CD45 (0.2% *w*/*v*), and 20 µL of FITC-conjugated antibody CD71 (0.2% *w*/*v*) in the dark at room temperature for 30 min. The cells were nucleus-stained with Hoechst (0.1 mg/mL in deionized water) for 10 min and then washed with deionized water.

2.4. ε-Polylysine Coating

The SACA chip was treated with O_2 plasma for 5 min, followed by the addition of 20 µL of ε-polylysine into its wells. The system was left undisturbed for 10 min and subsequently baked in an oven at 60 °C for 2 h.

2.5. FnRBC Screening, Isolation, and Release

After staining the cells, all cells were loaded into the SACA chip and incubate at room temperature for 5 min until the cells settled [18]. The automated cell imaging and sorting system began to perform three types of fluorescent and visible light scans. The operator performs FnRBC identification, location, and counting based on the three types of fluorescent image signals (Hoechst+/CD71+/HbF+/CD45−) after scanning and the cell types are observed by bright field. The cell isolation device moves to the FnRBCs' position according to the coordinates selected by the user and is fitted with a high-precision micron-level syringe pump for the selection and storage of cells in the corresponding container for subsequent detection and analysis. The complete process flow is shown in Figure S1.

2.6. Whole Genome Amplification (WGA)

Whole genome amplification (WGA) was performed using the QIAGEN REPLI-G Kit (Qiagen Inc., Redwood City, CA, USA) following the manufacturer's protocol with slight modifications. The sorted target cells were transferred into a PCR tube and centrifuged to collect the cell pellet. After centrifugation, the supernatant was carefully removed, leaving approximately 4 µL of liquid in the tube. Subsequently, 3 µL of Buffer D2 was added to the sample, mixed thoroughly, and incubated on a heating block at 65 °C for 10 min to facilitate cell lysis. After incubation, 3 µL of stop solution was added to the reaction, and the sample was placed on ice to halt the lysis process. Next, 40 µL of a master mix, consisting of an amplification buffer and polymerase, was added to the sample and mixed thoroughly to initiate the WGA reaction. The sample was incubated at 30 °C for 8 h on a heating block to allow for DNA amplification. Following the amplification step, the reaction was terminated by incubating the sample at 65 °C for 3 min. After the WGA process, the amplified DNA was analyzed for quality and concentration. This was achieved by measuring the DNA concentration and absorbance values (OD260/280) using an enzyme-linked immunosorbent assay (ELISA) reader. These steps ensured the accuracy and efficiency of the DNA amplification process for subsequent analyses.

2.7. Short Tandem Repeats (STR)

This experiment was conducted by the Genelabs Life Science Corp., New Taipei City, Taiwan, using their specialized testing services. The short tandem repeat (STR) analysis was performed using the Promega GenePrint® 24 System kit, and the raw data was analyzed with GeneMapper® Software V3.7. DNA samples were extracted and prepared following standard protocols to ensure sufficient concentration and purity for analysis. PCR amplification was performed using the Promega GenePrint® 24 System kit specified reaction mixture and thermal cycling conditions to amplify the 24 STR loci. The amplified products were then subjected to capillary electrophoresis to separate STR fragments by size, following the manufacturer's guidelines. The resulting raw electrophoresis data were imported into GeneMapper® Software V3.7, where the STR alleles for each locus were identified and sized. Positive and negative controls were included to ensure accuracy and reliability. After the verification of the allele calls and consistency across loci, the STR profiles were compiled and documented for downstream applications, such as fetal–maternal comparisons or genetic analysis. All procedures were conducted under strict quality control to maintain the integrity of the results.

3. Results

3.1. Cell Isolation Device

The self-assembled cell array (SACA) chip, previously developed and published by our team [19], is an innovative microfluidic platform designed for the efficient enrichment and immunolabeling of rare cells from biological samples. By leveraging gravity and lateral flow dynamics, the SACA chip enables the rapid formation of a monolayer of cells without requiring external fluid control equipment, significantly simplifying the cell detection process. The resulting single-cell layer is illustrated in Figure S2.

The cell sorting device comprises a high-precision syringe pump (SP101 was purchased by Yotec Instruments Co., Ltd., Taipei, Taiwan), a three-axis motor (Micromanipulator-MP285 was purchased by the Sutter Instrument Company, Novato, CA, USA), a cell needle holder, and a cell needle (Figure 1a). The syringe pump precisely controls the flow rate and volume during cell picking and releasing, which can be adjusted based on the number of target cells. For single-cell isolation, the flow rate is typically set at approximately 40 μL/min, while higher rates are required for cell clusters. The three-axis motor, with a movement accuracy of 1 μm, ensures the precise positioning of the target cell. The cell needle, fabricated from a glass capillary tube, features inner and outer diameters determined by the length of the truncated needle tip, ranging from a minimum inner diameter of 7 μm to a maximum of 50 μm (Figure 1b).

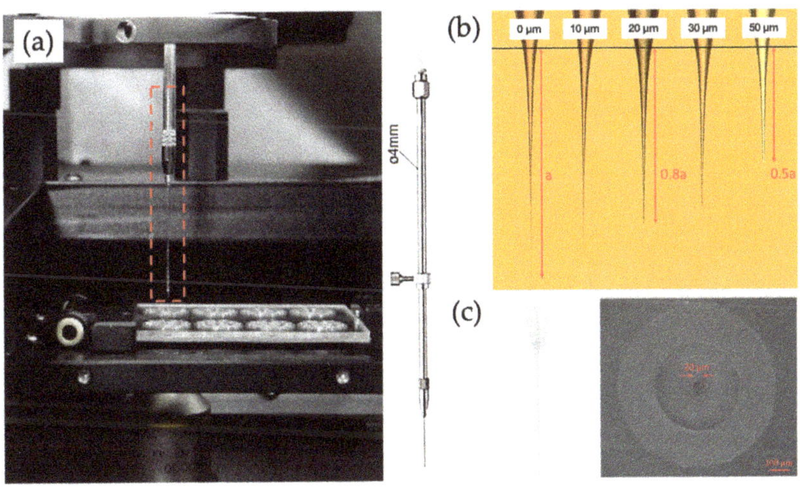

Figure 1. Cell isolation device. (**a**) The view inside the automatic cell image scanning and isolation system. The red cube represents the cell needle holder and the cell needle. (**b**) The relationship between the length of the needle tip and the inner diameter of the drawn glass capillary tube. (**c**) Structure of the plastic needle enclosing the glass needle and the microscopic field of view. Scale bar: 100 μm.

To integrate the newly developed automated image recognition software, we have further enhanced the system. The upgraded design incorporates a plastic needle to prevent damage to the chip while stabilizing the lateral and vertical positioning of the glass needle. This modification reduces the needle diameter from 400 μm to 20 μm—a 20-fold reduction—making it ideal for isolating target cells (FnRBCs) with diameters of approximately 12–18 μm, as shown in Figure 1c.

The system first identifies the target cell's position and uses a high-precision three-axis motor to maneuver the cell isolation device to the designated location. The device then employs precise control of the volume and flow rate to accurately aspirate the cell and release it into the corresponding storage container. The entire process of cell uptake and release is illustrated in Figure 2. This sequence of images demonstrates the isolation of a single FnRBC from a specific well of the SACA chip and its subsequent transfer to an empty well, ensuring precise and contamination-free manipulation.

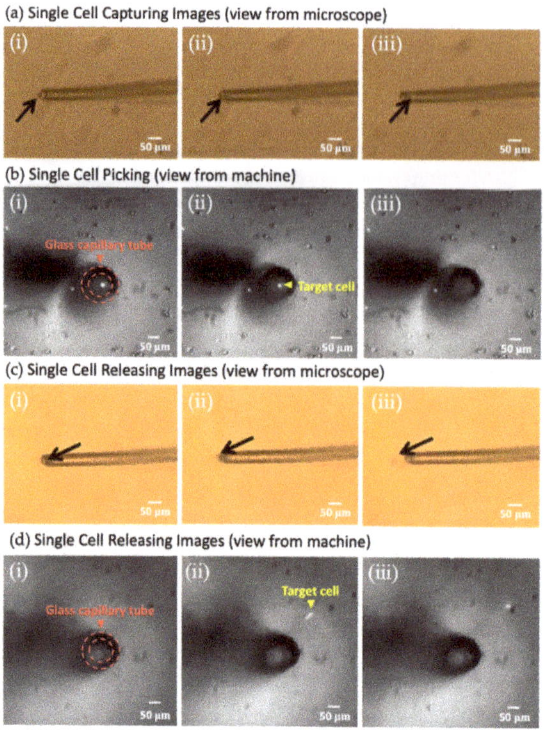

Figure 2. The process of single-cell isolation. (**a**,**b**) The process of single-cell picking. (**c**,**d**) The process of single-cell releasing. (i), (ii), and (iii) represent the processes of cells being aspirated or released, respectively. In Figures (**a**,**c**), the black arrows indicate the positions of cell movement within the microscopic field of view, while Figures (**b**,**d**) depict the top-down view of cell movement within the device.

In order to improve the accuracy of identifying FnRBCs, we incorporated ε-Polylysine (ε-PL) molecules synthesized by the JNC Corporation into our device. The ε-PL demonstrates a unique capability of forming NH_2 functional groups upon binding to polycarbonate (PC) surfaces, imparting a positive charge to the surface. Given that cells exhibit a net negative charge in phosphate-buffered saline (PBS), the positively charged NH_2 functional groups on the modified PC surface facilitate effective cell adhesion. As illustrated in Figure S3a, this modification allows for the efficient immobilization of cells on the chip surface, enabling straightforward retrieval by gently tapping the cells with a micropipette. As shown in Figure S3b, compared to the previously used F127, ε-PL coating effectively immobilizes most cells. Additionally, it reduces the number of non-target cells captured, resulting in a 31.2% decrease in the number of displaced cells during the cell-capturing process.

3.2. Precision of Single-Cell Isolation

To evaluate the sorting capability of the device, we conducted a precision quantification test for single-cell sorting. We specified the target cell number (1–8 cells) of the well of the SACA chip and isolated the cells from the well containing immunofluorescence-stained cells until we achieved the number of the target cell. After sorting, we performed automatic image scanning to calculate the actual number of cells in the well. The accuracy is evaluated through the relative error (Er), which is defined as $Er = (Xi - Xt)/Xt \times 100\%$, where Xi is the actual measured value and Xt is the target value. Figure 3a is the result after we have isolated the target cells in the new well, and Figure 3b is the statistical result of the number of specified target cells we have isolated five times. It can be seen from the graph that when the number of sorted cells is 1–4, the accuracy is 100%. When the sorting number is between 5–8 cells, there is a difference of ±2 cells between the actual number and the target number. According to the results of the experiments and calculations, the average relative error is about 2.15%, which means that this system has a high cell isolation ability with a sorting accuracy of 97.85%.

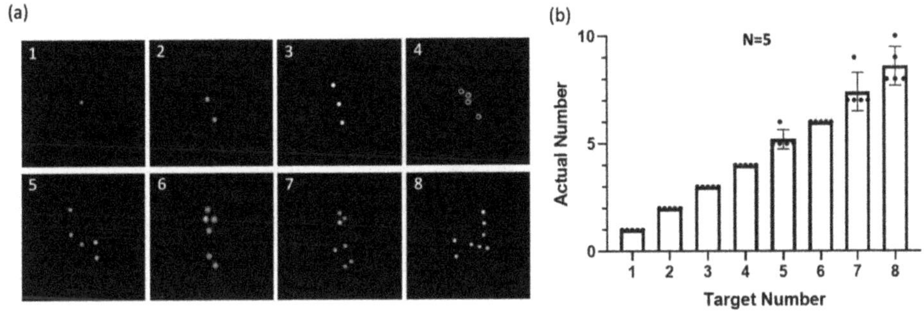

Figure 3. Precision test of single-cell isolation using the SACA system. (**a**) Representative images showing isolated cells from different target number groups (1–8 cells). (**b**) Statistical analysis of the number of FnRBCs isolated, repeated five times for each target group. The results demonstrate high precision, with isolated cell numbers closely matching the target numbers across all groups (n = 5). Error bars indicate the standard deviation.

In addition, to evaluate the detection limit of FnRBCs on the automated system, we conducted an experiment to simulate conditions for sorting rare target cells, enabling potential application for pregnant samples at earlier gestational stages. Specifically, 20 FITC-labeled white blood cells were mixed into a background of 500 million unlabeled white blood cells (Figure S4a) and subjected to sorting. As shown in Figure S4b, the system successfully identified and isolated all 20 fluorescently labeled white blood cells, demonstrating its capability to detect rare targets even in complex cellular environments. This result establishes the system's limit of detection (LOD) at 1 in 10^8, providing a robust foundation for future applications in NIPT.

3.3. FnRBCs Isolation

The immunofluorescence-stained cells are loaded into the well of the SACA chip, and each well can contain around 500,000 cells. The micro-structure and hydrophilic surface treatment drive the lateral flow field and the gravity field. The cells are flattened into a dense monolayer within five minutes. The single layer and tight cell arrangement effectively avoids cell stacking that can cause difficulty in image recognition or the sorting of non-target cells. After the cells have settled, four types of fluorescent and visible light automated image scanning are performed. FnRBCs can be identified by the cell types presented by fluorescent signals (Heochst+/CD71+/HbF+/CD45−) and visible light images.

In addition to recognizing the FnRBCs' signal from our automated machine, we also used a fluorescent inverted microscope for further confirmation. The FnRBCs' signal from the automated cell imaging scanner (Figure 4a) is the same as the signal from the fluorescence microscope (Figure 4b), which is Heochst+/CD71+/HbF+/CD45−. Therefore, the cell an FnRBC. The sorted FnRBCs should be stored in the corresponding container for subsequent analysis, such as STR and other tests, for the verification of FnRBCs.

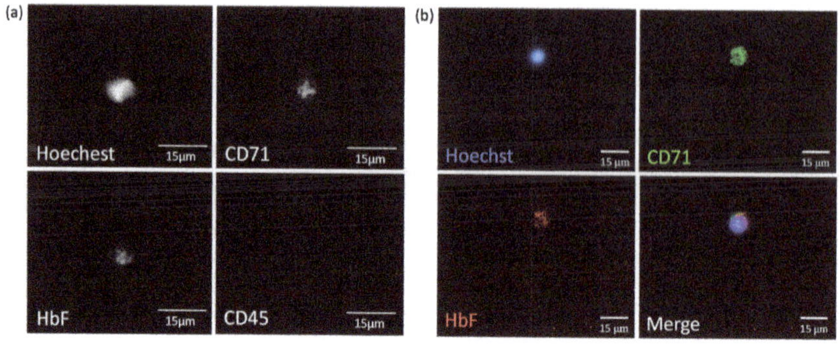

Figure 4. Images of FnRBCs on the SACA chip. (**a**) Signals (Heochst+/CD71+/HbF+/CD45−) from the cell imaging scanner before pickup. (**b**) Released FnRBC signals (Heochst+/CD71+/HbF+/CD45−) from a fluorescence microscope.

Additionally, we obtained blood samples from non-pregnant women and performed staining with different combinations of antibodies: (A) CD147+CD71+HbF+CD45−, (B) Hoechst+CD147+CD71+CD45−, and (C) Hoechst+CD71+HbF+CD45−, as shown in Figure S5. Statistical analysis revealed that the p-values for group B and group C compared to group A were 0.021079 and 0.03709, respectively. The results indicate that group B showed no significant difference compared to group A, suggesting that CD147 is less specific compared to HbF. Furthermore, the similarity between group B and group A highlights that the addition of CD147 does not provide significant advantages. Thus, HbF alone is sufficient for reliable identification. Based on these findings, we selected HbF and CD71 as the primary antibodies in our final staining protocol.

3.4. The Volume and the Background Noise of Single-Cell Isolation

Cell sorting is a critical step for downstream applications such as short tandem repeat (STR) analysis, which imposes an upper limit on the volume of the sorted sample. When a single cell is stored in a container with an excessive liquid volume, it can generate significant noise during subsequent cell signal amplification, complicating detection and analysis. Therefore, achieving cell sorting in an extremely small volume is essential to improve sorting purity and analytical precision.

To address this, we conducted 20 single-cell sorting experiments and measured the volume of each using a contact angle microdroplet volume measurement technique. The results showed that the average volume of a single cell sorted by our device is approximately 0.304 μL, as illustrated in Figure 5a. This volume is approximately 1/100 of the upper detection volume limit for STR analysis, demonstrating its suitability for downstream applications. Using this average volume as a parameter, we configured the high-precision syringe pump for optimized single-cell sorting. While only a single cell is sorted, maternal free DNA in the solution may still be inadvertently transferred into the storage container during aspiration and release. To quantify the potential impact of background contamination, we used PBS containing uniformly distributed 0.37 μm fluorescent polystyrene (PS) beads at a defined concentration (100% background noise) and introduced the solution into the SACA chip wells. Six different sorting volumes (5, 10, 15, 20, 25, and 30 μL) were tested, and the absorbance values were measured using an ELISA reader (GloMax® Explorer Multimode Microplate Reader, Promega Corporation, Madison, WI, USA). The calibration curve was used to calculate the corresponding background noise concentrations, as shown in Figure 5b.

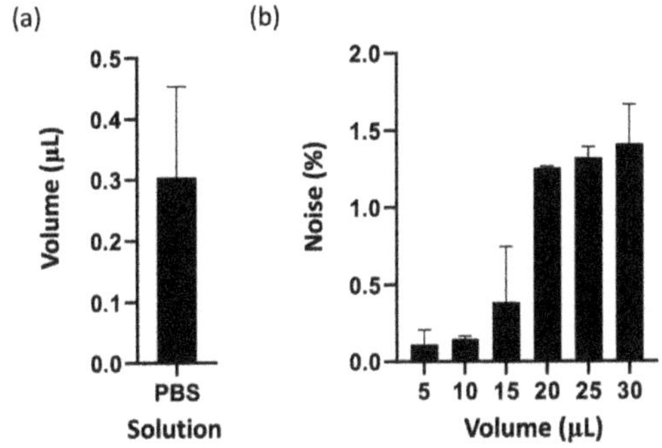

Figure 5. The average volume and background noise of single-cell isolation using PBS. (**a**) The average isolation volume for single cells is approximately 0.304 μL, demonstrating precise volume control during isolation ($n = 3$). Error bars indicate the standard deviation. (**b**) Background noise concentration at varying isolation volumes. Noise levels increase with larger isolation volumes, ranging from 0.11% at 5 μL to approximately 1.5% at 30 μL, highlighting the importance of minimizing isolation volume to reduce background interference ($n = 3$). Error bars indicate the standard deviation.

The results indicated that the background noise for a 5 μL sorting volume was approximately 0.11%, with a positive correlation observed between sorting volume and background noise concentration. At a sorting volume of 20 μL, the background noise concentration approached saturation, and at 30 μL, the maximum background noise reached around 1.41%. However, given the average sorting volume of 0.304 μL in our single-cell sorting experiments, the background noise concentration is estimated to be less than 0.11%, specifically around 0.0067%, based on proportional volume percentage calculations.

This remarkably low sorting volume significantly minimizes background contamination, ensuring high sorting purity. Based on the results of these quantitative experiments, achieving a lower single-cell sorting volume not only complies with the upper volume limits required for subsequent analysis but also effectively reduces background noise, enhancing the reliability and accuracy of downstream applications.

3.5. Maternal Blood Sample for STR Test

To validate the potential application of our approach in NIPT, we designed an STR experiment to assess the accuracy of FnRBC isolation from maternal blood. Using the SACA chip and fluorescent antibody staining with Hoechst+CD71+HbF+CD45−, we isolated FnRBCs and evaluated their suitability for downstream DNA analysis.

Five groups of samples (labeled A to E) were subjected to whole genome amplification (WGA) following cell capture. The number of FnRBCs captured from each group was eight, five, three, three, and three, respectively, as shown in Figure S6a. The OD260/OD280 ratios of the extracted DNA ranged between 1.5 and 1.7, slightly below the ideal range of 1.7 to 1.9, indicating suboptimal purity. This deviation suggests minor protein or contaminant residues in the samples, likely due to the limited DNA quantity obtained from microvolume samples. Nevertheless, the observed purity is within an acceptable range for amplified microcell samples, demonstrating the feasibility of using the SACA chip and our antibody combination for isolating FnRBCs suitable for genetic analysis.

As shown in Figure S6b, a positive correlation was observed between the number of cells and DNA concentration, with fewer cells yielding lower DNA concentrations. Furthermore, Figure 6a illustrates that our SACA system successfully captured fetal cells from maternal blood. In these results, 'fetal' refers to alleles where one copy originates from the father and the other from the mother, while 'maternal' refers to alleles identical to those of the mother. Across 24 loci, the fetal cells exhibited a higher number of signals compared to maternal cells. Figure 6b demonstrates a positive correlation between DNA quality and loci number, and Figure 6c shows that higher DNA concentrations further enhance the detection of additional loci.

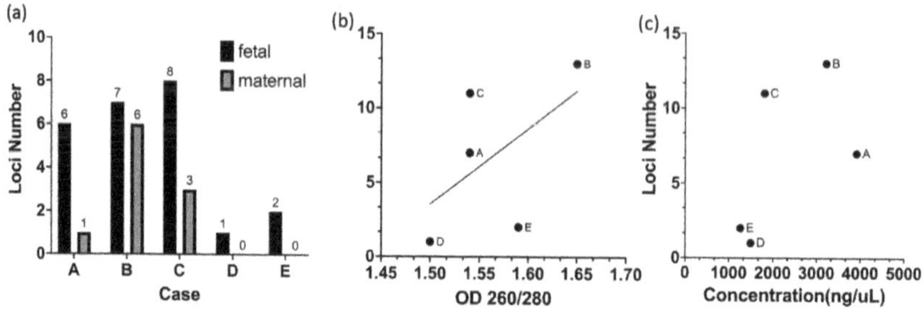

Figure 6. The STR results for FnRBCs isolated using the SACA chip and a Hoechst+CD71+HbF+CD45− antibody combination. (**a**) Comparison of detected loci numbers in fetal and maternal cells across five sample groups (A–E). (**b**) Correlation between OD260/280 ratios and the number of detected loci. (**c**) Relationship between DNA concentration and loci number.

These results emphasize the importance of capturing an adequate number of FnRBCs and maintaining high DNA quality to ensure sufficient DNA yield for downstream analysis. High-quality DNA contributes to the improved detection of genetic markers, supporting the reliability and accuracy of NIPT. This study provides a foundation for further optimization to enhance DNA quality and ensure robust downstream applications in NIPT.

4. Discussion

The precision and reliability of single-cell isolation demonstrated by the SACA chip, automated imaging, and sorting system are exceptional. With an average sorting accuracy of 97.85% and a relative error of approximately 2.15%, the system provides a reliable platform for isolating target cells, including rare populations such as FnRBCs. This high degree of precision is crucial for ensuring the purity and integrity of isolated cells, which is paramount for downstream analyses such as STR testing. Compared to conventional methods that often suffer from lower throughput and increased contamination risk, this technology represents a significant advancement in the field of single-cell sorting [20,21].

Although significant efforts have been made by various research groups to isolate circulating fetal cells (CFCs), particularly FnRBCs, and apply the technology in clinical settings, very few have reported successful results [22,23]. Many existing methods, such as fluorescence-activated cell sorting (FACS) and magnetic-activated cell sorting (MACS), involve labor-intensive and time-consuming experimental steps, limiting their practicality in clinical cytogenetics laboratories. In contrast, the SACA system achieves near automation in both cell capture and recovery processes, significantly reducing manual intervention and improving workflow efficiency. Our SACA system integrates a high-precision syringe pump, a three-axis motor, and an automated imaging scanning function, enabling the automatic localization and separation of FnRBCs. Its level of automation reduces the need for manual operations, minimizing errors caused by human intervention and enhancing the accuracy and efficiency of cell capture and recovery.

The system's low detection limits and background noise reduction further highlight its efficacy. The detection limit (LOD) of 1 in 10^8, established by spike-in experiments using FITC-labeled white blood cells, underscores the platform's ability to detect and isolate rare target cells even in highly complex biological samples. Additionally, the low average sorting volume of 0.304 μL minimizes background noise to less than 0.11%, making it suitable for sensitive analyses such as STR. These findings demonstrate that the platform can provide clean, high-quality cell samples, essential for applications like non-invasive prenatal testing (NIPT) [24].

The successful isolation of FnRBCs from maternal blood using Hoechst+/CD71+/HbF+/CD45− antibodies and subsequent STR analysis validates the FnRBC isolation efficiency and clinical feasibility of the SACA system. The positive correlation between DNA concentration and loci detection highlights the importance of capturing sufficient FnRBCs and ensuring high DNA quality for robust genetic analysis. Observed purity and signal fidelity provide a reliable basis for distinguishing fetal and maternal DNA, supporting the platform's utility in prenatal diagnostics.

Moreover, the SACA chip's surface coating enhancements for specificity significantly improve its performance. The application of ε-polylysine (ε-PL) coating increases friction between the chip surface and target cells, reducing the displacement of non-target cells by 31.2%. This enhanced cell retention improves sorting specificity and minimizes contamination. Such surface engineering not only strengthens the platform's performance but also demonstrates its adaptability for a wide range of biological applications.

Currently, many emerging platforms target trophoblasts for cbNIPD, as these cells are larger and easier to differentiate from the maternal white blood cell background. However, while trophoblasts provide some genetic information, their fragmented state often limits the analysis [15,25]. FnRBCs, on the other hand, represent the true fetal genome and therefore offer superior fidelity for genetic analysis. Trophoblasts and FnRBCs each have distinct advantages and limitations when used as targets for NIPT. Trophoblasts are more abundant in maternal blood and larger in size, making them easier to isolate and process. These factors contribute to the relatively low cost and established nature of trophoblast-targeting platforms [26]. However, trophoblasts often undergo DNA fragmentation and may exhibit placental mosaicism, which can compromise the accuracy of genetic analysis. In contrast, FnRBCs represent the true fetal genome, providing complete and high-quality DNA for detailed genetic analysis [11]. They also have minimal risk of contamination, ensuring reliable results. However, FnRBCs are extremely rare in maternal blood, requiring advanced isolation technologies like the SACA system to efficiently capture them. While this makes the process more technically demanding and costly, FnRBCs offer unparalleled accuracy and reliability for advanced genetic diagnostics. Our SACA system directly addresses this gap, providing a robust method for isolating FnRBCs with minimal contamination and high purity. This capability makes the SACA system a more reliable and accurate platform for cbNIPD compared to conventional cffDNA-based tests, which are prone to confounding factors such as maternal body mass index (BMI), fetoplacental mosaicism, and vanishing twin syndrome [27,28].

Finally, the clinical implications and future directions of this technology are promising. The SACA chip-based platform offers a reliable solution for NIPT by isolating FnRBCs from maternal blood with high purity and precision. Its ability to achieve accurate STR analysis makes it an invaluable tool for fetal genetic screening. Beyond NIPT, this technology has potential applications in other clinical fields, such as isolating circulating tumor cells (CTCs) for cancer diagnostics and monitoring. Continued refinements, particularly in integrating the system into sterile environments, will further expand its clinical utility, bridging the gap between research innovations and practical medical applications [29,30].

5. Conclusions

This study establishes the SACA chip as a highly effective and reliable platform for isolating FnRBCs from maternal blood. Its advanced design, precision sorting capabilities, and low detection limits make it a valuable tool for NIPT and other clinical applications. By providing high-quality, contamination-free samples, this technology lays the groundwork for accurate and efficient prenatal diagnostics, ensuring confidence in clinical outcomes and paving the way for future innovations.

6. Patents

R.O.C. patent number: TWI825620B.

Supplementary Materials: The following supporting information can be downloaded at: https://www.mdpi.com/article/10.3390/mi15121515/s1, Figure S1. Workflow for blood sample processing and the SACA system experiment. Figure S2. Principle of the SACA chip. Figure S3. Comparison of the effects of JNC coating and F127 surface modifications on cell capture and retention. Figure S4. White blood cell spike-in experiment using the SACA chip and automated system. Figure S5. False-positive testing using three different antibody combinations in non-pregnant female blood samples. Figure S6. DNA purity and concentration analysis of FnRBC samples obtained from five cases (A–E).

Author Contributions: Conceptualization, J.-K.W. and F.-G.T.; data curation, H.-Y.Y.; formal analysis, H.-Y.Y.; funding acquisition, F.-G.T.; investigation, H.-Y.Y.; methodology, H.-Y.Y., C.-H.L. (Che-Hsien Lin), Y.-W.H., C.-H.C., M.-C.H. and C.-H.L. (Chun-Hao Lai); project administration, F.-G.T.; resources, F.-G.T.; supervision, F.-G.T.; writing—original draft, Y.-W.H.; writing—review and editing, H.-Y.Y. All authors have read and agreed to the published version of the manuscript.

Funding: This research was funded by the National Science and Technology Council (project number: NSTC 113-2221-E-007 -044 -MY3) and Emerging Technology Application program (project number: 111AT17A).

Institutional Review Board Statement: The study was conducted according to the guidelines of the Declaration of Helsinki and approved by the Institutional Review Board, Taipei Medical University Hospital (IRB number: N202201034).

Informed Consent Statement: Informed consent was obtained from all subjects involved in the study.

Data Availability Statement: The original contributions presented in this study are included in the article. Further inquiries can be directed to the corresponding author.

Acknowledgments: Thanks to Cheng-Hsuan, Chu, an undergraduate student from Tseng's lab who assisted in the needle experiment.

Conflicts of Interest: The funders had no role in the design of the study; in the collection, analyses, or interpretation of data; in the writing of the manuscript, or in the decision to publish the results.

References

1. Hurley, P.A.; Ward, R.H.; Teisner, B.; Iles, R.K.; Lucas, M.; Grudzinskas, J.G. Serum PAPP-A measurements in first-trimester screening for Down syndrome. *Prenat. Diagn.* **1993**, *13*, 903–908. [CrossRef]
2. Falcon, O.; Auer, M.; Gerovassili, A.; Spencer, K.; Nicolaides, K.H. Screening for trisomy 21 by fetal tricuspid regurgitation, nuchal translucency and maternal serum free beta-hCG and PAPP-A at 11 + 0 to 13 + 6 weeks. *Ultrasound Obstet. Gynecol.* **2006**, *27*, 151–155. [CrossRef] [PubMed]
3. Bianchi, D.W.; Chiu, R.W.K. Sequencing of Circulating Cell-free DNA during Pregnancy. *N. Engl. J. Med.* **2018**, *379*, 464–473. [CrossRef] [PubMed]
4. Taylor-Phillips, S.; Freeman, K.; Geppert, J.; Agbebiyi, A.; Uthman, O.A.; Madan, J.; Clarke, A.; Quenby, S.; Clarke, A. Accuracy of non-invasive prenatal testing using cell-free DNA for detection of Down, Edwards and Patau syndromes: A systematic review and meta-analysis. *BMJ Open* **2016**, *6*, e010002. [CrossRef] [PubMed]
5. Bianchi, D.W.; Flint, A.F.; Pizzimenti, M.F.; Knoll, J.H.; Latt, S.A. Isolation of fetal DNA from nucleated erythrocytes in maternal blood. *Proc. Natl. Acad. Sci. USA* **1990**, *87*, 3279–3283. [CrossRef] [PubMed]
6. Bianchi, D.W.; Simpson, J.L.; Jackson, L.G.; Elias, S.; Holzgreve, W.; Evans, M.I.; Dukes, K.A.; Sullivan, L.M.; Klinger, K.W.; Bischoff, F.Z.; et al. Fetal gender and aneuploidy detection using fetal cells in maternal blood: Analysis of NIFTY I data. National Institute of Child Health and Development Fetal Cell Isolation Study. *Prenat. Diagn.* **2002**, *22*, 609–615. [CrossRef]
7. Wang, L.; Asghar, W.; Demirci, U.; Wan, Y. Nanostructured substrates for isolation of circulating tumor cells. *Nano Today* **2013**, *8*, 347–387. [CrossRef]
8. Feng, C.; He, Z.; Cai, B.; Peng, J.; Song, J.; Yu, X.; Sun, Y.; Yuan, J.; Zhao, X.; Zhang, Y. Non-invasive Prenatal Diagnosis of Chromosomal Aneuploidies and Microdeletion Syndrome Using Fetal Nucleated Red Blood Cells Isolated by Nanostructure Microchips. *Theranostics* **2018**, *8*, 1301–1311. [CrossRef]
9. Hoshino, K.; Huang, Y.Y.; Lane, N.; Huebschman, M.; Uhr, J.W.; Frenkel, E.P.; Zhang, X. Microchip-based immunomagnetic detection of circulating tumor cells. *Lab Chip* **2011**, *11*, 3449–3457. [CrossRef] [PubMed]
10. Nemescu, D.; Constantinescu, D.; Gorduza, V.; Carauleanu, A.; Caba, L.; Navolan, D.B. Comparison between paramagnetic and CD71 magnetic activated cell sorting of fetal nucleated red blood cells from the maternal blood. *J. Clin. Lab. Anal.* **2020**, *34*, e23420. [CrossRef]
11. Wei, X.; Ao, Z.; Cheng, L.; He, Z.; Huang, Q.; Cai, B.; Rao, L.; Meng, Q.; Wang, Z.; Sun, Y.; et al. Highly sensitive and rapid isolation of fetal nucleated red blood cells with microbead-based selective sedimentation for non-invasive prenatal diagnostics. *Nanotechnology* **2018**, *29*, 434001. [CrossRef] [PubMed]

12. Grzywa, T.M.; Nowis, D.; Golab, J. The role of CD71+ erythroid cells in the regulation of the immune response. *Pharmacol. Ther.* **2021**, *228*, 107927. [CrossRef] [PubMed]
13. Solomonia, N.; Playforth, K.; Reynolds, E.W. Fetal-maternal hemorrhage: A case and literature review. *AJP Rep.* **2012**, *2*, 7–14. [CrossRef]
14. Bischoff, F.Z.; Lewis, D.E.; Nguyen, D.D.; Murrell, S.; Schober, W.; Scott, J.; Simpson, J.L.; Elias, S. Prenatal diagnosis with use of fetal cells isolated from maternal blood: Five-color fluorescent in situ hybridization analysis on flow-sorted cells for chromosomes X, Y, 13, 18, and 21. *Am. J. Obstet. Gynecol.* **1998**, *179*, 203–209. [CrossRef] [PubMed]
15. Huang, C.-E.; Ma, G.-C.; Jou, H.-J.; Lin, W.-H.; Lee, D.-J.; Lin, Y.-S.; Ginsberg, N.A.; Chen, H.-F.; Chang, F.M.-C.; Chen, M. Noninvasive prenatal diagnosis of fetal aneuploidy by circulating fetal nucleated red blood cells and extravillous trophoblasts using silicon-based nanostructured microfluidics. *Mol. Cytogenet.* **2017**, *10*, 44. [CrossRef]
16. Zhu, P.; Ding, J.; Zhou, J.; Dong, W.J.; Fan, C.M.; Chen, Z.N. Expression of CD147 on monocytes/macrophages in rheumatoid arthritis: Its potential role in monocyte accumulation and matrix metalloproteinase production. *Arthritis Res. Ther.* **2005**, *7*, R1023–R1033. [CrossRef]
17. Yurchenko, V.; Constant, S.; Bukrinsky, M. Dealing with the family: CD147 interactions with cyclophilins. *Immunology* **2006**, *117*, 301–309. [CrossRef]
18. Chen, T.J.; Wu, J.K.; Chang, Y.C.; Fu, C.Y.; Wang, T.P.; Lin, C.Y.; Chang, H.Y.; Chieng, C.C.; Tzeng, C.Y.; Tseng, F.G. High-efficiency rare cell identification on a high-density self-assembled cell arrangement chip. *Biomicrofluidics* **2014**, *8*, 036501. [CrossRef] [PubMed]
19. Chu, H.-Y.; Lu, L.-S.; Cho, W.; Wu, S.-Y.; Chang, Y.-C.; Lin, C.-P.; Yang, C.-Y.; Lin, C.-H.; Jiang, J.-K.; Tseng, F.-G. Enumerating Circulating Tumor Cells with a Self-Assembled Cell Array (SACA) Chip: A Feasibility Study in Patients with Colorectal Cancer. *Cancers* **2019**, *11*, 56. [CrossRef] [PubMed]
20. Pratt, V.M.; Del Tredici, A.L.; Hachad, H.; Ji, Y.; Kalman, L.V.; Scott, S.A.; Weck, K.E. Recommendations for Clinical CYP2C19 Genotyping Allele Selection: A Report of the Association for Molecular Pathology. *J. Mol. Diagn.* **2018**, *20*, 269–276. [CrossRef]
21. Wu, Y.; Lu, Y.-C.; Jacobs, M.; Pradhan, S.; Kapse, K.; Zhao, L.; Niforatos-Andescavage, N.; Vezina, G.; du Plessis, A.J.; Limperopoulos, C. Association of Prenatal Maternal Psychological Distress With Fetal Brain Growth, Metabolism, and Cortical Maturation. *JAMA Netw. Open* **2020**, *3*, e1919940. [CrossRef] [PubMed]
22. Lu, W.; Huang, T.; Wang, X.R.; Zhou, J.H.; Yuan, H.Z.; Yang, Y.; Huang, T.T.; Liu, D.P.; Liu, Y.Q. Next-generation sequencing: A follow-up of 36,913 singleton pregnancies with noninvasive prenatal testing in central China. *J. Assist. Reprod. Genet.* **2020**, *37*, 3143–3150. [CrossRef] [PubMed]
23. Xiang, J.; Ding, Y.; Song, X.; Mao, J.; Liu, M.; Liu, Y.; Huang, C.; Zhang, Q.; Wang, T. Clinical Utility of SNP Array Analysis in Prenatal Diagnosis: A Cohort Study of 5000 Pregnancies. *Front. Genet.* **2020**, *11*, 571219. [CrossRef] [PubMed]
24. Pinhas-Hamiel, O.; Zalel, Y.; Smith, E.; Mazkereth, R.; Aviram, A.; Lipitz, S.; Achiron, R. Prenatal diagnosis of sex differentiation disorders: The role of fetal ultrasound. *J. Clin. Endocrinol. Metab.* **2002**, *87*, 4547–4553. [CrossRef]
25. Pfeifer, I.; Benachi, A.; Saker, A.; Bonnefont, J.P.; Mouawia, H.; Broncy, L.; Frydman, R.; Brival, M.L.; Lacour, B.; Dachez, R.; et al. Cervical trophoblasts for non-invasive single-cell genotyping and prenatal diagnosis. *Placenta* **2016**, *37*, 56–60. [CrossRef] [PubMed]
26. Vossaert, L.; Chakchouk, I.; Zemet, R.; Van den Veyver, I.B. Overview and recent developments in cell-based noninvasive prenatal testing. *Prenat. Diagn.* **2021**, *41*, 1202–1214. [CrossRef] [PubMed]
27. Jayamohan, H.; Lambert, C.J.; Sant, H.J.; Jafek, A.; Patel, D.; Feng, H.; Beeman, M.; Mahmood, T.; Nze, U.; Gale, B.K. SARS-CoV-2 pandemic: A review of molecular diagnostic tools including sample collection and commercial response with associated advantages and limitations. *Anal. Bioanal. Chem.* **2021**, *413*, 49–71. [CrossRef] [PubMed]
28. Hsiao, C.H.; Chen, C.H.; Cheng, P.J.; Shaw, S.W.; Chu, W.C.; Chen, R.C. The impact of prenatal screening tests on prenatal diagnosis in Taiwan from 2006 to 2019: A regional cohort study. *BMC Pregnancy Childbirth* **2022**, *22*, 23. [CrossRef] [PubMed]
29. Bu, X.; Zhou, S.; Li, X.; Li, S.; Li, H.; Ding, S.; He, J.; Linpeng, S. A retrospective single-center analysis of prenatal diagnosis and follow-up of 626 chinese patients with positive non-invasive prenatal screening results. *Front. Genet.* **2022**, *13*, 965106. [CrossRef]
30. Li, C.; Xiong, M.; Zhan, Y.; Zhang, J.; Qiao, G.; Li, J.; Yang, H. Clinical Potential of Expanded Noninvasive Prenatal Testing for Detection of Aneuploidies and Microdeletion/Microduplication Syndromes. *Mol. Diagn. Ther.* **2023**, *27*, 769–779. [CrossRef] [PubMed]

micromachines

MDPI

Article

Elasticity of Carrier Fluid: A Key Factor Affecting Mechanical Phenotyping in Deformability Cytometry

Hassan Pouraria and Jessica P. Houston *

Department of Chemical and Materials Engineering, New Mexico State University, Las Cruces, NM 88003, USA; pouraria1@gmail.com

* Correspondence: jph@nmsu.edu

Abstract: Recently, microfluidics deformability cytometry has emerged as a powerful tool for high-throughput mechanical phenotyping of large populations of cells. These methods characterize cells by their mechanical fingerprints by exerting hydrodynamic forces and monitoring the resulting deformation. These devices have shown great promise for label-free cytometry, yet there is a critical need to improve their accuracy and reconcile any discrepancies with other methods, such as atomic force microscopy. In this study, we employ computational fluid dynamics simulations and uncover how the elasticity of frequently used carrier fluids, such as methylcellulose dissolved in phosphate-buffered saline, is significantly influential to the resulting cellular deformation. We conducted CFD simulations conventionally used within the deformability cytometry field, which neglect fluid elasticity. Subsequently, we incorporated a more comprehensive model that simulates the viscoelastic nature of the carrier fluid. A comparison of the predicted stresses between these two approaches underscores the significance of the emerging elastic stresses in addition to the well-recognized viscous stresses along the channel. Furthermore, we utilize a two-phase flow model to predict the deformation of a promyelocyte (i.e., HL-60 cell type) within a hydrodynamic constriction channel. The obtained results highlight a substantial impact of the elasticity of carrier fluid on cellular deformation and raise questions about the accuracy of mechanical property estimates derived by neglecting elastic stresses.

Keywords: deformability cytometry; microfluidics; elasticity; carrier fluid

1. Introduction

Single-cell, high-throughput analyses generally involve the labeling of molecular biomarkers with exogenous fluorophores, many protocols of which require cell preparation, calibration, and standardization protocols before clinical diagnosis. Accordingly, emerging technologies are now focusing on how mechanical properties, akin to cellular antigens and proteins, are reporters of cellular function and disease. Notably, alterations in cellular mechanical properties have for decades been correlated to cellular function owing to the characterization of cytoskeletal and nuclear organization [1,2]. In general, mechanical traits, often dubbed "mechanical biomarkers", encompass the deformability of cells when subjected to external loads. Comparable to the intrinsic properties, such as gene and protein expression, these mechanical biomarkers contribute to the phenotyping of cell populations. Several studies have established connections between alteration in cellular mechanical properties and various processes, including cell cycle progression [3], cancer malignancy [4–6], leukocyte activation [7,8], and stem cell differentiation [9,10]. Hence, by assessing cell mechanics, the requirement for external markers, such as fluorescent dyes, is eliminated, making it a compelling and noninvasive biomarker for cell identification.

Conventional methodologies for mechanical phenotyping of cells include optical stretching [11], atomic force microscopy [12], micropipette aspiration [13], and parallel plate rheology [14]. Such methods evaluate the response of the cell under the applied forces and estimate the mechanical properties of the cells, such as stiffness and viscous modulus.

Despite their reliable results, these approaches suffer from a considerable drawback, which is low throughput, thus limiting their applicability for characterizing large populations of cells.

Accordingly, microfluidically driven cellular deformation approaches have become prominent owing to their high-throughput capabilities. In general, there are two main types of microfluidic deformation methods. The first method—constriction deformable cytometry—involves the use of a constriction channel narrower than the cell size [15,16]. Cells pass through the constriction channel, and their mechanical properties are evaluated based on passage times. The second approach—hydrodynamic forced deformation—employs either a constriction channel slightly larger than the size of cells [3] or a cross-slot channel generating a high extensional flow region [17]. Typical throughputs for constriction deformability, hydrodynamic constriction deformability, and hydrodynamic cross-slot deformability are 1, 100, and 1000 cells per second, respectively [3,18]. The hydrodynamic constriction channel, which allows real-time deformability cytometry (RT-DC), is compatible with active sorting devices [19,20]. Operating at a low Reynolds numbers ($Re < 1$), generally, inertia is neglected in RT-DC devices [3]. Furthermore, using a narrow channel slightly larger than the cells, very high shear rates and stress intensities can be achieved while allowing for contactless deformation. Significant effort has been devoted to developing a theoretical framework for extracting the mechanical properties of cells [21–23]. Deformability cytometry using cross-slot geometry typically operates at higher throughputs, where inertial effects come into play. The emerging inertial forces can be leveraged for the pre-alignment of particles, ensuring that cells experience identical path lines and stresses.

These microfluidic devices not only employ different channel geometries, but also utilize different probing time scales [18]. Generally, shortened measurement times are associated with rapid force application, which can significantly influence the response of the cell. Previous studies indicate that the high strain rate in cross-slot deformability cytometry can fail to detect responses to actin cytoskeleton perturbations [17]. However, such changes could be detected by using cross-slot geometries at lower flow rates [24,25].

For inducing sufficiently high hydrodynamic forces in the hydrodynamic constriction channel, high-viscosity fluids are employed [3,18]. However, in the cross-slot channel, mechanical phenotyping can be performed using both high- and low-viscosity fluids [24,25]. A commonly used high-viscosity fluid in such devices is methylcellulose dissolved in phosphate-buffered saline (MC-PBS) [3,18,25]. Due to the biocompatibility of methylcellulose and the capability of increasing the viscosity of carrier fluid, MC-PBS is widely used in both hydrodynamic constriction channels and cross-slot deformation cytometry. The calculation of the exerted hydrodynamic stresses in these devices hinges on the flow field and measured fluid viscosity, considering the fluid as a generalized Newtonian fluid (GNF) with shear thinning behavior. In fact, it is assumed that the observed deformation of the cell is mainly due to the applied shear stress from MC-PBS [22,23,26]. Nonetheless, polymer solutions, such as MC-PBS, usually exhibit viscoelastic behavior, and their elastic properties may lead to significant changes in fluid flow. A recent rheological investigation of MC-PBS solutions with different concentrations revealed the elasticity of these solutions [27]. Hence, it is crucial to investigate the behavior of MC-PBS solutions when flowing in the hydrodynamic constriction channel, especially in the regions close to the deforming cells where the fluid flow exhibits complex patterns with mixed kinematics. Therefore, disregarding the viscoelastic nature of carrier fluid may lead to significant errors in estimating the exerted stresses on the cells. Since cell stiffness is proportionate to the exerted stresses for a given deformation during experiments, neglecting a portion of exerted stresses could lead to a wrong estimation of cell stiffness. It is worth mentioning that the estimated cell stiffness in hydrodynamic constriction cytometry typically shows a discrepancy when compared to the measured results by using AFM [18,28]. It should be mentioned that, due to the distinct probing methods and operating conditions, individual methods for measuring the cell's mechanical properties often yield significantly different results. This disparity was highlighted in a recent work that compared elastic and viscous moduli obtained through an array of methods—atomic force microscopy, magnetic twisting, particle tracking microrheology, parallel-plate rheometry, and optical stretching—for the same cell line maintained in the standardized condition [29]. The resulting values exhibited a range spanning two orders of magnitude for an elastic modulus and three orders of magnitude for a viscous modulus. These disparities are attributed to the differences in the applied stress and strain rates, probe size, probing lengths scale, and time scale, and whether cells are attached to a surface or suspended in a fluid [28]. Hence, it is imperative to comprehend the temporal and spatial distributions of stress/strain applied to the cells in each method.

To investigate the significance of incorporating the complete viscoelastic behavior of the MC-PBS solution, we conduct computational fluid dynamics (CFD) simulations of single-phase MC-PBS flow within a microchannel with dimensions similar to those used in hydrodynamic constriction cytometry. In parallel, simulations are conducted for MC-PBS flow utilizing a GNF model that accounts for shear thinning behavior, but disregards the elasticity of fluid. A comparative analysis of the predicted flow fields employing a viscoelastic model and GNF model is presented. Subsequently, we model the deformation of a deforming particle representing a promyeloblast (i.e., HL-60 cell) navigating through the hydrodynamic constriction microchannel. Two MC-PBS fluid models are employed: one incorporating fluid elasticity and the other excluding it. A comparative evaluation of the flow fields, generated stress, and predicted shape of the suspended cell in these two fluid models elucidates the impact of fluid elasticity on the cell's deformation within the microchannel.

2. Numerical Method

Numerical simulations were conducted to investigate the behavior of MC-PBS flows within a narrow microchannel, with dimensions typically employed in hydrodynamics constriction deformability cytometry. Two CFD models were employed to predict the flow field of the MC-PBS solution: a simplified GNF model and a viscoelastic fluid model. The GNF approach employs the Carreau–Yasuda model, which accounts for shear thinning behavior, but ignores the elasticity of fluid. In contrast, the viscoelastic fluid model incorporated the linear form of Phan–Thien–Tanner (PTT) model, which is suitable for polymer solutions [30]. This model considers both viscous and elastic stresses as well as the shear thinning effect, providing a more comprehensive representation of fluid behavior [30,31]. The results obtained from these two models will be compared to assess the reliability of employing the GNF approach.

Governing equations of the flow field are fully resolved to predict the velocity, pressure, and stress distributions. The governing equations of isothermal incompressible fluid flow in the laminar regime are as follows:

$$\nabla \cdot u = 0 \tag{1}$$

$$\rho\left(\frac{\partial u}{\partial t} + u \cdot \nabla u\right) = -\nabla p + \nabla \cdot \tau \tag{2}$$

$$\tau = \tau_s + \tau_p = \mu_s\left(\nabla u + \nabla u^T\right) + \tau_P \tag{3}$$

$$\lambda\left[\frac{\partial \tau_P}{\partial t} + \nabla \cdot u \tau_P\right] + f(Tr\tau_P)\tau_P = \mu_p\left(\nabla u + \nabla u^T\right) + \lambda\left(\tau_P \cdot \nabla u + \nabla u^T \cdot \tau_P\right) \tag{4}$$

$$f(Tr\tau_P) = 1 + \frac{\lambda \varepsilon}{\mu_p} Tr(\tau_P) \tag{5}$$

Equations (1) and (2) are continuity and momentum equations, respectively, where u represents the velocity, p is pressure, and τ indicates the stress tensor. In the viscoelastic fluid context, the stress tensor τ is the sum of solvent stress (τ_s) and polymer stress (τ_p) (Equation (3)). Polymer stress was modeled using the PTT model (Equations (4) and (5)), where λ and μ_p represent the polymer relaxation time and polymer viscosity, respectively, and ε is a material parameter related to its extensional property. Generally, ε may range from 0 to 1, and a lower ε value in the PTT model corresponds to a higher predicted uniaxial extensional viscosity. In fact, when ε tends to zero, the PTT model reduces to the well-known Oldroyd-B model. However, contrary to the Oldroyd-B model, the PTT model predicts the shear-thinning behavior of fluid when $\varepsilon \neq 0$. The shear viscosity function in the PTT model is obtained by determining the appropriate value of ε [32,33]. On the contrary to the viscoelastic models, in the context of the GNF approach, the stress tensor includes only the first term on the right-hand side of Equation (3), representing the shear stress. The total shear viscosity, μ, is modeled using the Carreau–Yasuda model that follows Equation (6).

$$\mu(\dot{\gamma}) = \mu_\infty + (\mu_0 - \mu_\infty)\cdot\left(1 + (\lambda\dot{\gamma})^{\alpha}\right)^{\frac{v-1}{\alpha}} \tag{6}$$

where μ_0 and μ_∞ are zero shear and infinite shear viscosities, respectively. Furthermore, v, λ, and α are the power law index, relaxation time, and Carreau–Yasuda model constant, respectively. CFD simulations were conducted for the 0.5% MC-PBS solution. Experimental data regarding measured viscosity were obtained from Büyükurgancı et al. [27], where concentric cylinder (CC), cone–plate (CP), plate–plate (PP), and parallel-disk (PP) measurements were performed to cover very low to high shear rates. The viscosity of the solvent was assumed to be 1 mPa·S. Table 1 shows the parameters used to describe the viscosity of the 0.5% MC-PBS solution using the Carreau–Yasuda model [27]. Furthermore, Figure 1 compares the viscosity of the 0.5% MC-PBS solution as predicted by the PTT model ($\varepsilon = 0.013$) and Carreau–Yasuda model with the experimental measurements. As depicted, the PTT and Carreau–Yasuda models reasonably captured the shear thinning behavior of the fluid observed in the experimental data [27].

Table 1. Carreau–Yasuda model constants.

Fluid	μ_0	μ_∞	λ	v	α
0.5% MC-PBS	20 mPa.s	1 mPa.s	0.0012	0.65	1.02

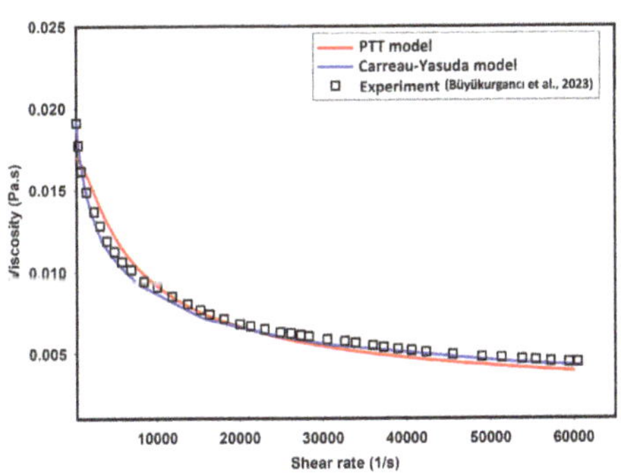

Figure 1. Experimental shear viscosity vs. PTT and Carreau–Yasuda models [27].

Numerical modeling was initially conducted to analyze a single-phase flow of the 0.5% MC-PBS solution within a 3D square channel. The primary aim was to compare the predicted stresses utilizing the Carreau–Yasuda model and the more comprehensive viscoelastic model know as PTT. Additionally, CFD simulations were performed to replicate the deformation of an HL-60 cell suspended in 0.5% MC-PBS solution when passing the channel. To this end, a viscous droplet model was employed, which takes into account the apparent cytoplasmic viscosity and membrane tension of cells. We used the level set method to model the two-phase flow and track the interface of the primary phase (0.5%

MC-PBS solution) flowing in the microchannel and a suspended droplet representing the cell. As such, the following equation was solved in conjunction with Equations (1)–(5):

$$\frac{\partial \varnothing}{\partial t} + u \cdot \nabla \varnothing = 0 \tag{7}$$

For the solution, volume fraction (α) and local material properties are updated based on $\varnothing$ and smoothed across the interface using a smooth Heaviside function. Furthermore, surface tension force is added to the right-hand side of the momentum equation and calculated as follows:

$$F_s = \sigma k n \delta_s \tag{8}$$

where σ represents the surface tension coefficient, k denotes the interface curvature, n signifies the normal vector, and δ_s represents the smoothed Dirac delta function centered at the interface. To restore its correct distributions near the interface, a re-distancing problem was solved [34–36].

Transport equations were solved by using the finite volume method (FVM) with the TransAT 5.7 code [37]. In the CFD model, pressure–velocity coupling was achieved by using the SIMPLEC algorithm. The spatial derivatives were discretized using a 2nd-order hybrid linear/parabolic approximation (HLPA) scheme. For the level set model, we used a 3rd-order WENO scheme for re-distancing [38]. The residual levels for convergence criteria were set to 1×10^{-6} for the velocity, level set, and stress components, and 1×10^{-5} for pressure.

3. Results and Discussions

3.1. Flow Field in the Microchannel Conveying the Single-Phase MC-PBS Solution

Numerical simulations were performed to analyze the flow characteristics of the 0.5% MC-PBS solution within a microchannel commonly utilized in hydrodynamic constriction deformability cytometry [18]. CFD simulations were initially employed to study the single-phase flow of the carrier fluid, utilizing two rheological models: the Carreau–Yasuda model and PTT model. Figure 2 shows a schematic of the microchannel and the adopted geometry and dimensions in the CFD model. The width, height, and length of the channel are 20 µm, 20 µm, and 400 µm, respectively. The dimensions of the microchannel and flow rate are identical to those reported experimentally [18]. A uniform grid size of 1 µm (total cell = 160,000) was adopted in the 3D CFD simulations.

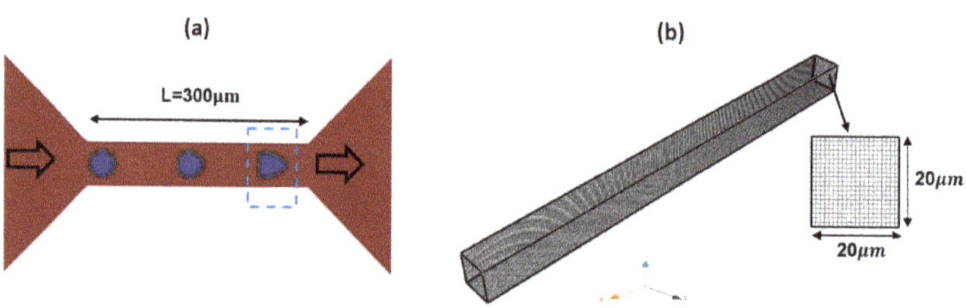

Figure 2. A schematic of typical microchannel geometry in hydrodynamic constriction deformability cytometry (**a**) and the adopted 3D grid in the CFD model (**b**).

Figure 3 illustrates the shear rate distribution as predicted by the Carreau–Yasuda model along the width of the channel, corresponding to an average flow velocity of 0.1 m/s under fully developed flow conditions. As observed, the predicted shear rates vary from nearly zero at the center of the channel to a maximum value of approximately 52,000 1/s at the walls.

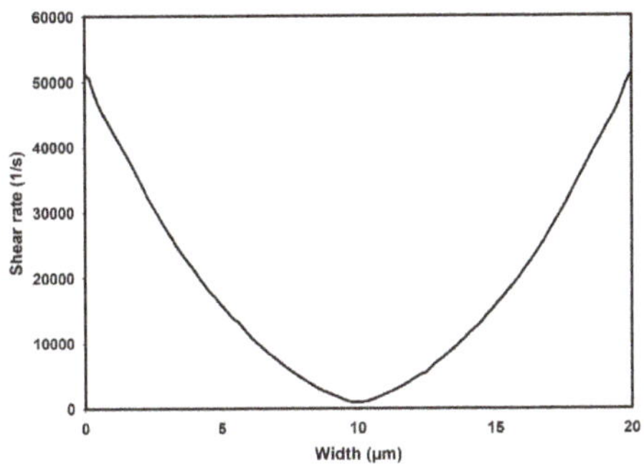

Figure 3. Shear rate distribution across the width of the channel as predicted by the Carreau–Yasuda model.

According to Figure 1, both the Carreau–Yasuda and PTT models can reasonably predict the shear viscosity of carrier fluid in the predicted shear range.

Figure 4a,b present contours illustrating the axial velocity within the cross-section of the channel, as predicted by the Carreau–Yasuda and PTT models upon reaching fully developed flow conditions. Notably, both models yield remarkably similar velocity distributions, showcasing a peak velocity of approximately 0.19 m/s at the channel center. In general, when dealing with shear-thinning fluids, one would anticipate flattened velocity profiles in comparison to their Newtonian counterparts.

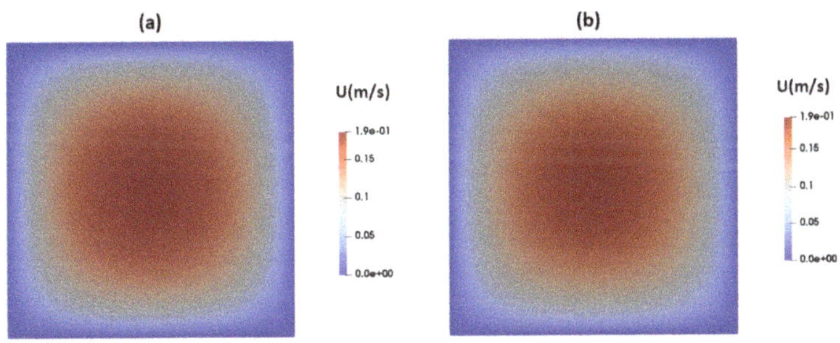

Figure 4. Distribution of axial velocity as predicted by (**a**) the PTT model and (**b**) Carreau–Yasuda model.

Figure 5 provides a more discernible representation of the contrast in axial velocity distribution across the channel width as determined by the PTT and Carreau–Yasuda models. As depicted in this figure, the PTT model predicts a slightly more flattened velocity profile at the center of the channel, suggesting slightly elevated shear-thinning behavior. The discrepancies observed in these velocity profiles could be attributed to the prediction of slightly different viscosities by the PTT and Carreau–Yasuda models, especially at high shear rates, as shown in Figure 1.

Figure 6 illustrates the contours of total shear stress as predicted by the Carreau–Yasuda model at the cross-section of the channel under fully developed flow conditions. As observed, the shear stress is at its maximum at the walls and gradually decreases to nearly zero when moving toward the center of the channel, where the velocity gradient is zero. In contrast to the Carreau–Yasuda model, the PTT model directly accounts for the contribution of polymers in the stress tensor. That is, the total stress is calculated as a summation of solvent and polymer stresses. Figure 7a,b display the contours of polymer shear stress and solvent shear stress as predicted by the PTT model within the channel cross-section. Figure 8 illustrates the predicted total shear stress obtained by the PTT model and

Carreau–Yasuda model across the width of the channel. As observed in this figure, there is good agreement between the predicted total shear stresses by the two models, except in regions close to the walls where the PTT model predicts slightly lower shear stresses when compared to the Carreau–Yasuda model. This small difference can be attributed to the accuracy of fitting the experimental shear viscosities by these two models.

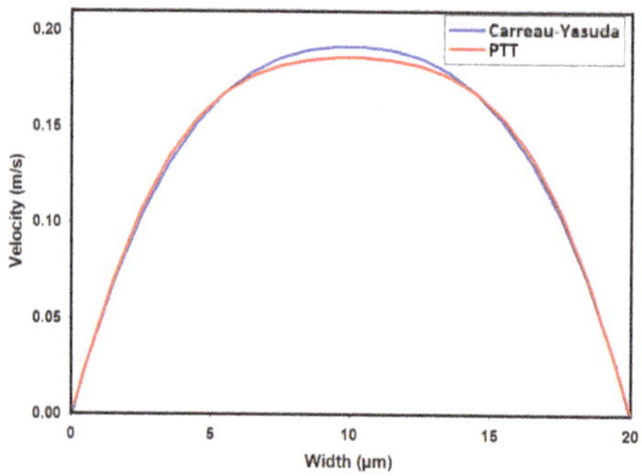

Figure 5. Velocity distribution across the width of the channel.

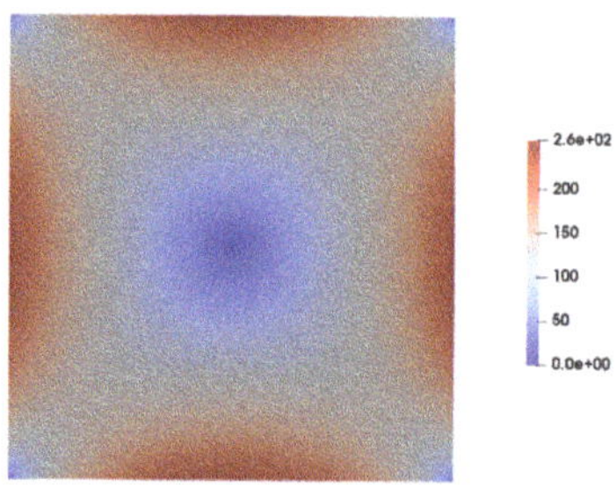

Figure 6. Contours of total shear stress at the channel cross-section as predicted by the Carreau–Yasuda model.

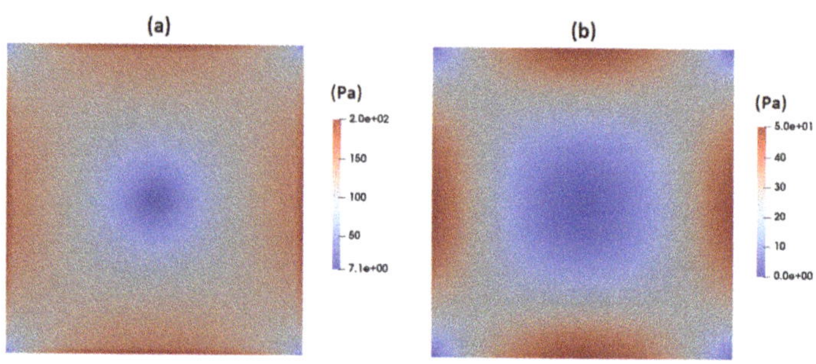

Figure 7. Contours of shear stress at the channel cross-section as predicted by the PTT model: (**a**) polymer shear stress and (**b**) solvent shear stress.

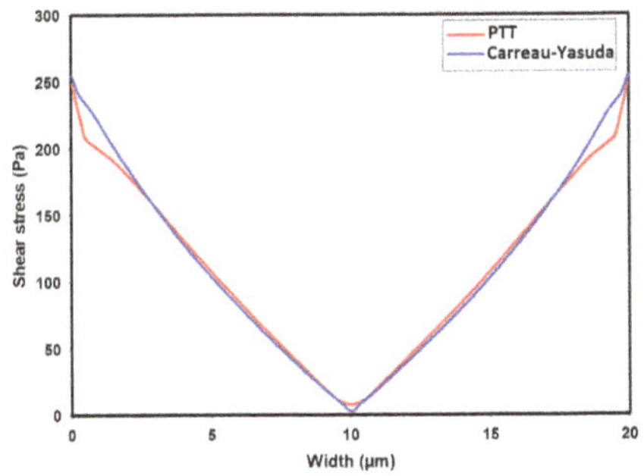

Figure 8. Distribution of total shear stress (τ_{xy}) across the width of the channel as predicted by the PTT and Carreau–Yasuda models.

Figure 9 shows the contour of normal stress (τ_{xx}) at the channel cross-section as predicted by the PTT model. This figure clearly shows significantly high elastic stress in the proximity of the channel walls, gradually diminishing to nearly zero at the channel center. Figure 10 presents the distribution of normal stress (τ_{xx}) across the channel width. A comparison of the predicted normal stress (τ_{xx}) in Figure 10 with the total shear stress in Figure 8 unequivocally reveals that the normal stress magnitude within the channel is an order of magnitude higher than that of the total shear stress, as predicted by the PTT and Carreau–Yasuda models. Previous studies underscore a substantial influence of such normal stresses on flow fields containing suspended particles. In fact, the observed stress gradients possess the potential to significantly influence the trajectory of suspended particles or induce deformations in the shape of pliable particles navigating through such flow fields [39–43].

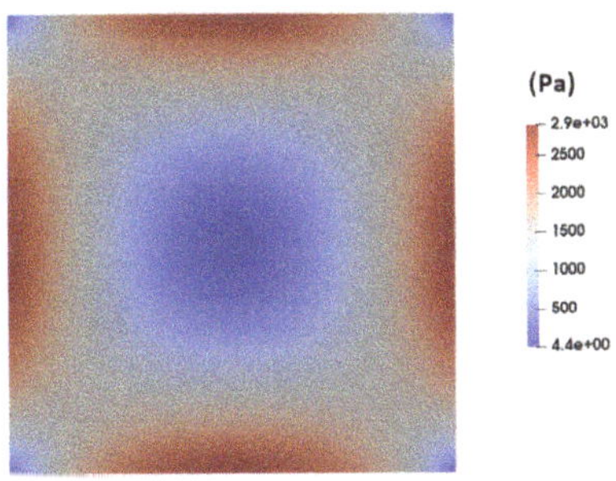

Figure 9. Distribution of normal stress (τ_{XX}) at the channel cross-section as predicted by the PTT model.

While the Carreau–Yasuda model can reasonably predict the velocity profile and total shear stresses in steady shear flows, it completely neglects the generation of normal stresses. Furthermore, its inability of directly accounting for the contribution of polymer chains in the stress tensor can be problematic in unsteady flows and flows with mixed kinematics [44,45]. Figure 11 shows the distribution of average shear stress at the wall of the channel as we move from the inlet toward the outlet. As seen in Figure 11, the predicted shear stresses by using the Carreau–Yasuda and PTT models in the developing region are different. According to this figure, for a single-phase flow of the 0.5% MC-PBS solution in a square channel, the flow reaches a fully developed condition within a short length in the entrance region. Hence, overall, using the Carreau–Yasuda model results in a reasonable estimation of the shear stress distribution in the channel.

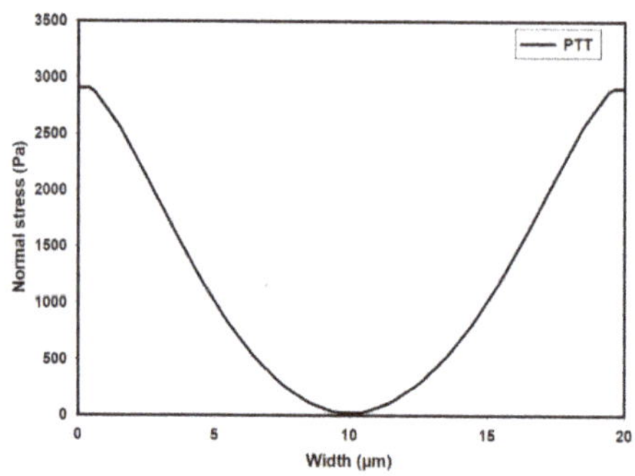

Figure 10. Distribution of normal stress (τ_{xx}) across the width of the channel as predicted by the PTT model.

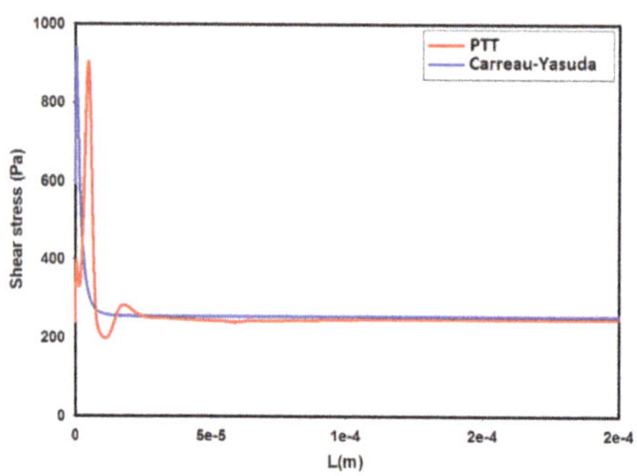

Figure 11. Distribution of average shear stress at the wall along the channel as predicted by the PTT and Carreau–Yasuda models.

3.2. Numerical Modeling of Cell Deformation in a Hydrodynamic Constriction Channel

The preceding numerical results pertaining to the single-phase flow of the 0.5% MC-PBS solution in the microchannel underscore the prevalence of normal stress over shear stress. In this section, we employ the level set model to investigate the impact of carrier fluid's elasticity on the exerted stress and deformation of suspended particles. In fact, due to the large size of the cells compared to the size of the channel, the flow field is substantially altered as the cells move within the channel. Hence, we model the deformation of a suspended viscous particle, representing an HL-60 cell, navigating through a hydrodynamic constriction microchannel using two MC-PBS fluid models—one incorporating elasticity and the other lacking it.

A method employed for modeling cell deformation is the viscous droplet model, where the cell is conceptualized as a viscous fluid-filled bag with constant surface tension. This model takes into account apparent membrane tension and cytoplasmic viscosity. While not appropriate for all cell types, it has demonstrated reasonable accuracy in describing the deformation of HL-60 cells [13,46,47]. It is crucial to emphasize that the main objective of this study is to explore the significance of the emerging elastic stress surrounding cells in motion within the channel and to investigate whether this elastic stress plays a crucial role in cell deformation compared to viscous stress. Consequently, irrespective of the accuracy of the viscous droplet model in replicating the exact behavior of HL-60 cells, any discernible differences in deformation predicted by two CFD models—one including the elastic effect and the other excluding it—would underscore the pivotal role of fluid elasticity in cell deformation.

We model the deformation of an HL-60 cell under an operating condition reported experimentally [18]. In order to reduce the computational cost, an axisymmetric CFD model is employed, which is compatible with the observed cell shape in the experiments. Hence, the predicted CFD results show the deformation of a single cell flowing in a tube with an identical hydraulic diameter (20 μm) rather than a 3D square channel. The accuracy of the CFD model in predicting the velocity profile of viscoelastic fluid within a circular tube was first validated by comparing the predicted results using the PTT model with an analytical solution [32]. Figure 12 shows the predicted velocity profile for a viscoelastic fluid, where the solvent viscosity ratio (β), Deborah number (De), and extensibility parameter (ε) are 0.1, 6.3, and 0.25, respectively. As observed in Figure 12, there is good agreement between the predicted numerical results and the analytical solution.

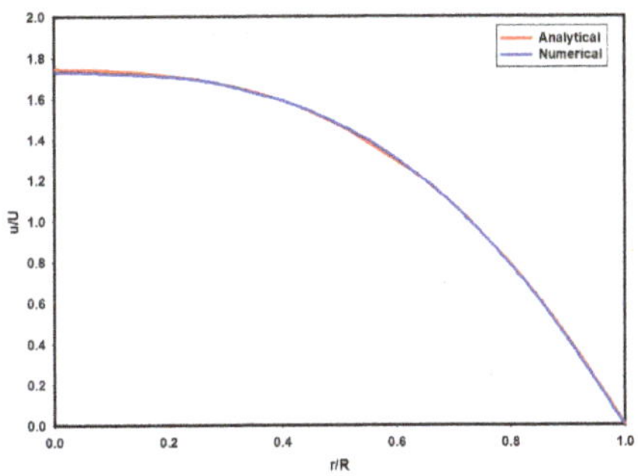

Figure 12. A comparison of the predicted velocity profiles using analytical and numerical models. De = 6.3, β = 0.1, and $\varepsilon = 0.25$.

Subsequently, CFD simulations were conducted for a two-phase flow of a suspended viscous droplet in the 0.5% MC-PBS solution and for a mean flow velocity of 0.1 m/s. Previous studies revealed that the apparent viscosity of HL-60 cells depends on the imposed shear rate. Using the empirical model proposed by Tsai et al. and considering an average shear rate of 20,000 1/s in the channel, the apparent viscosity was estimated to be 1.08 Pa·s [48]. Furthermore, the cortical membrane tension and density of HL-60 are 155 pN/μm and 1080 Kg/m^3, respectively [47,49]. The droplet representing an HL-60 cell was assumed to be initially spherical with a diameter of 14 μm at the inlet. Simulations were conducted using quad grids with a size of 0.5 μm (20 × 800 cells). The transient simulation was continued until the droplet traveled 300 μm downstream from the inlet. This distance mirrors the cell's journey to the measurement point within the lengths of the hydrodynamic constriction channel [18]. Furthermore, adaptive time stepping was employed to ensure that the Courant number remained below 0.25 in the solution.

Figure 13a,b show the shape of droplets 300 mm downstream from the inlet as predicted by the PTT and Carreau–Yasuda models, respectively. A discernible contrast in droplet deformation is evident, with the PTT model indicating greater deformation due to the consideration of fluid elasticity. Figure 14 compares the cell shape predicted by CFD simulations using the PTT and Carreau–Yasuda models with experimental data on cell shape. Aside from demonstrating a discernible difference in the predicted deformation between the two CFD models, the obtained results exhibit an excellent agreement between the predicted shape using the viscoelastic fluid model and the experimental observation under the specific experimental condition used in this study. However, employing more sophisticated cell models would result in more reliable predictions for different cells and across various time scales [26,50].

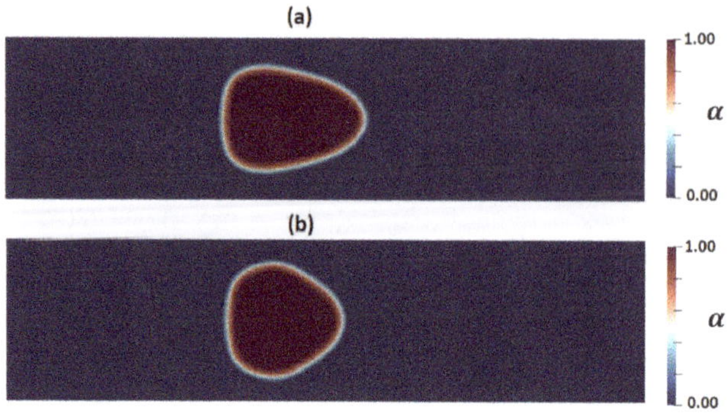

Figure 13. Contours of cell volume fraction (α) as predicted by (**a**) the PTT and (**b**) Carreau–Yasuda models.

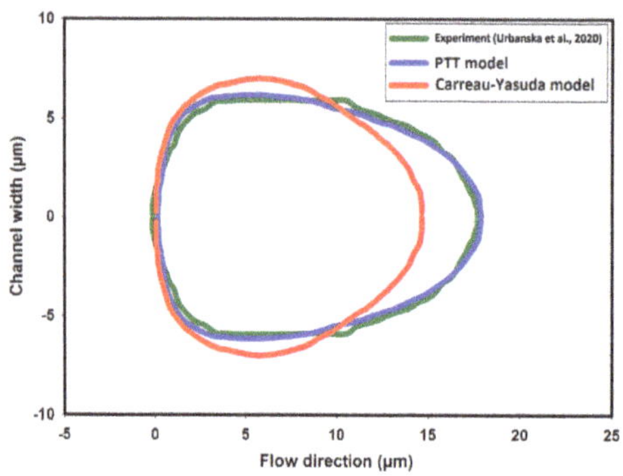

Figure 14. A comparison of cell deformations as measured in the experiment and predicted by the PTT and Carreau–Yasuda models [18].

Figure 15 shows the distribution of shear stress around the suspended droplets 300 μm downstream of the inlet as predicted by the PTT (Figure 15a) and Carreau–Yasuda (Figure 15b) models. According to these figures, the maximum shear stress near the droplet in the Carreau–Yasuda model is higher than that predicted by the PTT model. Moreover, the overall predicted distribution of shear stress is significantly different between the two.

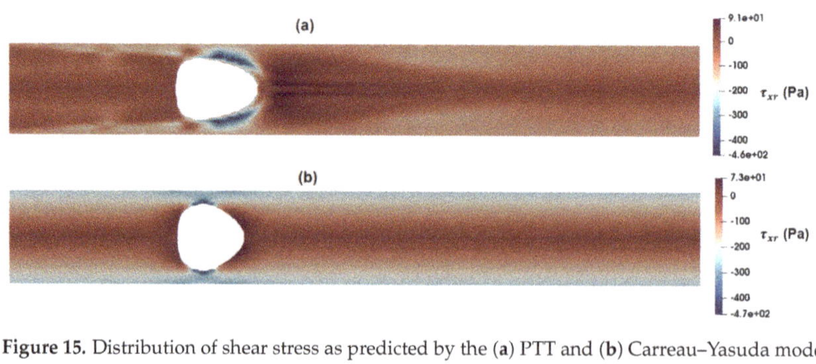

Figure 15. Distribution of shear stress as predicted by the (**a**) PTT and (**b**) Carreau–Yasuda models.

Figure 16a,b show the distribution of axial and radial normal stresses around the suspended droplets. A comparison of these figures with Figure 15a,b indicates that the intensity of the radial component of elastic stress, τ_{rr}, is less than that of shear stress, while the axial component of normal stress, τ_{xx}, reaches significantly high values around the droplet. According to the present results, the intensity of axial normal stress is significantly higher than that of shear stress around the droplet, which in turn results in exerting a compressive force and further stretching the deformable particle.

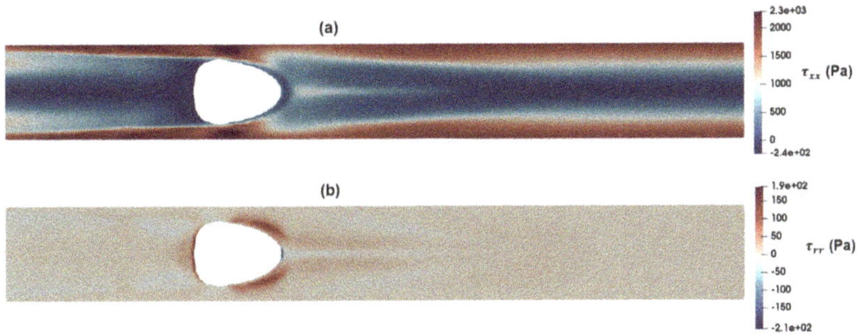

Figure 16. Distribution of normal stress components as predicted by the PTT model: (**a**) τ_{xx} and (**b**) τ_{rr}.

Figure 17a,b show the distribution of the axial velocity around the suspended droplets as predicted by the PTT and Carreau–Yasuda models, respectively. As observed in these figures, the PTT model predicts a high-velocity region at the front of the droplet, while the velocity distribution predicted by the Carreau–Yasuda model shows a velocity profile similar to that of fully developed flow. Figure 18a,b show the distribution of radial velocity around the suspended droplets as predicted by the PTT (Figure 18a) and Carreau–Yasuda (Figure 18b) models. According to these figures, the radial velocity at the rear and front parts of droplet are toward the wall and center of the channel, respectively. Such velocity distributions suggest the formation of a contraction flow regime in the front region of the droplet. Apart from their inability to predict elastic stress and time dependency, GNF models, such as Carreau–Yasuda, also predict a constant Trouton ratio, that is the ratio of extensional viscosity to shear viscosity, which remains constant at the Newtonian value of 3. Thus, their application for flow fields with mixed kinematics is more limited [44]. Figures 16a and 17a clearly show an extensional region predicted by the PTT model in the front region of the droplet, while the Carreau–Yasuda model fails to predict such an extensional flow. Moreover, the memory effect inherent in viscoelastic fluids necessitates a careful consideration of the minimum distance between two passing cells or the frequency of events. This ensures that the flow field at a specific point has adequate time to attain the equilibrium condition encountered by the proceeding cell. Without this attention to temporal and spatial dynamics, the observed deformation of distinct cells may not yield reliable results when estimating their mechanical properties or conducting comparative analysis.

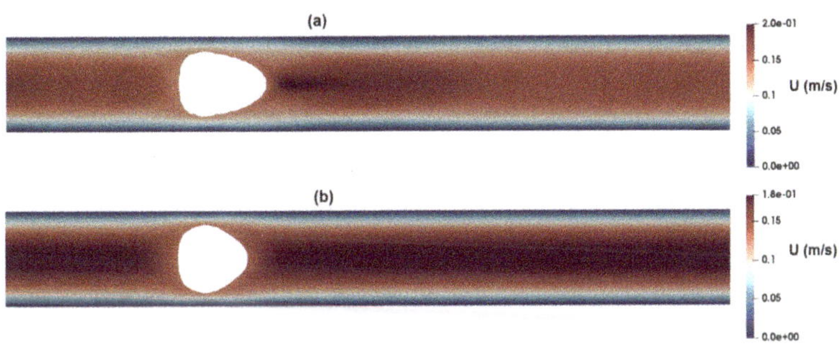

Figure 17. Distribution of axial velocity as predicted by the (**a**) PTT and (**b**) Carreau–Yasuda models.

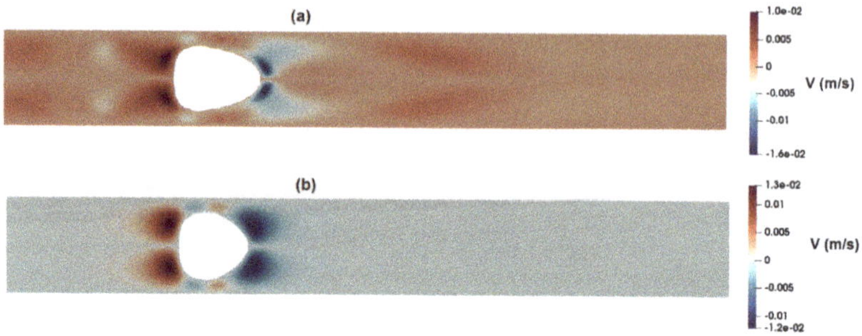

Figure 18. Distribution of radial velocity as predicted by the (**a**) PTT and (**b**) Carreau–Yasuda models.

4. Conclusions

The application of microfluidic systems for deformability cytometry has gained momentum, with numerous laboratories aiming to translate these technologies into commercially viable cytometers. Despite the common use of viscoelastic carrier fluids, the characterization of cells within these devices often relies on simplified computational frameworks that only account for viscous stresses while neglecting fluid elasticity.

The present study investigates the impact of the elasticity of the commonly used carrier fluid, the MC-PBS solution, on cell deformation in hydrodynamic constriction deformability cytometry. Three-dimensional computational analysis was conducted using two fluid models: a conventional GNF model that considers shear-thinning behavior but disregards fluid elasticity, and a comprehensive viscoelastic model, PTT, which incorporates both shear-thinning and elastic properties. The numerical results for the single-phase flow of a 0.5% MC-PBS solution indicate that both models yield nearly identical distributions of shear stress and velocity fields for fully developed steady shear flow. However, the incorporation of the viscoelastic model reveals significant normal stress generation along the channel, contradicting the assumption of zero normal stress in commonly used GNF fluid models. Importantly, CFD results highlight that the predicted intensity of normal stress surpasses that of shear stress.

The level set model was employed to predict the transient flow of suspended viscous particles representing HL-60 cells in a 0.5% MC-PBS solution. Initially neglecting fluid elasticity and subsequently considering the viscoelastic nature of the carrier fluid enabled a comprehensive comparison of predicted flow fields, stress distributions, and resulting deformations.

The obtained CFD results unveil substantial alterations in the flow field around moving particles within the channel. Accounting for the viscoelastic behavior of the carrier fluid led to slightly different predictions of shear stress distribution around the particles. However, the primary disparity lies in the prediction of significant normal stresses around the deforming particles, indicating a notable compressive force.

Comparing the predicted particle shapes using two CFD models—one incorporating elastic effects and the other excluding them—underscored the pivotal role of fluid elasticity in particle deformation. The predicted velocity fields around the moving particles also revealed a combination of shear and extensional flows in the polymer solution, necessitating the use of a CFD model that considers the viscoelastic behavior of the fluid.

These findings raise pertinent questions regarding the accuracy of mechanical property estimates derived from neglecting elastic stresses. The observed disparities underscore the importance of considering fluid elasticity in deformability cytometry analysis, particularly for chips with hydrodynamic constriction channels.

Author Contributions: Conceptualization, H.P.; Methodology, H.P.; Formal analysis, H.P.; Writing—original draft, H.P.; Writing—review & editing, J.P.H.; Supervision, J.P.H.; Project administration, J.P.H.; Funding acquisition, J.P.H. All authors have read and agreed to the published version of the manuscript.

Funding: This research was funded by the National Institutes of Health, National Institute for General Medical Sciences grant number 1R35GM152076. The APC was funded by 1R35GM152076.

Data Availability Statement: The original contributions presented in the study are included in the article, further inquiries can be directed to the corresponding author.

Conflicts of Interest: The authors declare no conflicts of interest.

References

1. Nematbakhsh, Y.; Lim, C.T. Cell biomechanics and its applications in human disease diagnosis. *Acta Mech. Sin.* **2015**, *31*, 268–273. [CrossRef]
2. Darling, E.M.; Di Carlo, D. High-throughput assessment of cellular mechanical properties. *Annu. Rev. Biomed. Eng.* **2015**, *17*, 35–62. [CrossRef] [PubMed]
3. Otto, O.; Rosendahl, P.; Mietke, A.; Golfier, S.; Herold, C.; Klaue, D.; Girardo, S.; Pagliara, S.; Ekpenyong, A.; Jacobi, A.; et al. Real-time deformability cytometry: On-the-fly cell mechanical phenotyping. *Nat. Methods* **2015**, *12*, 199–202. [CrossRef]
4. Guck, J.; Schinkinger, S.; Lincoln, B.; Wottawah, F.; Ebert, S.; Romeyke, M.; Lenz, D.; Erickson, H.M.; Ananthakrishnan, R.; Mitchell, D.; et al. Optical deformability as an inherent cell marker for testing malignant transformation and metastatic competence. *Biophys. J.* **2005**, *88*, 3689–3698. [CrossRef] [PubMed]
5. Swaminathan, V.; Mythreye, K.; O'Brien, E.T.; Berchuck, A.; Blobe, G.C.; Superfine, R. Mechanical stiffness grades metastatic potential in patient tumor cells and in cancer cell lines. *Cancer Res.* **2011**, *71*, 5075–5080. [CrossRef] [PubMed]
6. Tse, H.T.K.; Gossett, D.R.; Moon, Y.S.; Masaeli, M.; Sohsman, M.; Ying, Y.; Mislick, K.; Adams, R.P.; Rao, J.; Di Carlo, D. Quantitative diagnosis of malignant pleural effusions by single-cell mechanophenotyping. *Sci. Transl. Med.* **2013**, *5*, 212ra163. [CrossRef]
7. Bufi, N.; Saitakis, M.; Dogniaux, S.; Buschinger, O.; Bohineust, A.; Richert, A.; Maurin, M.; Hivroz, C.; Asnacios, A. Human primary immune cells exhibit distinct mechanical properties that are modified by inflammation. *Biophys. J.* **2015**, *108*, 2181–2190. [CrossRef]
8. Bashant, K.R.; Vassallo, A.; Herold, C.; Berner, R.; Menschner, L.; Subburayalu, J.; Kaplan, M.J.; Summers, C.; Guck, J.; Chilvers, E.R.; et al. Real-time deformability cytometry reveals sequential contraction and expansion during neutrophil priming. *J. Leukoc. Biol.* **2019**, *105*, 1143–1153. [CrossRef]
9. Ekpenyong, A.E.; Whyte, G.; Chalut, K.; Pagliara, S.; Lautenschläger, F.; Fiddler, C.; Paschke, S.; Keyser, U.F.; Chilvers, E.R.; Guck, J. Viscoelastic Properties of Differentiating Blood Cells Are Fate- and Function-Dependent. *PLoS ONE* **2012**, *7*, e45237. [CrossRef]
10. Lin, J.; Kim, D.; Tse, H.T.; Tseng, P.; Peng, L.; Dhar, M.; Karumbayaram, S.; Di Carlo, D. High-throughput physical phenotyping of cell differentiation. *Microsyst. Nanoeng.* **2017**, *3*, 17013. [CrossRef]
11. Guck, J.; Ananthakrishnan, R.; Mahmood, H.; Moon, T.J.; Cunningham, C.C.; Käs, J. The optical stretcher: A novel laser tool to micromanipulate cells. *Biophys. J.* **2001**, *81*, 767–784. [CrossRef] [PubMed]
12. Radmacher, M. Studying the mechanics of cellular processes by atomic force microscopy. *Methods Cell Biol.* **2007**, *83*, 347–372. [CrossRef] [PubMed]
13. Hochmuth, R.M. Micropipette aspiration of living cells. *J. Biomech.* **2000**, *33*, 15–22. [CrossRef] [PubMed]
14. Thoumine, O.; Ott, A.; Cardoso, O.; Meister, J.J. Microplates: A new tool for manipulation and mechanical perturbation of individual cells. *J. Biochem. Biophys. Methods* **1999**, *39*, 47–62. [CrossRef] [PubMed]
15. Lange, J.R.; Steinwachs, J.; Kolb, T.; Lautscham, L.A.; Harder, I.; Whyte, G.; Fabry, B. Microconstriction arrays for high-throughput quantitative measurements of cell mechanical properties. *Biophys. J.* **2015**, *109*, 26–34. [CrossRef] [PubMed]
16. Nyberg, K.D.; Hu, K.H.; Kleinman, S.H.; Khismatullin, D.B.; Butte, M.J.; Rowat, A.C. Quantitative deformability cytometry: Rapid, calibrated measurements of cell mechanical properties. *Biophys. J.* **2017**, *113*, 1574–1584. [CrossRef] [PubMed]
17. Gossett, D.R.; Tse, H.T.K.; Lee, S.A.; Ying, Y.; Lindgren, A.G.; Yang, O.O.; Rao, J.; Clark, A.T.; Di Carlo, D. Hydrodynamic stretching of single cells for large population mechanical phenotyping. *Proc. Natl. Acad. Sci. USA* **2012**, *109*, 7630–7635. [CrossRef]
18. Urbanska, M.; Muñoz, H.E.; Shaw Bagnall, J.; Otto, O.; Manalis, S.R.; Di Carlo, D.; Guck, J. A comparison of microfluidic methods for high-throughput cell deformability measurements. *Nat. Methods* **2020**, *17*, 587–593. [CrossRef] [PubMed]
19. Nawaz, A.A.; Urbanska, M.; Herbig, M.; Nötzel, M.; Kräter, M.; Rosendahl, P.; Herold, C.; Toepfner, N.; Kubankova, M.; Goswami, R.; et al. Using real-time fluorescence and deformability cytometry and deep learning to transfer molecular specificity to label-free sorting. *bioRxiv* **2019**. [CrossRef]
20. Rosendahl, P.; Plak, K.; Jacobi, A.; Kraeter, M.; Toepfner, N.; Otto, O.; Herold, C.; Winzi, M.; Herbig, M.; Ge, Y.; et al. Real-time fluorescence and deformability cytometry. *Nat. Methods* **2018**, *15*, 355–358. [CrossRef]

21. Fregin, B.; Czerwinski, F.; Biedenweg, D.; Girardo, S.; Gross, S.; Aurich, K.; Otto, O. High-throughput single-cell rheology in complex samples by dynamic real-time deformability cytometry. *Nat. Commun.* **2019**, *10*, 415. [CrossRef]
22. Fregin, B.; Czerwinski, F.; Biedenweg, D.; Girardo, S.; Gross, S.; Aurich, K.; Otto, O. Extracting cell stiffness from real-time deformability cytometry: Theory and experiment. *Biophys. J.* **2015**, *109*, 2023–2036. [CrossRef]
23. Mokbel, M.; Mokbel, D.; Mietke, A.; Träber, N.; Girardo, S.; Otto, O.; Guck, J.; Aland, S. Numerical simulation of real-time deformability cytometry to extract cell mechanical properties. *ACS Biomater. Sci. Eng.* **2017**, *3*, 2962–2973. [CrossRef] [PubMed]
24. Guillou, L.; Dahl, J.B.; Lin, J.M.G.; Barakat, A.I.; Husson, J.; Muller, S.J.; Kumar, S. Measuring cell viscoelastic properties using a microfluidic extensional flow device. *Biophys. J.* **2016**, *111*, 2039–2050. [CrossRef] [PubMed]
25. Armistead, F.J.; De Pablo, J.G.; Gadêlha, H.; Peyman, S.A.; Evans, S.D. Cells under stress: An inertial-shear microfluidic determination of cell behavior. *Biophys. J.* **2019**, *116*, 1127–1135. [CrossRef] [PubMed]
26. Wang, Z.; Lu, R.; Wang, W.; Tian, F.B.; Feng, J.J.; Sui, Y. A computational model for the transit of a cancer cell through a constricted microchannel. *Biomech. Model. Mechanobiol.* **2023**, *22*, 1129–1143. [CrossRef] [PubMed]
27. Büyükurgancı, B.; Basu, S.K.; Neuner, M.; Guck, J.; Wierschem, A.; Reichel, F. Shear rheology of methyl cellulose based solutions for cell mechanical measurements at high shear rates. *Soft Matter* **2023**, *19*, 1739–1748. [CrossRef] [PubMed]
28. Urbanska, M. Single-Cell Mechanical Phenotyping Across Timescales and Cell State Transitions. Doctor Dissertation, Technische Universität Dresden, Dresden, Germany, 2022. Available online: https://nbn-resolving.org/urn:nbn:de:bsz:14-qucosa2-775638 (accessed on 1 February 2024).
29. Wu, P.-H.; Raz-Ben Aroush, D.; Asnacios, A.; Chen, W.-C.; Dokukin, M.E.; Doss, B.L.; Durand-Smet, P.; Ekpenyong, A.; Guck, J.; Guz, N.V.; et al. A comparison of methods to assess cell mechanical properties. *Nat. Methods* **2018**, *15*, 491–498. [CrossRef] [PubMed]
30. Thien, N.P.; Tanner, R.I. A new constitutive equation derived from network theory. *J. Non-Newton. Fluid Mech.* **1977**, *2*, 353–365. [CrossRef]
31. Alves, M.A.; Oliveira, P.J.; Pinho, F.T. Numerical methods for viscoelastic fluid flows. *Annu. Rev. Fluid Mech.* **2021**, *53*, 509–541. [CrossRef]
32. Cruz, D.O.A.; Pinho, F.; Oliveira, P.J. Analytical solutions for fully developed laminar flow of some viscoelastic liquids with a Newtonian solvent contribution. *J. Non-Newton. Fluid Mech.* **2005**, *132*, 28–35. [CrossRef]
33. Azaiez, J.; Guénette, R.; Ait-Kadi, A. Numerical simulation of viscoelastic flows through a planar contraction. *J. Non-Newton. Fluid Mech.* **1996**, *62*, 253–277. [CrossRef]
34. Lizarraga-Garcia, E.; Buongiorno, J.; Al-Safran, E.; Lakehal, D. A broadly-applicable unified closure relation for Taylor bubble rise velocity in pipes with stagnant liquid. *Int. J. Multiphase Flow* **2017**, *89*, 345–358. [CrossRef]
35. Caviezel, D.; Narayanan, C.; Lakehal, D. Adherence and bouncing of liquid droplets impacting on dry surfaces. *Microfluid. Nanofluid.* **2008**, *5*, 469–478. [CrossRef]
36. Lakehal, D.; Larrignon, G.; Narayanan, C. Computational heat transfer and two-phase flow topology in miniature tubes. *Microfluid. Nanofluid.* **2008**, *4*, 261–271. [CrossRef]
37. Available online: https://transat-cfd.com/ (accessed on 1 February 2024).
38. Lakehal, D.; Meier, M.; Fulgosi, M. Interface tracking towards the direct simulation of heat and mass transfer in multiphase flows. *Int. J. Heat Fluid Flow* **2002**, *23*, 242–257. [CrossRef]
39. Suzuki, T.; Kalyan, S.; Berlinicke, C.; Yoseph, S.; Zack, D.J.; Hur, S.C. Deciphering viscoelastic cell manipulation in rectangular microchannels. *Phys. Fluids* **2023**, *35*, 103117. [CrossRef]
40. Pouraria, H.; Foudazi, R.; Houston, J.P. Exploitation of elasto-inertial fluid flow for the separation of nano-sized particles: Simulating the isolation of extracellular vesicles. *Cytom. Part A* **2023**, *103*, 786–795. [CrossRef] [PubMed]
41. Poryles, R.; Zenit, R. Encapsulation of droplets using cusp formation behind a drop rising in a non-Newtonian fluid. *Fluids* **2018**, *3*, 54. [CrossRef]
42. Ohta, M.; Kobayashi, N.; Shigekane, Y.; Yoshida, Y.; Iwata, S. The dynamic motion of single bubbles with unique shapes rising freely in hydrophobically modified alkali-soluble emulsion polymer solutions. *J. Rheol.* **2015**, *59*, 303–316. [CrossRef]
43. Esposito, G.; Dimakopoulos, Y.; Tsamopoulos, J. Buoyancy driven flow of a viscous drop in viscoelastic materials. *J. Non-Newton. Fluid Mech.* **2023**, *321*, 105124. [CrossRef]
44. Poole, R.J. Inelastic and flow-type parameter models for non-Newtonian fluids. *J. Non-Newton. Fluid Mech.* **2023**, *320*, 105106. [CrossRef]
45. Badami, M.M.; Alizadeh, P.; Almasi, S.; Riasi, A.; Sadeghy, K. Numerical study of blood hammer phenomenon considering blood viscoelastic effects. *Eur. J. Mech. B/Fluids* **2022**, *95*, 212–220. [CrossRef]
46. Mirzaaghaian, A.; Ramiar, A.; Ranjbar, A.A.; Warkiani, M.E. Application of level-set method in simulation of normal and cancer cells deformability within a microfluidic device. *J. Biomech.* **2020**, *112*, 110066. [CrossRef] [PubMed]
47. Nooranidoost, M.; Kumar, R. Improving viability of leukemia cells by tailoring shell fluid rheology in constricted microcapillary. *Sci. Rep.* **2020**, *10*, 11570. [CrossRef]
48. Tsai, M.A.; Waugh, R.E.; Keng, P.C. Cell cycle-dependence of HL-60 cell deformability. *Biophys. J.* **1996**, *70*, 2023–2029. [CrossRef] [PubMed]

49. Rosenbluth, M.J.; Lam, W.A.; Fletcher, D.A. Force microscopy of nonadherent cells: A comparison of leukemia cell deformability. *Biophys. J.* **2006**, *90*, 2994–3003. [CrossRef]
50. Wittwer, L.D.; Reichel, F.; Müller, P.; Guck, J.; Aland, S. A new hyperelastic lookup table for RT-DC. *Soft Matter* **2023**, *19*, 2064–2073. [CrossRef]

micromachines

MDPI

Article

A Microfluidic Approach for Assessing the Rheological Properties of Healthy Versus Thalassemic Red Blood Cells

Hao Jiang [1,2], Xueying Li [2], Zhuoyan Liu [2], Siyu Luo [2], Junbin Huang [3], Chun Chen [3], Rui Chen [1,*] and Fenfang Li [2,*]

1 Department of Electrical and Electronic Engineering, Southern University of Science and Technology, Shenzhen 518055, China; 12232176@mail.sustech.edu.cn

2 Shenzhen Bay Laboratory, Institute of Biomedical Engineering, Shenzhen 518107, China; lixueying921@yeah.net (X.L.); liuzhuoy@szbl.ac.cn (Z.L.); m19852337009@163.com (S.L.)

3 Pediatric Hematology Laboratory, Division of Hematology/Oncology, Department of Pediatrics, The Seventh Affiliated Hospital of Sun Yat-Sen University, Shenzhen 518107, China; huangjb37@mail.sysu.edu.cn (J.H.); chenchun@mail.sysu.edu.cn (C.C.)

* Correspondence: chenr@sustech.edu.cn (R.C.); fenfang.li@szbl.ac.cn (F.L.)

Abstract

The deformability of red blood cells (RBCs) is critical for microvascular circulation and is impaired in hematological disorders such as thalassemia, a prevalent public health concern in Guangdong, China. While microfluidics enable high-precision deformability assessment, current studies lack standardization in deformation metrics and rarely investigate post-deformation recovery dynamics. This study introduces an automated microfluidic platform for systematically evaluating RBC deformability in healthy and thalassemic individuals. A biomimetic chip featuring 4 μm, 8 μm, and 16 μm wide channels (7 μm in height) was designed to simulate capillary dimensions, with COMSOL CFD numerical modeling validating shear stress profiles. RBC suspensions (10^7 cells/mL in DPBS) were hydrodynamically focused through constrictions while high-speed imaging (15,000 fps) captured deformation–recovery dynamics. Custom-built algorithms with deep-learning networks automated cell tracking, contour analysis, and multi-parametric quantification. Validation confirmed significantly reduced deformability in Paraformaldehyde (PFA)-treated RBCs compared to normal controls. Narrower channels and higher flow velocities amplified shear-induced deformations, with more deformable cells exhibiting faster post-constriction shape recovery. Crucially, the platform distinguished thalassemia patient-derived RBCs from healthy samples, revealing significantly lower deformability in diseased cells, particularly in 4 μm channels. These results establish a standardized, high-throughput framework for RBC mechanical characterization, uncovering previously unreported recovery dynamics and clinically relevant differences in deformability in thalassemia. The method's diagnostic sensitivity highlights its translational potential for screening hematological disorders.

Keywords: red blood cells; deformability assessment; shape recovery; microfluidic platform; thalassemia

1. Introduction

Normal human red blood cells (RBCs) have a distinctive biconcave disc shape, measuring approximately 8 μm in diameter and about 2 μm in thickness. During microcirculation, they can undergo passive deformation while maintaining mechanical stability [1]. The deformability of RBCs depends on the structural properties of the cytoskeletal components [2], the vertical interactions between the cytoskeleton and the integral transmembrane complexes [3], as well as intracellular viscosity, RBC hydration state, and surface area–volume interactions [4]. Healthy RBCs readily respond to shear stress in the microcirculation, allowing them to deform effectively and pass through capillaries. Changes in the deformability of RBCs directly influence cell movement, morphological maintenance, and intercellular interactions [5], thereby affecting functions such as cell migration [6], mechanosensing [7], and signal transduction [8]. These alterations significantly influence the progression and severity of cardiovascular diseases, metabolic disorders, and neurological conditions [9,10]. The restoration of RBC morphology following deformation is essential for maintaining normal physiological function [11]. Mechanistically, this restorative process is governed by two intrinsic biophysical parameters: (i) the surface area-to-volume ratio (SA/V), which dictates the geometric constraints for reversible shape transitions [12], and (ii) the viscoelastic properties of the erythrocyte membrane, whose dynamic behavior is modulated by membrane-associated proteins [13]. Understanding both the deformability and shape recovery of RBCs is essential for elucidating their biophysical effects.

To measure the deformability of RBCs, several experimental methodologies have been proposed, which can be broadly categorized into direct and indirect methods. Direct methods include micropipette aspiration, optical tweezer, twisting micromagnetic beads, atomic force microscopy, parallel plate rheometry, and controlled cavitation rheology. Micropipette aspiration technology [14,15] employs a pressure difference to aspirate cells into a micropipette in a quasi-static way, allowing for the assessment of cellular deformability and viscoelasticity. Optical tweezer technology [16,17] and twisting micromagnetic beads [18] utilize laser beams and a magnetic field, respectively, to capture microbeads attached to the cell surface, generating a controlled force to assess cell mechanics [19]. Atomic force microscopy (AFM) [20,21] measures the mechanical properties of cells by obtaining force–distance curves at the cell surface, enabling high-precision elasticity measurements of individual living cells under physiological conditions. Parallel plate rheometry [22] quantifies the deformation of single cells under external stress. These methods assess the time-resolved response to force and extract parameters related to cellular deformability. However, these techniques are technically demanding and time-consuming, limiting measurement efficiency and making high-throughput measurements difficult.

In contrast, microfluidic-based methods [23] represent a highly attractive alternative due to their potential for single-cell precision and high-throughput deformability analysis [24–26]. The controlled cavitation rheology method stretches a collection of RBCs using impulsive flow generated by a single laser-induced cavitation bubble in a microchannel [27]. It can stretch many cells simultaneously and allows one to obtain large initial cell deformations and yield strength [28–30]. However, this direct method generates heterogeneous forces and cell deformations and requires integration of optics with microfluidics. In recent years, indirect methods for measuring the deformability of cells based on microfluidic technology have demonstrated ease of operation, low cost, and high throughput. They can be categorized into three types: constriction-based deformability cytometry (cDC), extensional flow deformability cytometry (xDC), and shear flow deformability cytometry (sDC) [31]. With the conventional cDC method, cells deform when passing through a narrow constriction channel with dimensions smaller than their diameter [32,33], where cell deformability is influenced by factors such as surface friction, stiffness, viscoelasticity, and adhesion properties. The Suspended Microchannel Resonator (SMR) quantifies cellular stiffness by measuring the time required for a cell to pass through a constriction channel comparable in size to the cell itself [34]. However, the measurement results provide a composite indicator of cellular deformability and do not characterize the deformation process of the cell within the channel or its recovery after exiting the constriction. The xDC methods quantify cellular stiffness by measuring the deformation ratio of the long axis to the short axis of cells in an extensional flow field, assuming an elliptical cell shape [35,36]. Multiparametric deformability cytometry (m-DC) [37] was developed based on this, integrating parameters such as cell size, deformability, deformation kinetics, and morphology. However, due to the anisotropic nature of RBCs [11], this method is not suitable for measuring their deformability.

The sDC method is a non-contact flow cytometry method that applies hydrodynamic flow shear to cells in a microchannel with dimensions slightly larger than the cell diameter at a constant flow velocity, inducing mechanical cell deformation under a strong shear stress gradient [38–40]. Recent studies have utilized high-speed charge-coupled device (CCD) cameras to capture real-time images of cells at various stages of deformation [41]. Additionally, some studies have integrated microscopic cell imaging with numerical simulations to further investigate the factors contributing to red blood cell deformation [42]. Moreover, due to their operational simplicity, these methods hold potential for applications in biological laboratories and clinical settings [43,44]. In recent years, the rapid advancement of microfluidic technology has provided an excellent platform for studying the biophysical properties of RBCs under both physiological and pathological conditions.

The deformability of RBCs is influenced by many pathological conditions, such as malaria [45], diabetes [46,47], sickle cell disease [48], thalassemia [49], hereditary spherocytosis [50], and hereditary xerocytosis [51]. The deformability of RBCs can be affected by metabolic processes that control ATP levels and redox state [52]. The inability to maintain deformability leads to a shortened RBC lifespan, and if not compensated for by the production of new RBCs, can result in hemolytic anemia [53]. Therefore, reliably estimating the deformability of RBCs and understanding the factors that control this process are crucial for assessing the severity of a patient's condition and selecting the best treatment strategies. For example, thalassemia, a globally significant hereditary blood disorder, is closely associated with the biophysical properties of red blood cells. Caused by autosomal gene defects, the disease results in an imbalance in the synthesis ratio of α- and β-globin chains in hemoglobin [54,55]. The unpaired globin chains precipitate and adhere to the red blood cell membrane skeleton, reducing the membrane's viscoelasticity and thereby impairing the deformability of red blood cells. This ultimately leads to hemolytic anemia [49].

In this study, we developed a high-throughput method for assessing RBC deformability based on microfluidic technology. The RBC suspension was introduced into a microfluidic channel with a microinjection pump. A high-speed camera was employed to record the complete motion of RBCs as they traversed a narrow channel at tens of thousands of frames per second (fps). A custom-developed program was utilized to analyze the deformation of RBCs throughout their motion, incorporating a deep learning algorithm to accurately extract RBC contours. Four parameters were established to characterize deformability. Additionally, we observed the morphological recovery of RBCs after exiting the narrow channel and compared several different RBC models, including healthy RBCs, RBCs fixed with paraformaldehyde (PFA), and those from patients with thalassemia. The results suggest that this method holds promise as a potential tool for disease diagnosis using label-free biophysical markers of RBCs.

2. Materials and Methods

2.1. Cell Sample Preparation

Whole blood samples (1–2 mL) were collected from the veins of healthy donors during medical checkups or from patients with thalassemia at the Seventh Affiliated Hospital of Sun Yat-Sen University. Heparin was used as an anticoagulant. Both healthy donors and patients with thalassemia were aged between 2 and 8 years old and resided in Guangdong province. The Seventh Affiliated Hospital of Sun Yat-Sen University's Ethics Committee approved the collection of patient samples, and the study complied with institutional norms, the Declaration of Helsinki, and guardians' informed consent. The collected blood samples were stored and transported at 4 °C to the Shenzhen Bay laboratory within 2 h of collection.

The fresh whole blood samples were washed 3 times and resuspended in 1X DPBS (Ca^{2+}+, Mg^{2+}+) after centrifugation at 1000 rpm for 3 min to achieve a final RBC concentration of approximately 10^7 cells/mL (the concentration of red blood cells in whole blood is approximately 4.7–6.1 $\times$ 10^9 cells/mL [56]).

For PFA treatment, the above samples were resuspended in 0.5% PFA in 1X DPBS (Ca^{2+}+, Mg^{2+}+) and incubated at room temperature for 30 min, after which the supernatants were discarded after centrifugation and the pellets resuspended in 1X DPBS (Ca^{2+}+, Mg^{2+}+) to achieve a final cell concentration of ~10^7 cells/mL.

2.2. *Microfluidic Device Preparation*

The capillary vessel-mimicking microchannel was designed using AutoCAD 2021 software. We used standard soft lithography techniques to fabricate the patterned SU-8-based master mold, from which the polydimethylsiloxane (PDMS, Sylgard 184 Silicone Elastomer Kit (Dow Corning (Dowsil), Midland, MI, USA), 10:1 mix ratio, cured at 60 °C for 4 h) microchannel was cast. The microchannel was then bonded to a glass slide (50 $\times$ 75 mm^2) immediately after 50 s of treatment in a plasma cleaner (Zepto one, Diener, Ebhausen, Germany).

2.3. *Experimental Operation for Cell Deformation and Recovery*

The microfluidic chip was placed on an inverted microscope (BDS500, CNoptec, Chongqing, China). A syringe pump (R462, KDS, Issy-les-Moulineaux, France) was used to introduce the cell suspensions into the microfluidic chip at a flow rate of 0.01 μL/min, and the RBCs were focused on the channel center by a sheath flow of 1 $\times$ PBS driven by another syringe pump (R462, RWD, Shenzen, China) at 0.02 μL/min. RBCs experienced elevated shear stress and became deformed upon entering the narrow channel, but started to recover their shape after exiting the channel. A high-speed camera (Nova S12, Photron, Tokyo, Japan) was used to capture the deformation and shape recovery of RBCs through a 63$\times$ oil immersion objective at a frame rate of 10,000 fps and exposure time of 10 μs to minimize motion blur. A custom-built script was employed for automatic cell tracking and contour analysis, from which the velocity and deformation of RBCs were extracted. The number of pixels per RBC is critical for tracking cells and analyzing cell contours. In our study, each red blood cell occupied approximately 510 pixels in total, which we found to be sufficient to ensure accuracy of cellular contour detection.

2.4. *Numerical Simulation*

Numerical simulation of the flow field in the microchannel was performed with a finite-element solver (COMSOL Multiphysics 6.0 trial, CFD module, COMSOL, Burlington, MA, USA). The geometry of the microchannel was imported into COMSOL from AutoCAD. Non-slip boundary conditions were employed, and the initial velocity at the inlets was calculated from experimental conditions. Though the flow demonstrated a Poiseuille profile, we were more interested in the velocity along the channel centerline in the flow direction (x) as we used sheath flow and focused RBCs to the channel center. Based on both experimental recordings and numerical simulations, the steep velocity gradients along the flow direction (x) and the channel centerline at the entrance and exit of the narrow channel expose RBCs to markedly higher shear stress, resulting in notable extensional deformation. The shear stress σ herein refers to that related to this velocity gradient.

2.5. Image Processing and Data Analysis

2.5.1. Cell Tracking

Custom-built MATLAB 2024b and Python 3.0 scripts were used to analyze each image frame of recorded cell motion and deformation. Image preprocessing was performed to improve the image contrast and enhance the edge areas by image filtering and adjusting the grayscale histogram. Subsequently, the current image was processed by subtracting a pre-saved background image. Then, grayscale thresholding and binarization operations were employed to extract the cellular regions in the images, from which the approximate location and area occupied by the cell in the current frame can be determined. The motion state of the cellular region in the current image was predicted using the Kalman filter algorithm. The Hungarian algorithm was utilized to match the cellular regions across different image frames [57,58]. This approach facilitated the acquisition of the complete motion trajectories of individual cells within the microfluidic channel.

2.5.2. Cell Contour Analysis Based on Deep Learning Methodologies

The extracted cellular regions were cropped from the original images. Then, the Canny algorithm combined with image erosion and dilation was applied to extract the cell contours [59], from which cell deformability was calculated. After exiting the narrow channels, due to imaging conditions, artefacts from a discernible white rim of the cells rendered the accurate extraction of cell contours using the Canny algorithm, which was based on image gradient operations. To solve this problem, a deep learning approach was employed with the open-source YOLOv8 convolutional neural network for the semantic segmentation of cell images [60,61], see Figure 1.

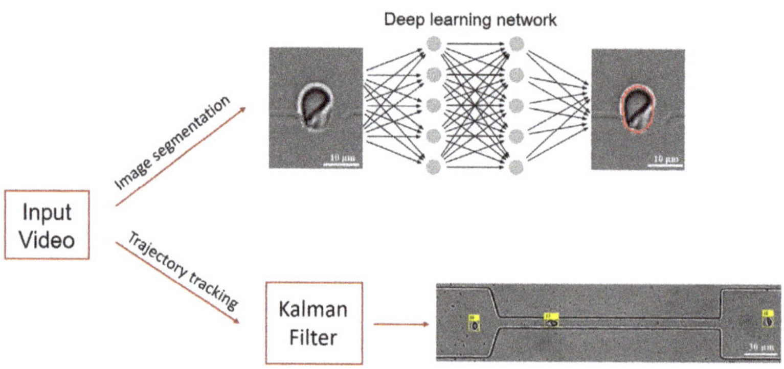

Figure 1. A schematic diagram of the core workflow for image processing of microfluidic experimental videos. It consists of two main steps: image segmentation and multi-object tracking. The YOLOv8 deep learning network model is employed to achieve accurate segmentation of the contour of red blood cells, and the Kalman filter algorithm is utilized to track the trajectories of multiple red blood cells in the video images. The red curve delineates the contour of the RBC, while the yellow square indicates the index of the RBC identified.

A dataset for red blood cell segmentation was constructed, comprising 60 images each for the narrow channels of three different widths that cells pass through. In this training set, half of the cells were normal healthy controls, and the other half were red blood cells from patients with thalassemia. LabelMe was used for data annotation. The pre-trained YOLOv8 deep learning network model was subsequently applied to the annotated dataset to perform model training. This was followed by testing and validation using 20 new cell images. Then, the trained network model was called in a MATLAB program to perform inference, yielding precise cell contours for subsequent analysis of cell deformation and recovery. The integration of traditional image processing techniques with contemporary deep learning methodologies enhanced the accuracy and reliability of cell contour extraction, providing a robust foundation for analyzing cell deformation and recovery features.

2.5.3. Cell Deformation Analysis

RBCs have a disc shape that can be distorted under deformation. Both our studies and those of others found that RBCs travelling through a microfluidic constriction channel typically adopt a limited set of morphologies [62]. We restricted our analysis to cells exhibiting the parachute shape, which was the most common in our study.

We detected the cell contour accurately, obtained the projected cell area A and perimeter P, and used four definitions of cell deformation. The centroid of an individual red blood cell was extracted from the obtained cell contours, from which the moving speed of the cell can be calculated between image frames. We assume an ellipse fit to the cell contour for aspect ratio and Taylor deformation, using the major and minor axes a and b. For the non-circularity index and changes in the major axis, we did not fit the cell contour to an ellipse. Based on these obtained parameters, four distinct definitions of cell deformation were generated:

Aspect ratio:

$$\mathrm{a/b}$$

Taylor deformation:

$$Ta = (\mathrm{a} - \mathrm{b})/(\mathrm{a} + \mathrm{b})$$

Non-circularity:

$$D = 1 - 2(\pi A)^{\frac{1}{2}}/P$$

Changes in the long axis of the cells:

$$a/a_0$$

where a_0 is the initial long axis of the cell before entering the narrow stretching channel.

2.5.4. Cell Shape Recovery Analysis

Aspect ratios *a/b* were extracted from cell contours after the cells exited the narrow channel. The dataset of a/b vs. time for each cell was first smoothed and then fitted with an exponential function $f = C_1 + C_2 \times exp(-t/\tau)$ with MATLAB using its built-in Non-Linear-Least-Squares and the robust method of fitting Bisquare and Trust-Region. Only data with goodness of fit (R-squared) above 0.95 were retained for subsequent statistical analysis of cell shape recovery.

The shear elastic moduli and viscous moduli for individual RBCs were estimated using methods described by Hocumuth et al. [63,64] and recently demonstrated in a similar experimental protocol to this study by Mancuso et al. [65]. We first use the classic Kelvin–Voigt model:

$$T = \frac{E}{2}\left(\lambda^2 - \lambda^{-2}\right) + \frac{2\eta}{\lambda} \cdot \frac{\partial \lambda}{\partial t} \tag{1}$$

where T is the uniaxial tension force, $\lambda = a/a_0$ is the stretch ratio, E is the elastic shear modulus, and η is the viscous modulus. The tension per unit length from purely uniaxial flow (valid along the centerline of our channels) is thus approximated as follows:

$$T = \frac{\sigma A}{L_0} = \frac{3A\mu\frac{du}{dx}}{L_0} \tag{2}$$

where $\sigma = 3\mu\frac{du}{dx}$ is the shear stress from a uniaxial extensional flow along the centerline of the channel, μ is the suspending fluid viscosity, $A = 136\ \mu m^2$ is the average area of the RBC membrane, and $L_0 = 8\ \mu m$ is the average resting length of a human RBC membrane. Settings (1) and (2) are equal to each other, and solving for $\frac{\partial\lambda}{\partial t}$ yields a first-order nonlinear ordinary differential equation for λ,

$$\frac{\partial\lambda}{\partial t} = \frac{1}{2\eta}\cdot\left(\frac{3A\mu\frac{du}{dx}}{L_0}\cdot\lambda - \frac{E}{2}(\lambda^3 - \lambda^{-1})\right) \quad (3)$$

2.6. Statistical Analysis

If only two groups were compared, a two-tailed Student's *t*-test was used for the statistical analysis of the deformability parameters. If more than two groups were compared, a one-way ANOVA test was performed first to determine whether one or more treatment groups were significantly different, followed by a post-hoc Tukey HSD multiple comparison test to identify which of the treatment pairs were significantly different from each other.

3. Results

3.1. Microfluidic Device Design, Experimental Setup, and Deformability Characterization

A microfluidic chip was designed to mimic small capillary vessels, where RBCs were stretched in a 200 μm long narrow part of the channel with widths of 4, 8, or 16 μm. Cell motion and deformation were imaged by a high-speed camera on an inverted microscope (Figure 2A). Each microfluidic unit had two inlets and one outlet. RBCs were focused by the sheath flow of 1× DPBS to the channel center before entering the narrow stretching region to ensure the reproducibility of the experiment (Figure 2A inset and Figure 2B). Numerical simulation of the flow field in the microchannel revealed the focusing effect of the sheath flow (dashed black box in Figure 2C) and the elevating velocity in the narrow stretching channel (red box in Figure 2C). The flow velocity in the narrow channel accelerated to around 0.07 m/s in the channel center, while the channel walls acted as a non-slip boundary, and the flow velocity before the entry and after the exit of the narrow channel varied between 0.01 and 0.02 m/s. These velocity gradients generated shear stress on cells entering and exiting the narrow channel, leading to cell deformation and shape recovery at relevant moments.

As shown in Figure 3A, an RBC was stretched when entering the narrow channel at 1.7 ms, it remained stretched from 3.0–5.0 ms until it exited the narrow channel at 6.3 ms and gradually recovered its shape (11.4 ms). Custom-built MATLAB and Python scripts were used for cell tracking (Figure 3B left). Traditional image processing techniques were integrated with contemporary deep learning methodologies to enhance the accuracy and reliability of cell contour extraction (Figure 3B right). Based on the obtained cell trajectories and cell contours, cell velocities and different definitions of cell deformation, including aspect ratio (*a/b*), Taylor deformation *Ta*, non-circularity, and change in the major axis of the cells (a/a_0), were calculated. Figure 3C shows that both the cell velocity and deformation increased when the cell entered the narrow stretching channel, remained at a high level, and declined when the cell exited the narrow channel. It is interesting to note that the deformation reached its peak value at the entrance of the narrow channel (e.g., ~1.7 ms), which can be explained by the velocity gradient along the cell flow direction, e.g., larger flow velocity on the right side of the cell compared to the left side. Similarly, when the cell exited the narrow channel and encountered the velocity gradient, its elongated shape was compressed in the flow direction due to the larger flow velocity on the left side of the cell.

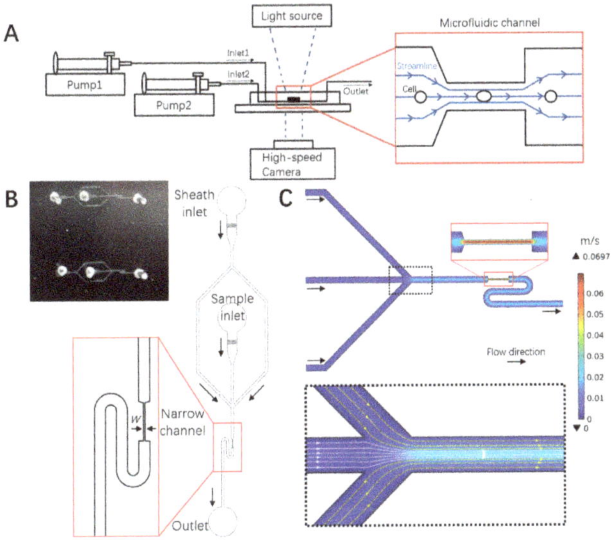

Figure 2. Microfluidic deformability cytometry of RBCs and flow field characterization. (**A**) Schematic diagram of the experimental setup, cell deformation, and flow field in the microfluidic chip. (**B**) Photograph (top left) and schematic of the microfluidic chip with sheath flow structure, where w is the narrow channel width (4, 8, 16 µm). The turning S shape following the exit of the narrow channel accommodated the size of the imaging field of the high-speed camera, allowing for the tracking of shape recovery. (**C**) Numerical simulation of the flow field in the microchannel.

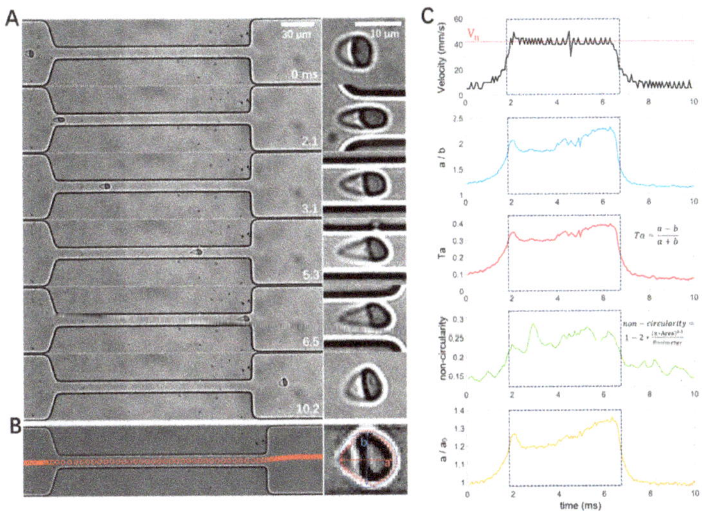

Figure 3. Illustration of cell motion, deformation, image processing, and data analysis with an example RBC from a normal healthy control. (**A**) Image sequence showing the deformation of the RBC passing through the narrow channel. (**B**) Cell trajectory and contour detection, with a and b depicting the major and minor axes, respectively. (**C**) Temporal evolution of cell velocity and deformability in different definitions measured from the same recording as in panel A. The dashed box denotes the time window when the cell was flowing inside the narrow channel. The width of the narrow channel is 16 µm.

To validate the operation and measurements of the microfluidic devices, we performed a control experiment comparing the deformability of non-treated healthy RBCs vs. PFA-treated RBCs (Figure 4). Differential deformability was observed between the two groups based on microfluidic measurements. The PFA-treated cells have a significantly lower deformability among the four different definitions, consistent with the fixation and stiffening effects of PFA. These results suggest that our measurements and analysis of the cell deformability are valid.

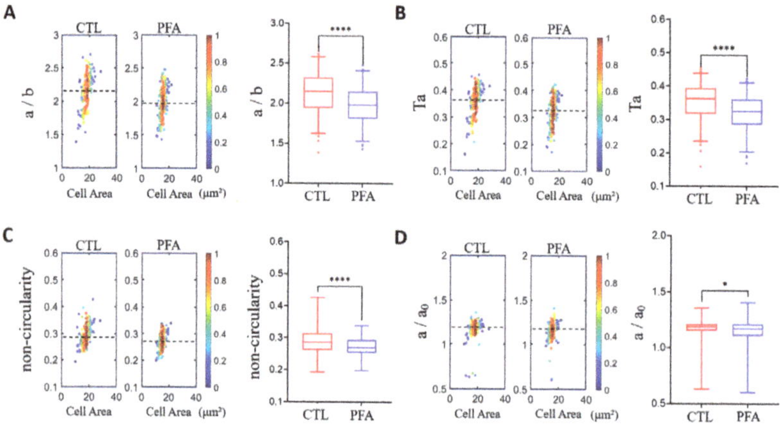

Figure 4. Validation of measurements by comparing the cell deformability of normal healthy RBCs and paraformaldehyde (PFA)-treated RBCs using density scatter plots and box plots. (**A**) Aspect ratio a/b. (**B**) Taylor deformation *Ta*. (**C**) Non-circularity. (**D**) Relative change in the major axis a/a_0. The dashed lines indicate median deformability. A hotter color indicates a higher data density. * $p < 0.05$, **** $p < 0.0001$. The narrow channel width is 8 μm, with an average cell velocity of 100 mm/s. $n = 165$ for normal RBCs and 144 for PFA-treated RBCs.

3.2. Influence of Narrow Channel Width and Flow Velocities on Cell Deformation

Under physiological conditions, RBCs need to squeeze through small capillary vessels of various sizes for circulation, where they may experience different flow velocities, shear stress, and deformation; therefore, it is necessary to investigate the influence of narrow channel width and flow velocities on the deformation of RBCs. As shown in Figure 5A, cell deformability increased significantly when the narrow channel width decreased across all four definitions of cell deformability, while the average flow rate was kept constant at 0.03 μL/min. Similarly, significantly higher cell deformability was observed when the average flow velocity increased while the narrow channel widths were kept constant at 8 μm (Figure 5B). These results are consistent with numerical modeling, which reveals that the flow shear stress increases as a result of elevated velocity gradient in a smaller channel width or larger average flow velocity (Figure 5C–E). For example, the instant shear stress from the uniaxial extensional flow along the centerline of the channel is defined as $\sigma = 3\mu\frac{du}{dx}$ [66] and the average flow shear stress along the flow direction at the entrance of the narrow channel and along the center line can be calculated as $\sigma_{ave} = 3\mu\frac{\Delta u}{\Delta x}$, where μ is the dynamic viscosity of the medium (998 kg·m^{-3}) and $\frac{\Delta u}{\Delta x}$ is the average velocity gradient along the flow direction (along the centre axis). The average flow shear stress is estimated to be 12.0, 27.9, and 60.3 dynes/cm^2. The max flow shear stress $\sigma_{max} = 3\mu\ (du/dx)_{max}$ along the center streamline is calculated from the maximum value of the first-order derivative of the position dependent velocity curves in Figure 5D, and estimated to be 25.8, 51.6, and 138.9 dynes/cm^2 for narrow channel widths of 16, 8, and 4 μm at a constant flowrate of 0.03 μL/min (Figure 5F). The average flow shear stress is estimated to 2.61, 8.1, 12.9, and 17.7 dynes/cm^2, and the max flow shear stress is estimated to be 10.2, 17.1, 27.0, and 36.9 dynes/cm^2 for average flow velocities of 15, 25, 40, and 55 mm/s at a narrow channel width of 8 μm, respectively (Figure 5G). Therefore, we choose 8 μm narrow channel width and 55 mm/s average flow velocity for all subsequent measurements of cell deformability, as these conditions are within the physiological range (15 to 55 mm/s) [67], from which larger cell deformation can be obtained.

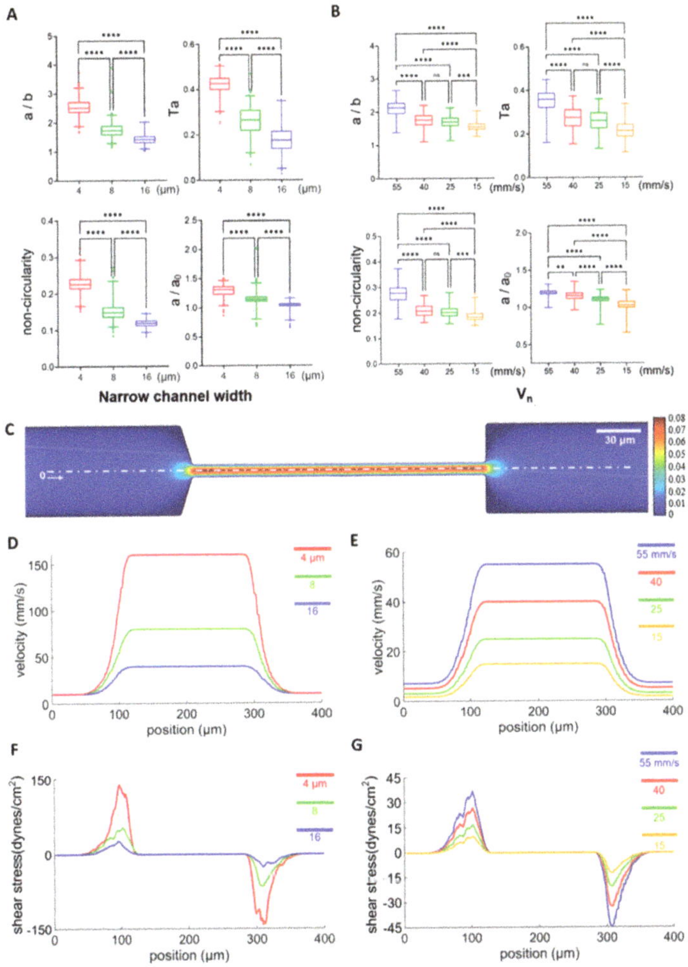

Figure 5. The influence of narrow channel width and flow rate on cell deformation. (**A**) Cell deformability from 4 definitions measured from narrow channel widths of 4, 8, 16 μm under a flow rate of 0.03 μL/min, and their statistical analysis. n = 182, 194, 265 for narrow channel widths of 4, 8, 16 μm, respectively. (**B**) Cell deformability from 4 definitions measured with maximum flow velocities (V_n) of 55, 40, 25, 15 mm/s with a narrow channel width of 8 μm. The corresponding n values are 64, 98, 125, and 58, respectively. (**C**) Representative results of numerical modeling of the flow field across the arrow channel with a channel width of 8 μm and V_n of 40 mm/s. (**D–G**) Flow velocity or shear stress distribution along the white dashed line (x) in panel C at narrow channel widths of 4, 8, and 16 μm with the flow rate fixed at 0.03 μL/min or maximum flow velocities (V_n) of 55, 40, 25, and 15 mm/s with the narrow channel width fixed at 8 μm. ** $p < 0.01$, *** $p < 0.001$, **** $p < 0.0001$, one-way ANOVA test followed by a post-hoc Tukey HSD multiple comparison test.

3.3. Label-Free Mechanical Phenotyping of RBCs from Patients with Thalassemia and Healthy Controls

Next, we employed the microfluidic approach to screen the deformability of RBCs in healthy controls and pathological conditions of thalassemia. Figure 6A shows the transit of an exemplary RBC from a patient with thalassemia through the narrow channel with the 8 μm width. The cell was stretched to an elongated shape when entering the narrow channel at 1.3 ms, it remained stretched from 1.3–4.6 ms until it exited the narrow channel at 5.3 ms and gradually recovered its shape (5.3–8.5 ms). Statistical analysis shows significantly lower cell deformability in the thalassemia group compared to the healthy control group, as observed using a narrow channel with widths of 4, 8, and 16 μm (Figure 6B–D). It is interesting to note that there is a more pronounced difference in aspect ratio between the two groups with smaller channel widths and larger shear stresses, indicating that the aspect ratio *a/b* may serve as a label-free biomarker to distinguish the deformability between RBCs from healthy donors and patients with thalassemia.

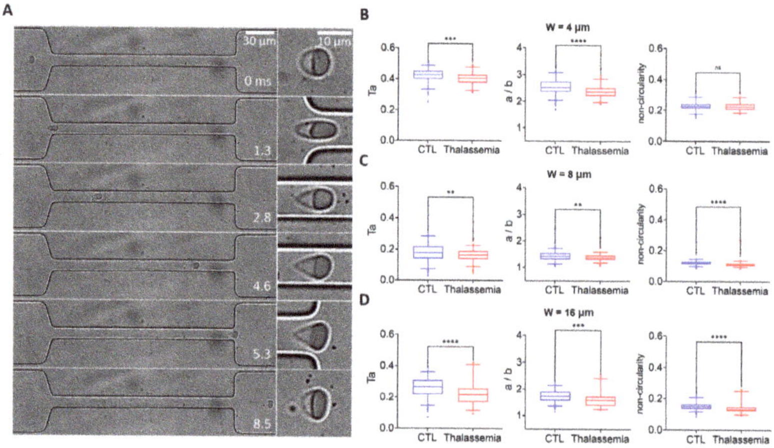

Figure 6. Deformability screening of RBCs in healthy and pathological conditions of thalassemia. (**A**) Transit and deformation of an example RBC from the patient with thalassemia in the microfluidic channel. An enlarged picture of the cell is shown on the right. Comparison of the deformability of RBCs between healthy donors and patients with thalassemia measured from narrow channels with widths of (**B**) 4 μm, (**C**) 8 μm, and (**D**) 16 μm. Deformability is defined by *Ta*, aspect ratio *a/b*, and non-circularity. The flow rate is 0.03 μL/min. ** $p < 0.01$, *** $p < 0.001$, **** $p < 0.0001$, one-way ANOVA test followed by a post-hoc Tukey HSD multiple comparison test.

3.4. Shape Recovery of RBCs from Patients with Thalassemia and Healthy Controls

We further investigated the dynamics of cell shape recovery when RBCs exited the narrow stretching region (Figure 7A), comparing representative individual RBCs from healthy control (non-treated and PFA-treated) and thalassemia groups. RBCs in all three cases gradually recovered their shape as they moved from the channel exit (0–267 s). Compared to RBCs from healthy controls and non-treated samples, PFA-treated RBCs exhibited rapid recovery from 0 to 267 μs, while little change in shape was observed thereafter. In contrast, it took longer for non-treated RBCs from healthy controls to recover their shapes (0–2935 s). Interestingly, RBCs in the pathological thalassemia condition seemed to recover their shape even more slowly. The aspect ratio *a/b* can be described using the following equation:

$$\gamma(t) = C1 + C2 \cdot e^{(-)t/\tau)}$$

where C1 and C2 are constants, τ denotes the characteristic relaxation time for cells to recover their shape, and t represents the time elapsed after the cells exit the narrow channel. The deformation data for each cell in each group are fitted using this equation (Figure 7B left) and a double logarithmic plot of $(\gamma(t) - C1)$ vs. time reveals that the shape recovery follows an exponential decay ($\ln(\gamma(t) - C1) = \ln(C2) - t/\tau$) (Figure 7B right), consistent with previous studies [64,68]. Statistical results demonstrate that RBCs treated with PFA exhibit the fastest recovery rate, while RBCs in the thalassemia group recover significantly slower than those from healthy controls.

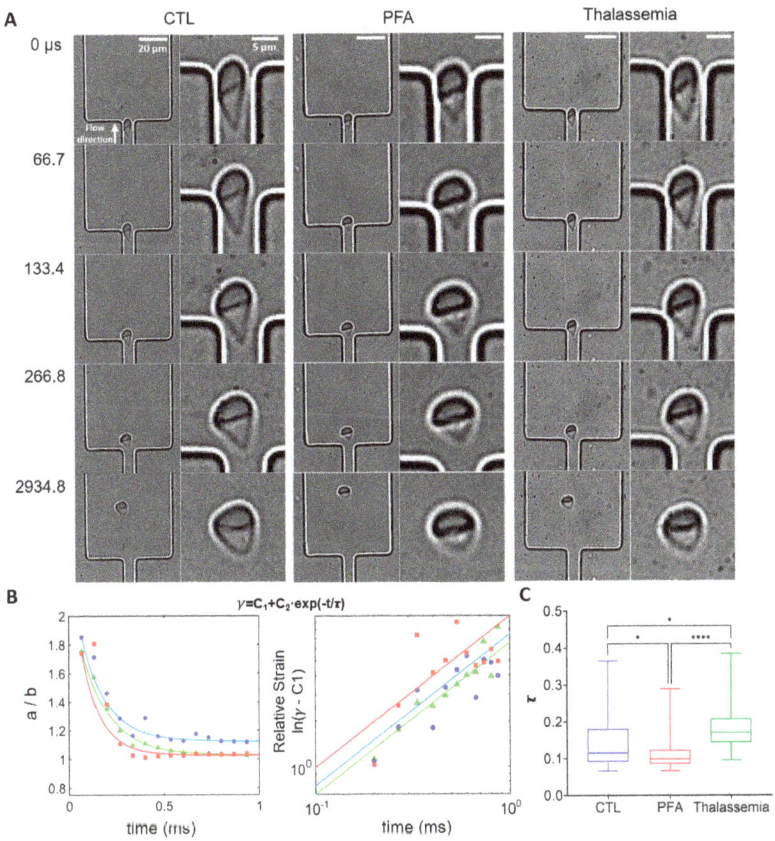

Figure 7. Analysis of shape recovery (strain decay) of RBCs from healthy control, PFA-treated healthy RBCs, and RBCs from patients with thalassemia. (**A**) Shape recovery of individual RBCs from different groups exiting the narrow stretching channel. The direction of flow is from bottom to top. The channel width is 8 μm, with V_n of 60 mm/s. (**B**) (**Left**) Representative strain γ (=a/b) for RBCs from healthy controls (blue), PFA-treated healthy RBCs (red), and RBCs from thalassemia patients (green) as a function of time. (**Right**) Double logarithmic plot of (γ − C1) as a function of time for the cells shown on the left. The solid lines represent the fits of the strain data to the exponential function $\gamma = C1 + C2 \times \exp(-t/\tau)$, where τ represents the characteristic relaxation time. τ for the cells are 0.15 s, 0.11 s, and 0.18 s for healthy RBCs, PFA-treated healthy RBCs, and RBCs from thalassemia patients, respectively. (**C**) Box plots of the characteristic relaxation time τ for the exponential decay fitting for each cell. The lines inside the boxes represent the median value of the data, while the edges of the boxes are the 25th and 75th percentiles. The goodness of the fit (average R2 value) is above 0.95 for all groups. n = 24, 47, and 48 for healthy controls (CTL), PFA-treated healthy RBCs, and RBCs from thalassemia patients. * $p < 0.05$, **** $p < 0.0001$, one-way ANOVA test followed by a post-hoc Tukey HSD multiple comparison test.

4. Discussion

To investigate differences in deformability between RBCs from thalassemia patients and healthy donors, we established a microfluidic flow cytometry based on shear flow deformation and high-speed optical imaging. To characterize the dynamic behavior of multiple RBCs in microfluidic flow cytometry, we developed a custom image processing program that enabled precise tracking of individual moving RBCs and accurate extraction of cell contours. From the extracted RBC contours, we quantified cellular deformability with multiple definitions by calculating key morphological parameters. This approach successfully differentiated between RBCs from thalassemia patients and healthy donors. We further investigated recovery dynamics following the exit from constricted microchannels. Statistical analysis revealed significant differences in the recovery rates between healthy and thalassemic RBC models.

The microfluidic system developed in this study enables precise control over RBCs' deformability and comprehensively records RBC motion before, during, and after transit through constricted microchannels. By implementing the sheath flow configuration, RBCs were hydrodynamically focused along the channel centerline, allowing cells to experience consistent flow shear with similar initial velocities when entering the stretching channel. To validate the reliability of our system, PFA-treated RBCs were compared with healthy RBCs, thereby establishing the platform's sensitivity in detecting biomechanical alterations. Furthermore, through precise modulation of flow velocity, RBC deformation was achieved across varying channel widths, revealing the influence of the flow shear on cell deformation. Our microchannels (4–60 μm in width) closely mimic physiological vascular scales [69], providing biomimetic relevance to in vivo microcirculation. In our study, the Reynolds number $Re < 0.72$, indicating a laminar flow regime without turbulent disturbances [70], enhancing the stability and controllability of cellular deformation processes. The system design improved experimental reproducibility, establishing a robust tool for investigating RBC rheological properties.

In this study, we have substantially improved image analysis by integrating a Kalman filter and Hungarian algorithm. It allowed continuous tracking of multiple individual RBC trajectories throughout microfluidic recordings, compared to only one cell tracking in previous studies [36], which eased the requirement for sample concentration. This framework was combined with a YOLOv8 deep learning network to achieve high-precision RBC contour extraction [71]. A curated RBC image dataset was constructed to train the model, yielding both exceptional contour detection accuracy (>95% mean average precision) and computational efficiency. Conventional edge detection algorithms [59] suffer significant limitations under suboptimal imaging conditions, particularly when RBC boundaries exhibit low contrast against the background. The deep learning approach effectively circumvents this constraint by learning discriminative morphological features. Collectively, this integrated tool provides a robust analytical framework for quantitative investigations of RBC dynamics.

Furthermore, we defined a suite of parameters to quantify RBC deformability, including the aspect ratio (a/b), Taylor deformation (Ta), non-circularity, and relative elongation change (a/a_0). These metrics collectively capture multidimensional aspects of cell morphological changes, enabling a comprehensive assessment of cellular deformability. While a/b and Ta directly reflect cellular elongation, the non-circularity characterizes contour irregularity. This multiparameter approach not only encompasses a broader spectrum of morphological features during deformation but also enhances assessment accuracy through integrated analysis.

Our results demonstrated significantly impaired deformability in thalassemic RBCs compared to healthy controls under the same experimental conditions. This deformation deficiency likely stems from structural and functional abnormalities in the RBC membrane [72]. Reduced deformability compromises microcirculatory perfusion efficiency, potentially exacerbating tissue hypoxia and triggering pathophysiological sequelae such as endothelial damage and inflammatory cascades [73,74].

To investigate recovery kinetics post deformation, we analyzed three RBC cohorts using our microfluidic platform: (1) healthy controls, (2) paraformaldehyde (PFA)-fixed RBCs (modeling rigidity), and (3) thalassemic RBCs. PFA-fixed RBCs exhibited the fastest post-constriction recovery, attributable to membrane stiffening and stabilized cytoskeletal architecture [75]. Conversely, thalassemic RBCs showed significantly slower recovery than healthy cells, consistent with disease-induced membrane defects and cytoskeletal dysfunction [76]. The elastic shear modulus and viscous modulus of RBCs could be estimated using Equations (1)–(3) described above. The elastic shear modulus E for the representative RBC from the control group is calculated to be more than three times bigger than that of the thalassemia groups (with their shape characteristic relaxation time τ at around the mean value of their respective groups). This finding is consistent with previous studies, which report that the average elastic shear modulus was higher for softer cells [63]. The viscous modulus η for the same representative RBC from the control group is found to be more than five times larger than that of the thalassemia groups, further confirming the disease-induced defects in cell structure and dysfunction in viscoelastic properties. These findings in our study elucidate distinct biomechanical signatures in strain recovery, providing mechanistic insights into cellular impairments associated with hematological disorders.

To approximate physiological conditions, the dimensions of the microfluidic constriction channel (7 μm in depth and 4, 8, 16 μm in width), the resulting RBC velocities (on the order of 10–100 mm/s) and shear stress for extensional deformation (1–10^2 dynes/cm^2) fall within the ranges reported for human blood vessels and flow speed of human erythrocytes [77,78]. However, several limitations exist. First, the microchannel in our study is rectangular in shape, unlike the cylindrical shape found in vivo. Therefore, RBCs were exposed to a more heterogeneous shear stress in the experiment compared to homogeneous shear stress in the radial direction in vivo. Nevertheless, we were more interested in the velocity gradient in the flow direction along the channel centerline and the resulting shear stress. Second, we used a highly diluted and RBC-only suspension in DPBS buffer prepared from whole blood. Thus, the extracellular environment differs markedly from native blood plasma, e.g., the fluid viscosity/density, chemical environment, and cell–cell communications could differ significantly [79]. Third, our experiment was performed at room temperature, which is lower than the in vivo temperature, and may affect the viscoelastic properties of RBCs and their deformability [80]. Fourth, our studies have a limited sample size (<150 cells), and considerable variability is observed in the measured parameters across the cell population. Due to the large field of view and high frame rate required to capture complete cell trajectories and deformations, tens of thousands of high-resolution images were generated within seconds. This resulted in a limitation on the number of cells that could be recorded per unit time, owing to the constrained storage capacity of our high-speed imaging system. The variability in the measured parameters could arise from the inherent heterogeneity within the RBC population [81]. Nevertheless, by employing sheath flow to focus cells toward the channel center and precisely detecting their contours, we reliably measured and distinguished statistically significant differences in cell deformability across groups.

Overall, the microfluidic system and data analysis pipeline presented serve as a robust platform for assessing the rheological properties of RBCs in health and disease, demonstrating good potential for in vitro screening of thalassemia and other hematological diseases.

Author Contributions: Conceptualization, F.L. and R.C.; methodology, H.J., X.L., Z.L., S.L., J.H., C.C. and F.L.; software, H.J.; validation, H.J. and F.L.; formal analysis, H.J. and F.L.; investigation, H.J., X.L., S.L., R.C. and F.L.; resources, J.H., C.C. and R.C.; data curation, H.J., X.L. and F.L.; writing—H.J., S.L., R.C. and F.L.; supervision, R.C. and F.L.; funding acquisition, C.C., R.C. and F.L. All authors have read and agreed to the published version of the manuscript.

Funding: This work was supported by the National Natural Science Foundation of China, the Science Foundation for Youths (No. 12204322), the Natural Science Foundation of Guangdong province (No. 2023A1515010649), the Guangdong Provincial Pearl River Talents Program 2023QN10X235, the Sanming Project of Medicine in Shenzhen (SZSM202011004), the National Natural Science Foundation of China (NSFC, Grant No.82370164), and the Science, Technology and Innovation Commission of Shenzhen Municipality (JCYJ20210324123004011 and JCYJ20180307150419435).

Data Availability Statement: Data are contained within the article.

Acknowledgments: We would like to acknowledge Zhihui Liu for their technical support with data analysis. We also thank Yang Zhang from Shenzhen Bay Laboratory for helpful discussions.

Conflicts of Interest: The authors declare no conflicts of interest.

References

1. Darling, E.M.; Di Carlo, D. High-throughput assessment of cellular mechanical properties. *Annu. Rev. Biomed. Eng.* **2015**, *17*, 35–62. [CrossRef] [PubMed]
2. Burton, N.M.; Bruce, L.J. Modelling the structure of the red cell membrane. *Biochem. Cell Biol.* **2011**, *89*, 200–215. [CrossRef] [PubMed]
3. Gallagher, P.G. Hereditary elliptocytosis: Spectrin and protein 4.1 R. *Semin. Hematol.* **2004**, *41*, 142–164. [CrossRef]
4. Clark, M.R.; Mohandas, N.; Shohet, S.B. Osmotic gradient ektacytometry: Comprehensive characterization of red cell volume and surface maintenance. *Blood* **1983**, *61*, 899–910. [CrossRef]
5. Diez-Silva, M.; Dao, M.; Han, J.; Lim, C.-T.; Suresh, S. Shape and biomechanical characteristics of human red blood cells in health and disease. *MRS Bull.* **2010**, *35*, 382–388. [CrossRef]
6. Chan, C.J.; Heisenberg, C.-P.; Hiiragi, T. Coordination of morphogenesis and cell-fate specification in development. *Curr. Biol.* **2017**, *27*, R1024–R1035. [CrossRef]
7. Barshtein, G.; Pajic-Lijakovic, I.; Gural, A. Deformability of stored red blood cells. *Front. Physiol.* **2021**, *12*, 722896. [CrossRef]
8. Wan, J.; Ristenpart, W.D.; Stone, H.A. Dynamics of shear-induced ATP release from red blood cells. *Proc. Natl. Acad. Sci. USA* **2008**, *105*, 16432–16437. [CrossRef]
9. Mohandas, N.; Gallagher, P.G. Red cell membrane: Past, present, and future. *Blood* **2008**, *112*, 3939–3948. [CrossRef]
10. Koch, M.; Wright, K.E.; Otto, O.; Herbig, M.; Salinas, N.D.; Tolia, N.H.; Satchwell, T.J.; Guck, J.; Brooks, N.J.; Baum, J. Plasmodium falciparum erythrocyte-binding antigen 175 triggers a biophysical change in the red blood cell that facilitates invasion. *Proc. Natl. Acad. Sci. USA* **2017**, *114*, 4225–4230. [CrossRef]
11. Evans, E.A. Structure and deformation properties of red blood cells: Concepts and quantitative methods. *Methods Enzymol.* **1989**, *173*, 3–35.
12. Namvar, A.; Blanch, A.; Dixon, M.; Carmo, O.; Liu, B.; Tiash, S.; Looker, O.; Andrew, D.; Chan, L.-J.; Tham, W.-H. Surface area-to-volume ratio, not cellular viscoelasticity is the major determinant of red blood cell traversal through small channels. *Cell Microbiol.* **2021**, *120*, 170a.
13. Takakuwa, Y. Regulation of red cell membrane protein interactions: Implications for red cell function. *Curr. Opin. Hematol.* **2001**, *8*, 80–84. [CrossRef]
14. Hochmuth, R.M. Micropipette aspiration of living cells. *J. Biomech.* **2000**, *33*, 15–22. [CrossRef] [PubMed]
15. Wang, H.; Zhou, F.; Guo, Y.; Ju, L.A. Micropipette-based biomechanical nanotools on living cells. *Eur. Biophys. J.* **2022**, *51*, 119–133. [CrossRef] [PubMed]
16. Smith, A.S.; Nowak, R.B.; Zhou, S.; Giannetto, M.; Gokhin, D.S.; Papoin, J.; Ghiran, I.C.; Blanc, L.; Wan, J.; Fowler, V.M. Myosin IIA interacts with the spectrin-actin membrane skeleton to control red blood cell membrane curvature and deformability. *Proc. Natl. Acad. Sci. USA* **2018**, *115*, E4377–E4385. [CrossRef]
17. Stroka, K.M.; Jiang, H.; Chen, S.-H.; Tong, Z.; Wirtz, D.; Sun, S.X.; Konstantopoulos, K. Water permeation drives tumor cell migration in confined microenvironments. *Cell* **2014**, *157*, 611–623. [CrossRef]
18. Puig-de-Morales-Marinkovic, M.; Turner, K.T.; Butler, J.P.; Fredberg, J.J.; Suresh, S. Viscoelasticity of the human red blood cell. *Am. J. Physiol.-Cell Physiol.* **2007**, *293*, C597–C605. [CrossRef]

19. Song, H.; Liu, Y.; Zhang, B.; Tian, K.; Zhu, P.; Lu, H.; Tang, Q. Study of in vitro RBCs membrane elasticity with AOD scanning optical tweezers. *Biomed. Opt. Express* **2016**, *8*, 384–394. [CrossRef]
20. Lim HW, G.; Wortis, M.; Mukhopadhyay, R. Stomatocyte–discocyte–echinocyte sequence of the human red blood cell: Evidence for the bilayer–couple hypothesis from membrane mechanics. *Proc. Natl. Acad. Sci. USA* **2002**, *99*, 16766–16769. [CrossRef]
21. Kariuki, S.N.; Marin-Menendez, A.; Introini, V.; Ravenhill, B.J.; Lin, Y.-C.; Macharia, A.; Makale, J.; Tendwa, M.; Nyamu, W.; Kotar, J. Red blood cell tension protects against severe malaria in the Dantu blood group. *Nature* **2020**, *585*, 579–583. [CrossRef] [PubMed]
22. Thoumine, O.; Ott, A.; Cardoso, O.; Meister, J.-J. Microplates: A new tool for manipulation and mechanical perturbation of individual cells. *J. Biochem. Biophys. Methods* **1999**, *39*, 47–62. [CrossRef] [PubMed]
23. Chen, H.; Guo, J.; Bian, F.; Zhao, Y. Microfluidic technologies for cell deformability cytometry. *Smart Med.* **2022**, *1*, e20220001. [CrossRef] [PubMed]
24. Chen, Y.; Guo, K.; Jiang, L.; Zhu, S.; Ni, Z.; Xiang, N. Microfluidic deformability cytometry: A review. *Talanta* **2023**, *251*, 123815. [CrossRef]
25. An, L.; Ji, F.; Zhao, E.; Liu, Y.; Liu, Y. Measuring cell deformation by microfluidics. *Front. Bioeng. Biotechnol.* **2023**, *11*, 1214544. [CrossRef]
26. Fajdiga, L.; Zemljič, Š.; Kokalj, T.; Derganc, J. Shear flow deformability cytometry: A microfluidic method advancing towards clinical use—A review. *Anal. Chim. Acta* **2025**, *1355*, 343894. [CrossRef]
27. Quinto-Su, P.A.; Kuss, C.; Preiser, P.R.; Ohl, C.-D. Red blood cell rheology using single controlled laser-induced cavitation bubbles. *Lab Chip* **2011**, *11*, 672–678. [CrossRef]
28. Mohammadzadeh, M.; Li, F.; Ohl, C.-D. Shearing flow from transient bubble oscillations in narrow gaps. *Phys. Rev. Fluids* **2017**, *2*, 014301. [CrossRef]
29. Li, F.; Chan, C.U.; Ohl, C.D. Yield strength of human erythrocyte membranes to impulsive stretching. *Biophys. J.* **2013**, *105*, 872–879. [CrossRef]
30. Tandiono, T.; Klaseboer, E.; Ohl, S.-W.; Ow, D.S.-W.; Choo, A.B.-H.; Li, F.; Ohl, C.-D. Resonant stretching of cells and other elastic objects from transient cavitation. *Soft Matter* **2013**, *9*, 8687–8696. [CrossRef]
31. Urbanska, M.; Muñoz, H.E.; Shaw Bagnall, J.; Otto, O.; Manalis, S.R.; Di Carlo, D.; Guck, J. A comparison of microfluidic methods for high-throughput cell deformability measurements. *Nat. Methods* **2020**, *17*, 587–593. [CrossRef]
32. Mirzaaghaian, A.; Ramiar, A.; Ranjbar, A.A.; Warkiani, M.E. Application of level-set method in simulation of normal and cancer cells deformability within a microfluidic device. *J. Biomech.* **2020**, *112*, 110066. [CrossRef] [PubMed]
33. Nyberg, K.D.; Scott, M.B.; Bruce, S.L.; Gopinath, A.B.; Bikos, D.; Mason, T.G.; Kim, J.W.; Choi, H.S.; Rowat, A.C. The physical origins of transit time measurements for rapid, single cell mechanotyping. *Lab Chip* **2016**, *16*, 3330–3339. [CrossRef] [PubMed]
34. Byun, S.; Son, S.; Amodei, D.; Cermak, N.; Shaw, J.; Kang, J.H.; Hecht, V.C.; Winslow, M.M.; Jacks, T.; Mallick, P. Characterizing deformability and surface friction of cancer cells. *Proc. Natl. Acad. Sci. USA* **2013**, *110*, 7580–7585. [CrossRef] [PubMed]
35. Gossett, D.R.; Tse, H.T.; Lee, S.A.; Ying, Y.; Lindgren, A.G.; Yang, O.O.; Rao, J.; Clark, A.T.; Di Carlo, D. Hydrodynamic stretching of single cells for large population mechanical phenotyping. *Proc. Natl. Acad. Sci. USA* **2012**, *109*, 7630–7635. [CrossRef]
36. Li, F.; Cima, I.; Vo, J.H.; Tan, M.-H.; Ohl, C.D. Single cell hydrodynamic stretching and microsieve filtration reveal genetic, phenotypic and treatment-related links to cellular deformability. *Micromachines* **2020**, *11*, 486. [CrossRef]
37. Lin, J.; Kim, D.; Tse, H.T.; Tseng, P.; Peng, L.; Dhar, M.; Karumbayaram, S.; Di Carlo, D. High-throughput physical phenotyping of cell differentiation. *Microsyst. Nanoeng.* **2017**, *3*, 17013. [CrossRef]
38. Cinar, E.; Zhou, S.; DeCourcey, J.; Wang, Y.; Waugh, R.E.; Wan, J. Piezo1 regulates mechanotransductive release of ATP from human RBCs. *Proc. Natl. Acad. Sci. USA* **2015**, *112*, 11783–11788. [CrossRef]
39. Forsyth, A.M.; Wan, J.; Ristenpart, W.D.; Stone, H.A. The dynamic behavior of chemically "stiffened" red blood cells in microchannel flows. *Microvasc. Res.* **2010**, *80*, 37–43. [CrossRef]
40. Forsyth, A.M.; Wan, J.; Owrutsky, P.D.; Abkarian, M.; Stone, H.A. Multiscale approach to link red blood cell dynamics, shear viscosity, and ATP release. *Proc. Natl. Acad. Sci. USA* **2011**, *108*, 10986–10991. [CrossRef]
41. Otto, O.; Rosendahl, P.; Mietke, A.; Golfier, S.; Herold, C.; Klaue, D.; Girardo, S.; Pagliara, S.; Ekpenyong, A.; Jacobi, A. Real-time deformability cytometry: On-the-fly cell mechanical phenotyping. *Nat. Methods* **2015**, *12*, 199–202. [CrossRef]
42. Saadat, A.; Huyke, D.A.; Oyarzun, D.I.; Escobar, P.V.; Øvreeide, I.H.; Shaqfeh, E.S.; Santiago, J.G. A system for the high-throughput measurement of the shear modulus distribution of human red blood cells. *Lab Chip* **2020**, *20*, 2927–2936. [CrossRef] [PubMed]
43. Tomaiuolo, G. Biomechanical properties of red blood cells in health and disease towards microfluidics. *Biomicrofluidics* **2014**, *8*, 051501. [CrossRef] [PubMed]
44. Kubánková, M.; Hohberger, B.; Hoffmanns, J.; Fürst, J.; Herrmann, M.; Guck, J.; Kräter, M. Physical phenotype of blood cells is altered in COVID-19. *Biophys. J.* **2021**, *120*, 2838–2847. [CrossRef]
45. Guo, Q.; Reiling, S.J.; Rohrbach, P.; Ma, H. Microfluidic biomechanical assay for red blood cells parasitized by Plasmodium fal-ciparum. *Lab Chip* **2012**, *12*, 1143–1150. [CrossRef]

46. Agrawal, R.; Smart, T.; Nobre-Cardoso, J.; Richards, C.; Bhatnagar, R.; Tufail, A.; Shima, D.; Jones, P.H.; Pavesio, C. Assessment of red blood cell deformability in type 2 diabetes mellitus and diabetic retinopathy by dual optical tweezers stretching technique. *Sci. Rep.* **2016**, *6*, 15873. [CrossRef]
47. Eluru, G.; Srinivasan, R.; Gorthi, S.S. Deformability Measurement of Single-Cells at High-Throughput with Imaging Flow Cytometry. *J. Light. Technol.* **2015**, *33*, 3475–3480. [CrossRef]
48. Alapan, Y.; Little, J.A.; Gurkan, U.A. Heterogeneous red blood cell adhesion and deformability in sickle cell disease. *Sci. Rep.* **2014**, *4*, 7173. [CrossRef]
49. Vayá, A.; Alis, R.; Suescún, M.; Rivera, L.; Murado, J.; Romagnoli, M.; Solá, E.; Hernandez-Mijares, A. Association of erythrocyte deformability with red blood cell distribution width in met-abolic diseases and thalassemia trait. *Clin. Hemorheol. Microcirc.* **2016**, *61*, 407–415. [CrossRef]
50. Perrotta, S.; Gallagher, P.G.; Mohandas, N. Hereditary spherocytosis. *Lancet* **2008**, *372*, 1411–1426. [CrossRef]
51. Huisjes, R.; Makhro, A.; Llaudet-Planas, E.; Hertz, L.; Petkova-Kirova, P.; Verhagen, L.P.; Pignatelli, S.; Rab, M.A.; Schiffelers, R.M.; Seiler, E.; et al. Density, heterogeneity and deformability of red cells as markers of clinical severity in hereditary spherocytosis. *Haematologica* **2020**, *105*, 338–347. [CrossRef]
52. Bogdanova, A.; Petrushanko, I.Y.; Hernansanz-Agustín, P.; Martínez-Ruiz, A. "Oxygen sensing" by Na, K-ATPase: These mi-raculous thiols. *Front. Physiol.* **2016**, *7*, 314. [CrossRef]
53. Phillips, J.; Henderson, A.C. Hemolytic anemia: Evaluation and differential diagnosis. *Am. Fam. Physician* **2018**, *98*, 354–361. [PubMed]
54. Kattamis, A.; Kwiatkowski, J.L.; Aydinok, Y. Thalassaemia. *Lancet* **2022**, *399*, 2310–2324. [CrossRef] [PubMed]
55. Krishnevskaya, E.; Molero, M.; Ancochea, Á.; Hernández, I.; Vives-Corrons, J.-L. New-Generation Ektacytometry Study of Red Blood Cells in Different Hemoglobinopathies and Thalassemia. *Thalass. Rep.* **2023**, *13*, 70–76. [CrossRef]
56. Basu, D.; Kulkarni, R. Overview of blood components and their preparation. *Indian J. Anaesth.* **2014**, *58*, 529–537. [CrossRef]
57. Welch, G.; Bishop, G. *An Introduction to the Kalman Filter*; University of North Carolina at Chapel Hill: Chapel Hill, NC, USA, 1995.
58. Khodarahmi, M.; Maihami, V. A review on Kalman filter models. *Arch. Comput. Methods Eng.* **2023**, *30*, 727–747. [CrossRef]
59. Ding, L.; Goshtasby, A. On the Canny edge detector. *Pattern Recognit.* **2001**, *34*, 721–725. [CrossRef]
60. Redmon, J.; Divvala, S.K.; Girshick, R.B.; Farhadi, A. You Only Look Once: Unified, Real-Time Object Detection. In Proceedings of the 2016 IEEE Conference on Computer Vision and Pattern Recognition (CVPR), Las Vegas, NV, USA, 27–30 June 2016; pp. 779–788.
61. Kaur, S.; Kaur, L.; Lal, M. *A Review: YOLO and Its Advancements*; Springer Nature Singapore: Singapore, 2024; pp. 577–592.
62. Reichel, F.; Mauer, J.; Nawaz, A.A.; Gompper, G.; Guck, J.; Fedosov, D.A. High-Throughput Microfluidic Characterization of Erythrocyte Shapes and Me-chanical Variability. *Biophys. J.* **2019**, *117*, 14–24. [CrossRef]
63. Williams, D.C.; Wood, D.K. High-throughput quantification of red blood cell deformability and oxygen saturation to probe mechanisms of sickle cell disease. *Proc. Natl. Acad. Sci. USA* **2023**, *120*, e2313755120. [CrossRef]
64. Hochmuth, R.; Worthy, P.; Evans, E. Red cell extensional recovery and the determination of membrane viscosity. *Biophys. J.* **1979**, *26*, 101–114. [CrossRef] [PubMed]
65. Mancuso, J.E.; Ristenpart, W.D. Stretching of red blood cells at high strain rates. *Phys. Rev. Fluids* **2017**, *2*, 101101. [CrossRef]
66. Macosko, C.W. *Rheology: Principle, Measurement and Application*; VCH: New York, NY, USA, 1994.
67. Popel, A.S.; Johnson, P.C. Microcirculation and hemorheology. *Annu. Rev. Fluid. Mech.* **2005**, *37*, 43–69. [CrossRef] [PubMed]
68. Evans, E.; Mohandas, N.; Leung, A. Static and dynamic rigidities of normal and sickle erythrocytes. Major influence of cell hemoglobin concentration. *J. Clin. Investig.* **1984**, *73*, 477–488. [CrossRef]
69. Tropres, I.; Grimault, S.; Vaeth, A.; Grillon, E.; Julien, C.; Payen, J.F.; Lamalle, L.; Décorps, M. Vessel size imaging. *Magn. Reson. Med.* **2001**, *45*, 397–408. [CrossRef]
70. Wu, C.; Almuaalemi, H.Y.M.; Sohan, A.M.F.; Yin, B. Effect of flow velocity on laminar flow in microfluidic chips. *Micromachines* **2023**, *14*, 1277. [CrossRef]
71. Fakhrurroja, H.; Fashihullisan, A.A.; Bangkit, H.; Pramesti, D.; Ismail, N.; Mahardiono, N.A. A Vision-Based System: Detecting Traffic Law Violation Case Study of Red-Light Running Using Pre-Trained YOLOv8 Model and OpenCV. In Proceedings of the 2024 IEEE International Conference on Smart Mechatronics (ICSMech), Yogyakarta, Indonesia, 19–21 November 2024; IEEE: New York, NY, USA, 2024.
72. Cooper, R.A. Influence of increased membrane cholesterol on membrane fluidity and cell function in human red blood cells. *J. Supramol. Struct.* **1978**, *8*, 413–430. [CrossRef]
73. Fedosov, D.A.; Peltomäki, M.; Gompper, G. Deformation and dynamics of red blood cells in flow through cylindrical mi-crochannels. *Soft Matter* **2014**, *10*, 4258–4267. [CrossRef]
74. Dri, E.; Lampas, E.; Lazaros, G.; Lazarou, E.; Theofilis, P.; Tsioufis, C.; Tousoulis, D. Inflammatory Mediators of Endothelial Dysfunction. *Life* **2023**, *13*, 1420. [CrossRef]

75. Kim, S.-O.; Kim, J.; Okajima, T.; Cho, N.-J. Mechanical properties of paraformaldehyde-treated individual cells investigated by atomic force microscopy and scanning ion conductance microscopy. *Nano Converg.* **2017**, *4*, 5. [CrossRef]
76. Li, Z.; Yao, X.; Zhang, J.; Yang, J.; Ni, J.; Wang, Y. Exploring the bone marrow micro environment in thalassemia patients: Potential therapeutic alternatives. *Front. Immunol.* **2024**, *15*, 1403458. [CrossRef]
77. Besedina, N.A.; Skverchinskaya, E.A.; Ivanov, A.S.; Kotlyar, K.P.; Morozov, I.A.; Filatov, N.A.; Mindukshev, I.V.; Bukatin, A.S. Microfluidic Characterization of Red Blood Cells Microcir-culation under Oxidative Stress. *Cells* **2021**, *10*, 3552. [CrossRef]
78. Amirouche, A.; Esteves, J.; Lavoignat, A.; Picot, S.; Ferrigno, R.; Faivre, M. Dual shape recovery of red blood cells flowing out of a microfluidic con-striction. *Biomicrofluidics* **2020**, *14*, 024116. [CrossRef]
79. Young, E.W.K.; Beebe, D.J. Fundamentals of microfluidic cell culture in controlled microenvironments. *Chem. Soc. Rev.* **2010**, *39*, 1036–1048. [CrossRef] [PubMed]
80. Lecklin, T.; Egginton, S.; Nash, G.B. Effect of temperature on the resistance of individual red blood cells to flow through capillary-sized apertures. *Pflug. Arch.* **1996**, *432*, 753–759. [CrossRef] [PubMed]
81. Jain, V.; Yang, W.-H.; Wu, J.; Roback, J.D.; Gregory, S.G.; Chi, J.-T. Single Cell RNA-Seq Analysis of Human Red Cells. *Front. Physiol.* **2022**, *13*, 828700. [CrossRef] [PubMed]

micromachines

MDPI

Article

A Novel Microfluidic Platform for Circulating Tumor Cell Identification in Non-Small-Cell Lung Cancer

Tingting Tian [1,†], Shanni Ma [1,†], Yan Wang [2], He Yin [3], Tiantian Dang [3], Guangqi Li [3], Jiaming Li [3], Weijie Feng [1], Mei Tian [4], Jinbo Ma [1] and Zhijun Zhao [5,6,*]

[1] First Clinical Medical College, Ningxia Medical University, Yinchuan 750001, China; tiantt6130@163.com (T.T.); 13139514013@163.com (S.M.); 230120810451@nxmu.edu.cn (W.F.); mk_runner_high@163.com (J.M.)
[2] Department of Internal Oncology, Cancer Hospital, General Hospital of Ningxia Medical University, Yinchuan 750001, China; mgwy1974@163.com
[3] Center of Medical Laboratory, General Hospital of Ningxia Medical University, Yinchuan 750001, China; 18295016264@163.com (H.Y.); D080519D@163.com (T.D.); liguangqi77@163.com (G.L.); lemon6966@163.com (J.L.)
[4] School of Medical Laboratory, Ningxia Medical University, Yinchuan 750001, China; tff15202635152@163.com
[5] Central Laboratory, Peking University First Hospital Ningxia Women and Children's Hospital (Ningxia Hui Autonomous Region Maternal and Child Health Hospital), Yinchuan 750004, China
[6] Third Clinical Medical College, Ningxia Medical University, Yinchuan 750001, China
* Correspondence: z15815z@163.com; Tel.: +86-186-9511-2255
† These authors contributed equally to this work.

Abstract

Circulating tumor cells (CTCs) are crucial biomarkers for lung cancer metastasis and recurrence, garnering significant clinical attention. Despite this, efficient and cost-effective detection methods remain scarce. Consequently, there is an urgent demand for the development of highly sensitive CTC detection technologies to enhance lung cancer diagnosis and treatment. This study utilized microspheres and A549 cells to model CTCs, assessing the impact of acoustic field forces on cell viability and proliferation and confirming capture efficiency. Subsequently, CTCs from the peripheral blood of patients with lung cancer were captured and identified using fluorescence in situ hybridization, and the results were compared to the immunomagnetic bead method to evaluate the differences between the techniques. Finally, epidermal growth factor receptor (EGFR) mutation analysis was conducted on CTC-positive samples. The findings showed that acoustic microfluidic technology effectively captures microspheres, A549 cells, and CTCs without compromising cell viability or proliferation. Moreover, EGFR mutation analysis successfully identified mutation types in four samples, establishing a basis for personalized targeted therapy. In conclusion, acoustic microfluidic technology preserves cell viability while efficiently capturing CTCs. When integrated with EGFR mutation analysis, it provides robust support for the precise diagnosis and treatment of lung cancer as well as personalized drug therapy.

Keywords: non-small-cell lung cancer; circulating tumor cells; microfluidics; acoustic; amplification refractory mutation system

1. Introduction

Lung cancer is the foremost cause of cancer-related mortality globally, presenting a significant threat to human health [1]. The majority of patients receive a diagnosis at an advanced stage, resulting in a five-year survival rate of merely 5% [2]. Early detection and timely intervention in tumor cases substantially enhance patient survival rates. Consequently, early diagnosis and treatment are crucial for improving prognosis and extending patient lifespan.

Histopathological examination remains the "gold standard" for lung cancer diagnosis, permitting direct microscopic analysis of pathological tissues to determine tumor type and stage accurately, thus guiding clinical decisions [3,4]. Despite its accuracy, this invasive procedure carries risks such as pain, bleeding, and infection. Challenges are amplified in advanced lung cancer, where tumor puncture is particularly risky and obtaining adequate samples is difficult. Conversely, liquid biopsy offers a non-invasive alternative, minimizing patient harm and enabling repeated sampling for real-time tumor monitoring. It effectively captures the spatial heterogeneity and evolution of tumors, addressing the limitations of tissue biopsy, which only reveals localized characteristics [5].

Circulating tumor cells (CTCs), a key marker for liquid biopsies, have recently demonstrated significant promise in diagnosing and prognosticating lung cancer [6]. CTCs are tumor cells that detach from primary or metastatic sites and enter the bloodstream, potentially spreading to tissues or organs, such as lymph nodes, the liver, kidneys, the brain, and bones [7–10]. CTCs contain abundant biological information and can provide various types of molecular dynamic information, including DNA, RNA, and proteins, offering a unique perspective for in-depth analysis of the biological characteristics of tumor cells [11]. Leveraging this advantage, CTC analysis facilitates real-time monitoring of tumor progression and precise evaluation of treatment efficacy, thereby laying a critical foundation for personalized and timely treatment modifications. Though CTC detection provides benefits, such as minimal invasiveness, high specificity, and the ability to reflect tumor heterogeneity [12], clinical application remains challenging due to the extremely low concentration of CTCs (1–10 cells/mL) in blood and their inherent heterogeneity [13].

The isolation and capture of CTCs depend on their distinct physical attributes, such as size, deformability, dielectric properties, and hydrodynamics, as well as their biological markers, including epithelial cell adhesion molecules [14]. Recently, integrating various physical field technologies—acoustic, optical, electrical, and magnetic—into microfluidic systems has significantly enhanced the purity and efficiency of CTC sorting [15]. Acoustic fields enable contactless cell manipulation through precise acoustic radiation forces, thereby preserving cell viability [16]. Optical fields utilize optical tweezers to capture and precisely arrange individual cells [17]. Electrical fields utilize dielectrophoretic forces for high-throughput sorting based on cellular electrical properties, facilitating downstream molecular analysis while maintaining cell integrity [18]. Magnetic fields use functionalized magnetic beads to specifically identify and separate target CTCs [19]. Among these technologies, acoustic field technology is distinguished by its unique advantages. It non-invasively manipulates cells using a non-contact acoustic radiation force, preserving their intrinsic state and viability. This method offers convenient operation, high efficiency, and controllability. While bioaffinity methods are also widely employed, their limitations, such as the omission of epithelial–mesenchymal transition-type CTCs, are increasingly recognized. To address this, research is trending towards multimodal sorting strategies, like combining acoustic fields with microfluidic chips. By applying an acoustic field force within these chips, cell capture is enhanced [20], providing a more precise sample basis for lung cancer gene research.

Research indicates that lung cancer occurrence and progression are closely linked to multiple gene mutations [21]. Among these, mutations in the epidermal growth factor receptor (EGFR) gene are a key driver of non-small-cell lung cancer (NSCLC) pathogenesis and progression [22]. The amplification refractory mutation system polymerase chain reaction (ARMS-PCR) employs a unique primer design, offering high sensitivity and specificity in detecting gene mutations. This method has become a leading technology in personalized tumor molecular diagnostics [23]. Utilizing ARMS-PCR for detecting EGFR mutations in patients with lung cancer enables rapid and precise identification of specific EGFR mutation types [24,25].

This study integrates acoustic microfluidic technology with ARMS-PCR to develop a comprehensive platform for lung cancer diagnosis and treatment, advancing precision medicine. Utilizing acoustic field forces, the platform effectively captures microspheres, A549 cells, and CTCs within a microfluidic chip, ensuring cell viability and preserving biological characteristics. By incorporating ARMS-PCR, it detects EGFR gene mutations in captured CTCs, providing crucial molecular data for personalized treatment. This platform not only supports lung cancer diagnosis, treatment guidance, and efficacy assessment but also opens new avenues for clinical research and practice.

2. Materials and Methods

2.1. Specifications of Microfluidic Chips

The microfluidic chip used in this experiment has external dimensions ranging from 12 × 11 mm to 30 × 20 mm, with a thickness of 0.5 to 1 mm. The inlets and outlets have a diameter of 1 mm and feature through-holes. The flow channel measures 10 × 1 mm in a serpentine configuration, with a depth of 0.2 mm. Micropores have a diameter of 0.1 mm, spaced 0.2 mm apart center to center, and a depth of 80 μm. The glass coverslip has external dimensions of 50 × 40 mm and a thickness of 0.15 mm.

2.2. Construction of Microfluidic Instruments

The microfluidic chip features circular micropores and is constructed from transparent polydimethylsiloxane (PDMS) material. A piezoelectric ceramic sheet, functioning as a sensor, converts electrical energy into mechanical vibrations and is the component most intimately linked to the microfluidic chip within the sorting system. The disc-shaped piezoelectric ceramic transducer used in this study has a diameter of 20 ± 0.5 mm and a total thickness of 0.42 ± 0.03 mm, composed of FT-27T-4.0A1 hard piezoelectric material. Following the bonding of the PDMS chip to a glass coverslip, a piezoelectric ceramic plate is attached to the underside, and thin tubes are inserted at both ends to link the sample injection device. The chip is then positioned on a microscope, with wires connecting it to the amplifier, signal generator, and piezoelectric ceramic plate. The frequency and voltage are adjusted, and observations are made using a microscope (OLYMPUS-IX53/Japan/Tokyo/Olympus Corporation).

2.3. Experiment on Simulating Cells with Polystyrene Microspheres

To prepare mixed solutions of 8, 15, and 25 μm microspheres with phosphate-buffered saline (PBS), combine 970 μL of PBS with 30 μL of the microsphere solution, ensuring thorough mixing. If adhesion occurs, incorporate 10–30 μL of Tween 80 and mix again. Once the device is connected, rinse the chip with 75% ethanol and use PBS to remove air bubbles. Load the sample, then observe microsphere capture under a microscope. Adjust the voltage, flow rate, and frequency as needed, and document the data.

2.4. Study on the Capture Efficiency of A549 Cells by Acoustic Microfluidic Technology

Cultured A549 cells were prepared into suspensions with concentrations of 1×10^6, 1×10^5, 1×10^4, and 1×10^3 cells/mL and then loaded onto the machine. Cell capture was observed microscopically, and the capture efficiency was determined by recording the ratio of captured cells to those passing through the chip per unit time.

2.5. Labeling A549 Cells and Investigating Their Capture Utilizing an Acoustic Microfluidic Chip

A549 cells were stained and labeled with the Cell Plasma Membrane Staining Kit with DiI (Red Fluorescence) (DiI) following the manufacturer's instructions. The experiment consisted of three groups: a. PBS buffer + A549 cell group (10:1 ratio); b. PBS buffer + supernatant blood + A549 cell group, where whole blood from healthy individuals was allowed to stand at room temperature for 10 min and then centrifuged at 12,000 rpm for 5 min, collected supernatant blood, and the above A549 cells suspension, supernatant blood, and PBS buffer were mixed in a 1:1:8 ratio; c. PBS buffer + whole blood + A549 cell group, where the A549 cells suspension, whole blood, and PBS buffer were mixed in a ratio of 1:1:8. After being passed through the microfluidic chip, cell capture was observed across all groups. Captured cells from group C underwent Calcein Acetoxymethyl Ester (Calcein AM) viability staining, followed by centrifugation at 100× *g* for 5 min at room temperature, and were washed twice with PBS. The Calcein AM assay working solution was added, and cells were incubated at 37 °C in the dark for 30 min. Staining was observed using a fluorescence microscope (OLYMPUS-IX53/Japan).

2.6. Capture of CTCs from the Peripheral Blood of Patients with Lung Cancer

2.6.1. Inclusion and Exclusion Criteria for Patients with Lung Cancer

Inclusion criteria: a. diagnosed with NSCLC via clinical or pathological assessment; b. aged 20 to 80 years, any gender; c. predominantly at an advanced stage (Stage III/IV) with tumor metastasis. Exclusion criteria: a. presence of severe hematological disorders; b. diagnosis of multiple malignant tumors; c. presence of severe immune dysfunction. Informed consent from all participants were obtained in the study. The study was conducted in accordance with the Declaration of Helsinki, and the protocol was approved by the General Hospital of Ningxia Medical University Medical Research Ethics Review Committee (KYLL-2022-0782) on 27 October 2022.

2.6.2. Collection of Peripheral Blood from Patients with Lung Cancer for CTC Capture by Acoustic Microfluidic Technique

Five milliliters of peripheral blood were collected from patients with lung cancer. After lysing erythrocytes with RBC lysate, ultrasound was employed for acoustic microfluidic capture. The captured cells were stained, dried overnight as instructed, and then subjected to immunostaining and fluorescence in situ hybridization (iFISH). Finally, fluorescence microscopy (Nikon/Japan/Tokyo) was used to photograph the slides to determine whether the captured cells were CTCs.

2.6.3. Peripheral Blood Was Collected from the Same Patient, and CTCs Were Detected Using Microfluidics Combined with Immunomagnetic Bead Technology, and Comparison of the Differences Between the Two Methods

Execute the protocol by sequentially performing cell separation, incubation, magnetic bead separation, and staining on a 5 mL sample of peripheral blood from the same patient. After staining, examine the final slide under a microscope (AXIO Imager Z2, Zeiss/Germany/Oberkochen) to determine if the cells isolated using microfluidics with immunomagnetic beads are CTCs. Compare these findings with CTCs captured via acoustic microfluidic technology.

2.7. Detection of EGFR Genetic Mutations in Patients with Lung Cancer Using the ARMS-PCR Method

Collect 10 mL of peripheral blood from patients whose CTCs have been identified through the above operations. Obtain 10 mL of peripheral blood from patients with identified CTCs. Centrifuge the sample at 2000× *g* for 10 min, and collect the supernatant, which constitutes the serum. Follow the provided protocol to extract DNA and determine its concentration. Set up the ARMS-PCR reaction system according to the guidelines, conduct on-machine detection, and analyze the results.

2.8. Data Analysis

Data analysis was performed using SPSS 26, with results expressed as mean ± standard deviation. Each test was conducted in triplicate, and statistical significance was set at $p < 0.05$. Graphical analysis was executed using GraphPad Prism 9.5.0. Fluorescence staining quantification was performed using ImageJ 8, and image composition was completed with Adobe Illustrator 2022.

3. Results

3.1. Model of Acoustic Microfluidic Chip and Mechanism of Capture

Acoustic microfluidic technology integrates acoustic waves with microfluidics, utilizing acoustic radiation and Stokes drag forces from acoustic streaming to manipulate the movement of particles and cells for efficient and rapid cell separation and capture [26–28]. Studies indicate that these forces are closely linked to cell diameter; larger cells experience a stronger capture force under identical acoustic conditions than smaller cells [26]. By modifying parameters like voltage and frequency, the micropores in the acoustic microfluidic chip can selectively capture target cells, which is central to its capture mechanism.

The acoustic microfluidic device is depicted in Figure 1a, with its chip model and actual chip illustrated in Figure 1b and 1c, respectively. Upon activating the device, the device is adjusted to a specific voltage and frequency. The application of an acoustic field force amplifies the sine wave from the signal generator, causing the piezoelectric transducer to vibrate. This vibration is transmitted to the microporous array structure within the PDMS chip. Viscous dissipation generates a pressure differential around the micropores, creating acoustic flow inside and outside the pores. Figure 2a presents the schematic of cell capture. A micro-syringe pump injects the detection sample into the chip. Larger cells experience an acoustic flow capture force exceeding the thrust of the flow field, resulting in their capture, while smaller cells continue with the flow field. This entire process is observable in real time through an inverted microscope. By placing the chip on the microscope stage, captured cells in the micropores are visible in real time by adjusting the lens and focus knob.

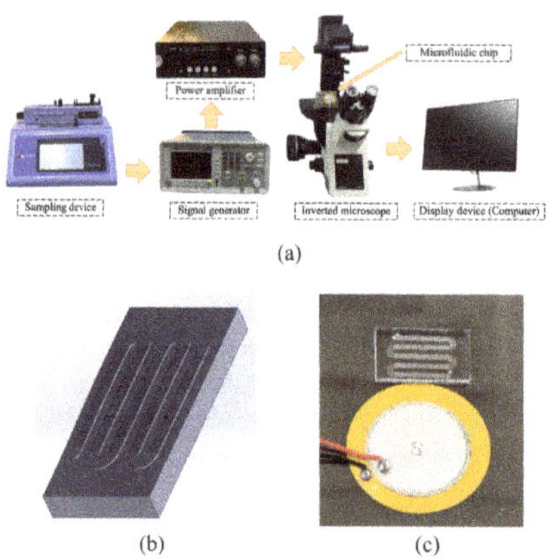

Figure 1. Diagram of materials for building acoustic microfluidic technology. (**a**) Diagram of the acoustic microfluidic device; (**b**) microfluidic chip model diagram; (**c**) microfluidic capture chip actual picture.

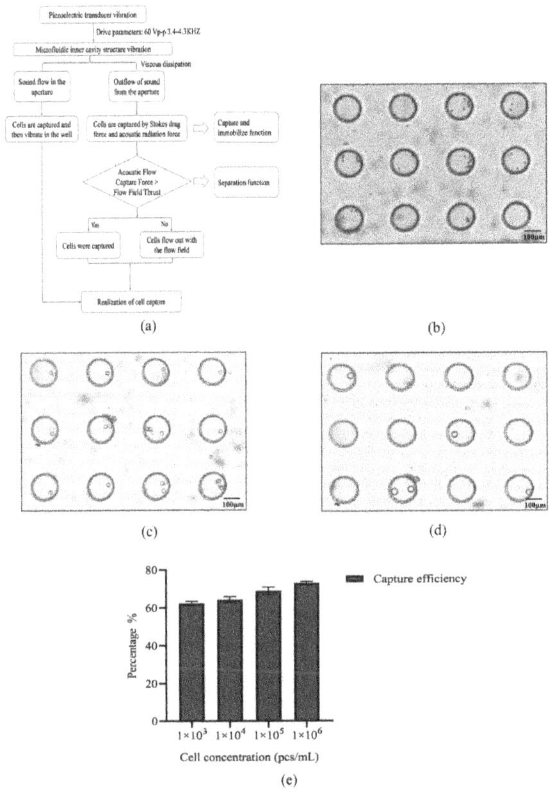

Figure 2. Schematic diagram of acoustic microfluidic technology for capture, and a schematic illustration of microsphere capture. (**a**) Overview of the capture principle of the acoustic microfluidic chip; (**b**) microscope image of an 8 μm microsphere captured by the chip; (**c**) microscope image of a 15 μm microsphere captured by the chip; (**d**) microscope image of a 25 μm microsphere captured by the chip; (**e**) graph of the capture efficiency of A549 cells under different concentration gradients.

Upon application of an acoustic field, the liquid advances under the influence of the acoustic force. Captured cells jump within the micropores, while uncaptured cells are carried away by the flow. The captured cells remain stable in the micropores during the acoustic field's action, enabling real-time observation or subsequent procedures. Once the acoustic field is deactivated, these cells resume flowing, allowing for cell sorting. By modifying parameter settings, cells of varying diameters and those in different solute conditions (such as PBS or whole blood) can be effectively captured.

3.2. Experiment on Simulating Cells with Polystyrene Microspheres by Using Acoustic Microfluidic Technology

In the acoustic microfluidic device, 8 μm polystyrene microspheres were trapped in the microchannels and exhibited a jumping motion at 2.5 V, 2 μL/min, and 3.5 kHz (Figure 2b). Adjusting the parameters to 3 V, 3 μL/min, and a frequency of 3.9–4.1 kHz resulted in the capture and similar jumping behavior of both 15 μm and 25 μm microspheres (Figure 2c,d).

3.3. Study on the Cell Capture Efficiency of Lung Cancer A549 Cells by Acoustic Microfluidic Technology

Under conditions of 3 V, 3 μL/min, and 4.0 kHz, experiments with varying concentrations of A549 cells demonstrated that capture efficiency increased significantly with higher concentrations, stabilizing above 70% at elevated levels (Figure 2e). This suggests that increasing the initial cell count is essential for optimizing CTC capture efficiency.

3.4. Investigation of the Capture of Pre-Labeled A549 Cells Using Acoustic Microfluidic Technology

DiI-labeled A549 cells were successfully captured in PBS at an applied voltage of 3 V, a flow rate of 0.6 μL/min, and a frequency of 3.2 kHz. Microscopic observation revealed the captured cells exhibiting beating behavior under bright-field illumination (Figure 3a) and fluorescent labeling (Figure 3b), as shown in a video in Supplementary Materials S1. Reducing the flow rate to 0.5 μL/min enabled the capture of DiI-labeled A549 cells in the supernatant of healthy human peripheral blood samples after centrifugation, with the captured cells visible under both bright-field and fluorescence microscopy and in a video (Figure 4a,b, Supplementary Materials S2). These results demonstrate the efficient capture of labeled A549 cells by the acoustic field, both in PBS and in the supernatant of human blood samples.

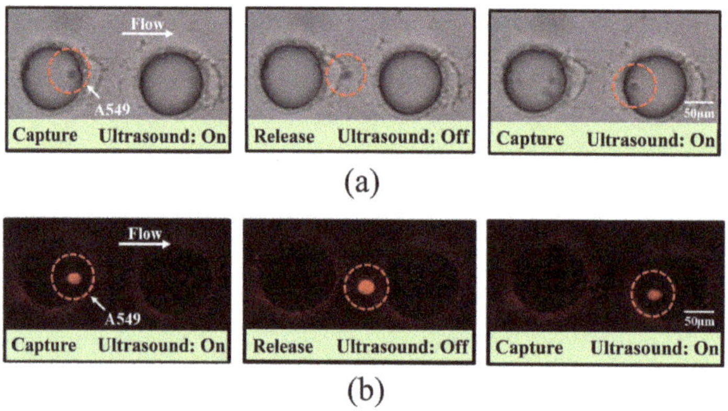

Figure 3. Acoustic microfluidic capture of pre-labeled A549 cells. (**a**) A549 cells in PBS captured by the acoustic microfluidic chip under bright-field microscopy; (**b**) A549 cells in PBS captured by the acoustic microfluidic chip under fluorescence microscopy.

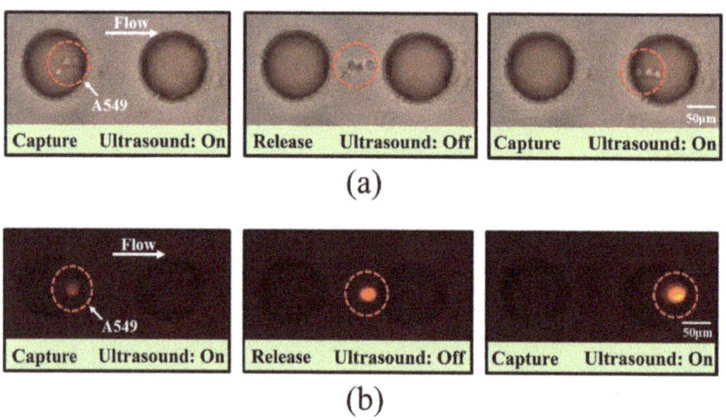

Figure 4. Acoustic microfluidic capture of pre-labeled A549 cells. (**a**) Bright-field image showing A549 cells captured by the acoustic microfluidic chip after mixing with the peripheral centrifuged supernatant blood of a healthy individual; (**b**) fluorescence image showing A549 cells captured by the acoustic microfluidic chip after mixing with the peripheral centrifuged supernatant blood of a healthy individual.

To assess the device's capture performance in complex environments, DiI-labeled A549 cells were introduced into whole blood samples. Successful cell capture was achieved at 3 V, 0.3 μL/min, and 3.2 kHz, as shown in the bright-field imaging (Figure 5a). However, due to the complexity of whole blood, cell types were indistinguishable in bright field alone. Fluorescence imaging confirmed the red cells as target A549 cells (Figure 5b), with a video provided in Supplementary Materials S3. Calcein-AM staining further differentiated cell types: live A549 cells were positive in both bright field and double fluorescence (red + green), dead A549 cells showed red fluorescence but not green, live leukocytes were positive in green fluorescence only, and dead leukocytes were visible only in bright field (Figure 5c). Table 1 demonstrates how DiI-labeled cells can be distinguished from other blood cells under a microscope based on the staining results after Calcein-AM staining. This experiment demonstrates that acoustic microfluidic technology can capture cells in complex blood samples and differentiate cell viability through staining.

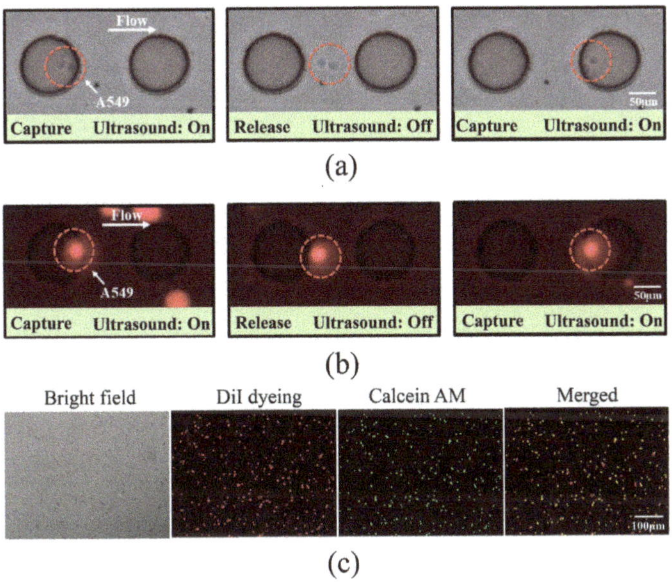

Figure 5. Acoustic microfluidic capture of pre-labeled A549 cells. (**a**) Bright-field view of A549 cell capture on an acoustic microfluidic chip following admixture with healthy human peripheral blood; (**b**) fluorescence view of A549 cell capture on an acoustic microfluidic chip under the same conditions; (**c**) cell viability staining of A549 cells after being captured in healthy human peripheral blood.

Table 1. The staining differences between DiI-labeled cells and other blood cells after Calcein-AM staining.

	Living A549 Cells	Dead A549 Cells	Living White Blood Cells	Dead White Blood Cells
Bright-field field of view	yes	yes	yes	yes
Red fluorescence field of view	yes	yes	no	no
Green fluorescence field of view	yes	no	yes	no

3.5. Utilizing Acoustic Microfluidics Technology to Isolate CTCs from Patients with Lung Cancer

Suitable cases were selected based on the specified inclusion and exclusion criteria, and the clinical data for these cases were systematically gathered and organized, as detailed in Table 2.

Table 2. Clinical data collected on patients with lung cancer.

Sample Number	Years	Sex	Histological Type	Metastasis	Tumor Staging	Personal History
Number 1	72	Female	Squamous Carcinoma	Bone	cT4N2M1 IV	No history of smoking
Number 2	74	Male	Adenocarcinoma	Head	cT1cN2M1b IVA	Previous history of smoking
Number 3	62	Male	Adenocarcinoma	None	cT1cN0M0 IA3	Previous history of smoking
Number 4	59	Male	Squamous Carcinoma	None	cT3N3M0 IIIc	No history of smoking
Number 5	74	Male	Squamous Carcinoma	Lymphatic Node	cT4N3M1 IV	No history of smoking
Number 6	64	Male	Squamous Carcinoma	None	cT4N3M1 IV	Previous history of smoking

Peripheral blood samples from patients with lung cancer were captured using an acoustic microfluidic device. Subsequent liquid cell staining and iFISH analyses were performed. The acoustic microfluidic chip successfully captured peripheral blood cells from six samples under the conditions of 3 V, 3 μL/min, and 4.0 kHz, as shown in Figure 6a. The captured cells were clearly visible under a microscope. Due to the similar cell sizes of CTCs and white blood cells [29], both cell types may be captured within a similar acoustic field frequency range, rendering morphological observation alone insufficient for distinguishing CTCs from white blood cells. To reliably identify the captured cells, the iFISH detection method was employed for verification. The iFISH analysis revealed the presence of one, six, and three CTCs in samples 1, 2, and 3, respectively (Figure 6b,c; Figure 7), four CTCs in sample 4 (Figure 8a), and one CTC each in samples 5 and 6 (Figure 8b).

CTC identification relies on the expression profiles of three markers: the platelet endothelial cell adhesion molecule (CD31), the leukocyte common antigen (CD45), and 4′,6-diamidino-2-phenylindole (DAPI) [30–33], alongside Chromosome 8 Centromere Probe (CEP8) labeling. CEP8 is used to label the chromosomal positions of the detected cells. CD31 distinguishes CTCs from circulating tumor vascular endothelial cells, enhancing specificity by excluding endothelial cell contamination, where endothelial cells express CD31. CD45 differentiates leukocytes from CTCs, ensuring the captured cells are not hematopoietic, as leukocytes are CD45-positive. Both CTCs and leukocytes exhibit positive DAPI staining because they have nuclei. Notably, in certain cases (Figure 6b), leukocytes may also express CD31. Figure 6b illustrates the CTCs detected in sample 1, with the CEP8 probe indicating chromosomal positions from iFISH. The marked cells are CTCs, while those above are leukocytes.

Immunofluorescent labeling analysis revealed that all cells in sample 2 were CTCs, characterized by double negativity for CD31 and CD45 and positivity for DAPI expression (Figure 6c). In sample 3, the labeled cells in field of view 1 were CTCs, while the unlabeled cells were CD45-positive leukocytes; the cells in fields of view 2 and 3 were all CTCs (Figure 7). Sample 4 contained CTCs in fields of view 1 and 2, as well as in the cells indicated by the dotted lines in fields of view 3 and 4, while the cell identified by the arrow in field of view 3 was a leukocyte. The cells within the box in field of view 5 were endothelial cells (Figure 8a). The cells in sample 5 were CTCs; the cells labeled in sample 6 were also identified as CTCs (Figure 8b).

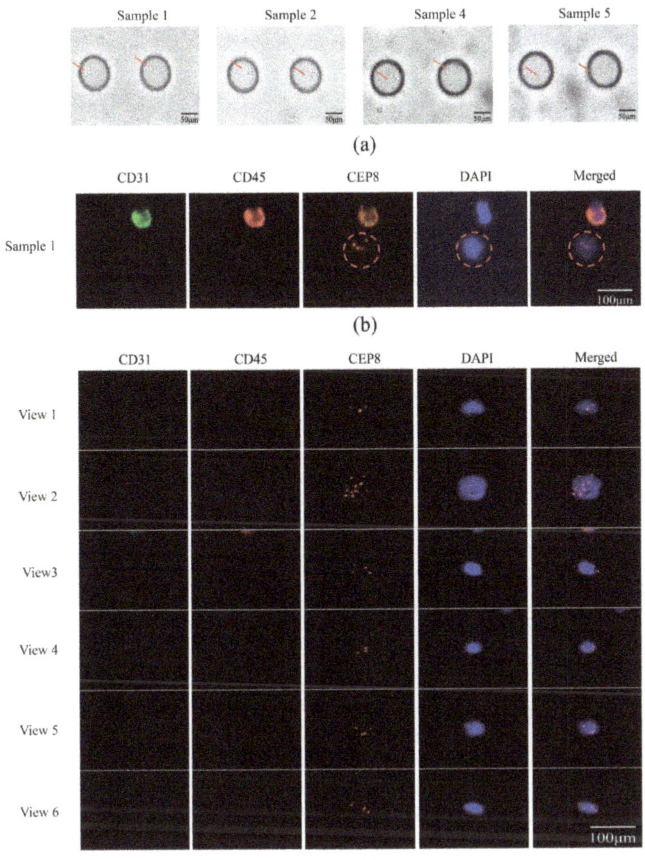

Figure 6. CTCs were captured using acoustic microfluidic technology from the peripheral blood of patients with lung cancer. (**a**) Cell capture of some samples (samples 1, 2, 4, 5) in the acoustic microfluidic device; (**b**) CTCs in sample 1 identified via iFISH fluorescent staining (red circle); (**c**) CTCs in sample 2 identified via iFISH fluorescent staining, the cells in fields of view 1–6 were all CTCs.

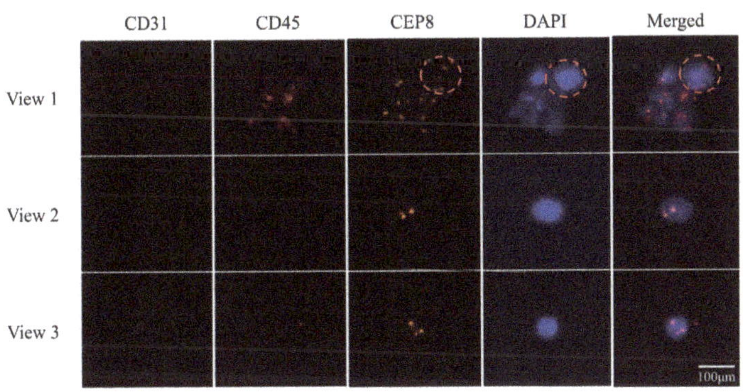

Figure 7. CTCs were captured using acoustic microfluidic technology from the peripheral blood of patients with lung cancer, with fluorescent staining of CTCs identified by iFISH in sample 3, the red circle cells in field of view 1 were CTCs, the cells in fields of view 2 and 3 were all CTCs.

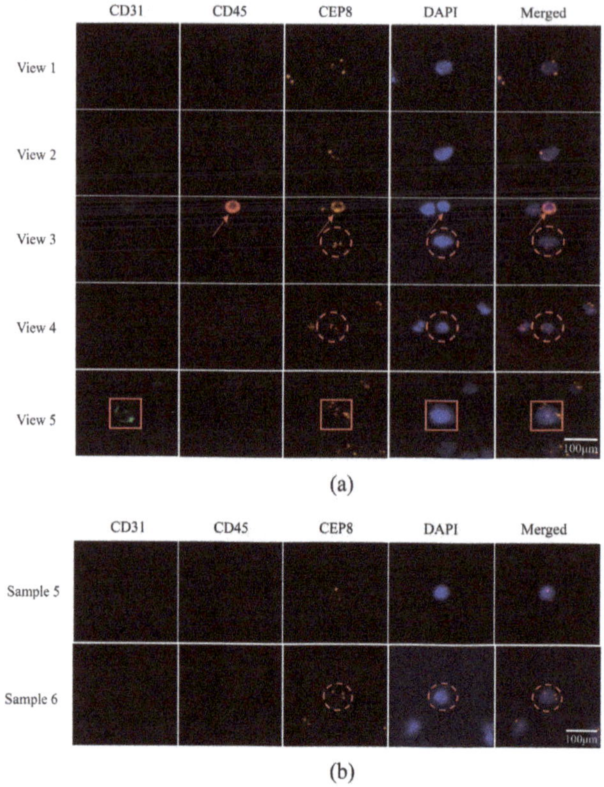

Figure 8. CTCs were captured using acoustic microfluidic technology from the peripheral blood of patients with lung cancer. (**a**) Fluorescent staining of CTCs, leukocytes, and endothelial cells identified by iFISH in sample 4, the cells in fields of view 1 and 2, the red circle cells in fields of view 3 and 4 were CTCs, the cell identified by the red arrow in field of view 3 was a leukocyte. The cells within the red box in field of view 5 were endothelial cells; (**b**) CTCs identified by iFISH fluorescent staining in samples 5 and 6, the cells in sample 5 were CTCs, the cells labeled with red circle in sample 6 were CTCs.

3.6. CTCs in the Peripheral Blood of Patients with Lung Cancer Were Detected Using a Microfluidic Technique Combined with Immunomagnetic Bead Separation and Compared to Acoustic Microfluidic Technology

CTCs were identified in all six samples using the microfluidics–immunomagnetic bead separation approach, which employs cytokeratin (CK), CD45, and DAPI as key markers. CK is a specific marker for CTCs, and its positive expression indicates the presence of CTCs [34]. CTCs were defined as CK-positive, CD45-negative, and DAPI-positive cells. The cells observed in samples 2 and 3 in Figure 9a were all identified as CTCs. Figure 9b shows the cell expression patterns in different fields of view of samples 5 and 6, where the cells marked with squares were identified as CTCs, and the unmarked cells were white blood cells. The numbers of CTCs detected in the six samples using the microfluidics–immunomagnetic bead separation and the acoustic microfluidic technologies were statistically analyzed, as presented in Table 3.

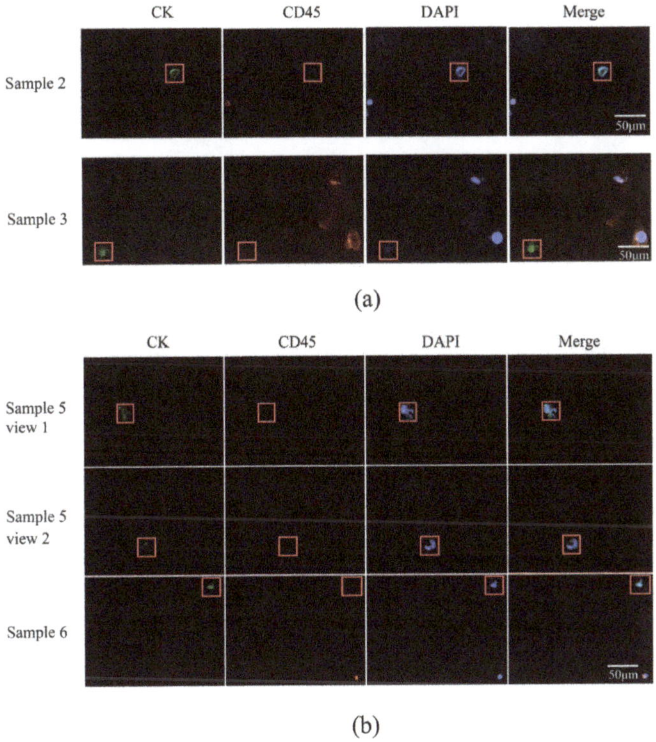

Figure 9. CTC fluorograms obtained using a microfluidic device with immunomagnetic bead separation. (**a**) Fluorescent staining-based CTC maps for samples 2 and 3, the cells with red box were CTCs. (**b**) Fluorescent staining-based CTC maps for samples 5 and 6, the cells with red box were CTCs.

Table 3. Statistics on CTC detections.

	Number of CTCs Captured by Acoustic Microfluidics Combined with iFISH	Number of CTCs Captured by Microfluidics Combined with the Immunomagnetic Bead Separation Technique
Number 1	1	52
Number 2	6	17
Number 3	3	13
Number 4	4	8
Number 5	1	9
Number 6	1	1

3.7. Detection of Genetic Mutations in Patients with Lung Cancer Using the ARMS-PCR Method

ARMS-PCR identified EGFR mutations in four out of six patients with lung cancer, as detailed in Table 4. The detected mutations comprised an exon 19 deletion (19-del), a leucine-to-arginine substitution at position 858 (L858R) in exon 21, a serine-to-isoleucine substitution at position 768 (S768I) in exon 20, and a threonine-to-methionine substitution at position 790 (T790M) in exon 20.

Table 4. Detection of EGFR gene mutations.

Sample Number	Type of EGFR Gene Mutation
Number 1	19-del, L858R mutation
Number 2	S768I mutation
Number 3	None
Number 4	T790M mutation
Number 5	None
Number 6	L858R mutation

4. Discussion

Lung cancer, a highly malignant tumor, imposes a profound burden on many families [35]. Conventional diagnostic and therapeutic approaches are costly and frequently ineffective. Recently, CTC detection technology has emerged as a promising avenue for early lung cancer screening and diagnosis. Nevertheless, existing CTC detection methods remain expensive and time-consuming.

This study employs acoustic microfluidic technology, integrating acoustic waves with microfluidics, to efficiently and rapidly separate and capture target cells. Initially, polystyrene microspheres were used due to their uniform size, ease of modification, chemical stability, low cost, and high capture efficiency [36]. The diameter of CTCs typically ranges from 15 μm to 25 μm [37], aligning with the microspheres used. Thus, initial capture experiments simulated cells with these microspheres. Subsequently, acoustic field parameters were optimized for capturing A549 cells. Short-term viability and long-term proliferation experiments confirmed that the acoustic field maintains cell viability and biological integrity during capture, a finding supported by multiple studies [38,39]. Detailed experimental data and results are provided in Supplementary Materials S4.

Fluorescently labeled A549 cells were utilized in subsequent experiments to demonstrate the capacity of acoustic microfluidic technology to reliably and efficiently capture cells across diverse solute environments, including PBS and whole blood. DiI was used for fluorescent labeling to specifically trace A549 cells, enabling clear identification in complex optical backgrounds, such as whole blood, and reducing interference from impurities or other cells, thus enhancing detection specificity and reliability. Additionally, to assess the impact of acoustic forces on cell viability, captured cells in peripheral blood were stained with Calcein AM/Propidium Iodide, allowing observation of the viability of DiI-labeled A549 cells post-capture. The capture selectivity of acoustic microfluidic technology for specific cell lines primarily arises from the interplay between the inherent physical properties of the cells (e.g., size, density, compressibility) and the external field parameters (frequency, voltage, flow rate) [40]. Due to their relatively large diameter (approximately 15–20 μm) and high acoustic impedance coefficient, A549 cells are readily captured by the acoustic radiation force under specific acoustic field conditions. In contrast, blood cells (such as red blood cells and lymphocytes) have smaller sizes or markedly different mechanical properties, rendering them less affected by the acoustic field under the same conditions, with slower velocities and, consequently, lower capture efficiency or even an inability to be captured [41,42]. This aligns with existing studies on acoustic microfluidic sorting, demonstrating that optimizing acoustic field conditions, such as frequency and voltage, can achieve efficient and selective cell capture [26].

Differences in CTC detection arise between microfluidics–immunomagnetic bead separation and acoustic microfluidic technologies, with the former identifying more CTCs. This discrepancy may be due to the acoustic microfluidic technology being in the research phase, lacking optimized acoustic field parameters, and exhibiting uneven acoustic force distribution in the capture area. Furthermore, these technologies vary in cell capture, staining, and identification criteria. Acoustic microfluidics employs CD31 and CD45 markers, whereas microfluidics–immunomagnetic bead separation uses CK and CD45. This approach may miss CTCs that have undergone epithelial-to-mesenchymal transition and do not express CK, reducing specificity and possibly detecting non-CTCs.

An acoustic microfluidic chip has enabled the capture of CTCs from the peripheral blood of patients with lung cancer. Unlike the CTC capture technique employing microfluidics combined with the immunomagnetic bead separation method, the acoustic microfluidic approach offers distinct advantages, including the elimination of antibody labeling [43], simplified and cost-effective operation, and the preservation of cell viability. Nonetheless, the microfluidics with immunomagnetic bead separation technique has shown high sensitivity and specificity in detecting various tumor cells and has been successfully applied in clinical CTC studies.

While the acoustic microfluidic technology employed in this study effectively captures CTCs, it detects fewer CTCs than microfluidics combined with the immunomagnetic bead separation method. Enhancing CTC detection sensitivity in the future may involve optimizing chip capture parameters, such as acoustic wave frequency, sound pressure intensity, and microfluidic channel design. Overall, the acoustic microfluidic technology efficiently completed CTC capture within approximately 2 h. When coupled with iFISH, the detection process can be completed within 48 h at a cost of only around CNY 1000, offering significant economic advantages over existing technologies (CNY 3000–4000) and demonstrating strong potential for clinical application.

In the first step of this study, CTCs from patients with lung cancer were captured using acoustic microfluidic technology. Patients with identified CTCs underwent EGFR gene mutation analysis. Personalized clinical treatment strategies were then developed based on the specific mutations detected. The ARMS–PCR technology, approved by the China National Medical Products Administration, is employed for gene mutation detection due to its simplicity, rapidity, and reliability. It is effective in identifying gene mutations at early tumor stages [44]. Details on the mutation sites and structural characteristics of EGFR gene mutations are provided in Figure S3 of Supplementary Materials S4. This study involved the extraction of cell-free DNA (cfDNA) from the peripheral blood plasma of patients with lung cancer for mutation analysis. CfDNA consists of non-cell-bound DNA fragments present in the bloodstream, primarily originating from DNA released during cell apoptosis or necrosis [45]. In patients with cancer, cfDNA includes circulating tumor DNA from tumor cells, which harbors tumor-specific genetic variations, such as insertion/deletion mutations, point mutations, and copy number variations [46]. The detection of EGFR gene mutations typically relies on liquid biopsy technology. This approach involves several key steps. First, the patient's peripheral blood sample is collected, which contains a variety of normal blood cells as well as, potentially, CTCs that have originated from the tumor and entered the bloodstream. Upon apoptosis or rupture of these CTCs, they release cfDNA into the blood. The blood sample is then centrifuged to isolate the serum, from which the DNA is extracted. Finally, a PCR-based technique, such as ARMS–PCR, is employed to detect the presence of EGFR gene mutations within the cfDNA. This information can then be utilized for tumor diagnosis, selection of targeted therapies, and monitoring of treatment effectiveness.

The 19-del and L858R point mutations are the predominant EGFR gene alterations in patients with NSCLC [47]. From a clinical perspective, patients with NSCLC harboring these two mutations have demonstrated favorable therapeutic responses to epidermal growth factor receptor tyrosine kinase inhibitors (EGFR-TKIs) [48,49], establishing them as an important first-line targeted treatment indication.

Notably, the T790M acquired resistance mutation was identified in the assay [50–52]. This mutation commonly emerges after 6–13 months of EGFR-TKI treatment and is a primary mechanism of resistance to EGFR-TKI therapy [53–55]. In response, third-generation EGFR-TKIs, such as osimertinib, have been developed that can effectively overcome T790M-mediated resistance [52,56]. Additionally, a relatively rare S768I mutation was detected [57], with an incidence of approximately 0.5–1% in patients with NSCLC. Due to the limited number of cases, the clinical characteristics, prognostic significance, and therapeutic response to EGFR-TKIs of this mutation type remain to be fully elucidated, necessitating further expansion of the sample size for in-depth investigation.

In this study, samples 1, 2, 4, and 6 were positive for EGFR gene mutations; however, due to financial or clinical constraints, these patients did not undergo routine genetic testing. Patient 2 received radiotherapy, while patients 1, 4, and 6 were treated with an immunotherapy-based chemotherapy regimen consisting of tislelizumab, albumin-bound paclitaxel, and carboplatin. Although this regimen is standard for advanced NSCLC, patients identified with EGFR mutations are often prioritized for EGFR-TKI therapies, such as osimertinib or gefitinib, which offer superior efficacy and survival benefits. This highlights the critical role of genetic testing in precision lung cancer treatment. Promoting genetic testing in clinical practice is essential to enhance treatment precision and improve patient survival outcomes. To enhance the accuracy of lung cancer diagnosis and treatment while maximizing clinical benefits, we advocate for the adoption of acoustic microfluidic technology in CTC detection. This technology is vital for evaluating metastasis and recurrence risks. Patients testing positive for CTCs should subsequently undergo EGFR mutation analysis via cell-free DNA. In metastatic lung cancer cases, the ARMS method yields superior accuracy for gene mutation detection. This sequential testing approach can significantly lower patient costs, improve test precision, boost diagnostic efficiency, and optimize medical resource allocation. Additionally, it provides comprehensive molecular data, offering robust diagnostic evidence to aid clinicians in formulating the most effective treatment strategies.

The detection of EGFR gene mutations using cfDNA still faces certain limitations. A major challenge is that cfDNA samples may contain DNA from various sources in the circulating blood, and the presence of this non-tumor-derived DNA can interfere with the accuracy of mutation detection. While current EGFR mutation detection kits demonstrate high sensitivity, directly isolating circulating tumor DNA from the peripheral blood of patients with lung cancer could further improve the detection accuracy and specificity. Future research should focus on optimizing the application of acoustic microfluidic technology for the capture of CTCs, emphasizing enhanced capture efficiency and cell purity. This approach would yield more reliable ctDNA for gene mutation analysis and provide high-quality samples for subsequent studies. Detecting EGFR gene mutations has become a crucial molecular diagnostic tool in NSCLC. Identifying specific mutations—such as 19-del, L858R, and T790M—enables clinicians to select targeted therapies more effectively. This precision medicine approach enhances treatment efficacy while reducing unnecessary drug exposure and potential side effects, thus advancing personalized lung cancer treatment.

In this study, we developed a comprehensive technology transformation pathway, progressing from basic parameter optimization (microspheres to cell lines) to clinical application (CTC detection to drug guidance), confirming the feasibility of combining acoustic microfluidic technology with EGFR detection. This stepwise detection strategy not only validated the effectiveness of acoustic microfluidics in capturing CTCs but also integrated CTC detection with EGFR gene mutation analysis. This method enables dynamic monitoring of patient conditions, facilitating precise lung cancer diagnosis and treatment, thereby holding significant clinical value for improving patient prognosis.

The integration of acoustic microfluidic technology, because of its non-destructive, efficient, and cost-effective CTC capture, with ARMS-PCR for precise genetic testing, facilitates comprehensive precision diagnosis and treatment of lung cancer—from initial diagnosis through treatment monitoring and personalized solution optimization. Despite current limitations, such as small sample sizes, advancements and the integration of additional technologies are anticipated to enhance early screening and dynamic monitoring levels of lung cancer, ushering in a new era of precision in its diagnosis and treatment.

5. Conclusions

The acoustic field force in acoustic microfluidics, as a non-contact force, excels in preserving cell integrity and viability in vitro. This technology enables efficient capture of microspheres and lung cancer A549 cells, as well as CTCs from the peripheral blood of patients with advanced lung cancer. This capability not only provides a more robust foundation for clinical disease evaluation but also offers crucial technical support for the diagnosis and treatment of lung cancer. When integrated with ARMS-PCR technology, the platform precisely detects EGFR gene mutations, guides targeted therapies, and advances precision medicine applications.

Supplementary Materials: The following supporting information can be downloaded at: https://www.mdpi.com/article/10.3390/mi16101136/s1, Supplementary Materials S1: The capture video of diI-labeled A549 in bright field and fluorescent field (in PBS); Supplementary Materials S2: The capture video of diI-labeled A549 in bright field and fluorescent field (in the supernatant of healthy human peripheral blood samples after centrifugation); Supplementary Materials S3: The capture video of diI-labeled A549 in bright field and fluorescent field (in whole blood samples); Supplementary Materials S4: The research of short-term viability and long-term proliferation experiments and the analyze of EGFR mutation sites.

Author Contributions: Conceptualization, T.T. and S.M.; methodology, Y.W.; software, H.Y.; validation, T.T., S.M., and H.Y.; formal analysis, T.D. and Y.W.; investigation, J.L. and Y.W.; resources, M.T. and T.D.; data curation, G.L. and W.F.; writing—original draft preparation, T.T. and S.M.; writing—review and editing, Z.Z.; visualization, J.M., M.T., and Z.Z.; supervision, J.L., W.F., and Z.Z.; project administration, T.T. and H.Y.; funding acquisition, T.T., S.M., and G.L. All authors have read and agreed to the published version of the manuscript.

Funding: This research was funded by the Key R&D Program Project of Ningxia Hui Autonomous Region, grant number 2023BEG02002, and the APC was funded by the Key R&D Program Project of Ningxia Hui Autonomous Region (No.2023BEG02002).

Data Availability Statement: The data that support the findings of this study are available from the corresponding author upon reasonable request.

Acknowledgments: We would like to thank the participants who provided samples for this study.

Conflicts of Interest: The authors declare no conflicts of interest.

Abbreviations

The following abbreviations are used in this manuscript:

CTCs	Circulating Tumor Cells
EGFR	Epidermal Growth Factor Receptor
NSCLC	Non-Small-Cell Lung Cancer
ARMS-PCR	Amplification Refractory Mutation System Polymerase Chain Reaction

PDMS	Polydimethylsiloxane
PBS	Phosphate-buffered Saline
DiI	Cell Plasma Membrane Staining Kit with DiI (Red Fluorescence)
Calcein AM	Calcein Acetoxymethyl Ester
iFISH	Immunostaining and Fluorescence in Situ Hybridization
CD31	Cluster of Differentiation 31
CD45	Cluster of Differentiation 45
DAPI	4′,6-diamidino-2-phenylindole
CEP8	Chromosome 8 Centromere Probe
CK	Cytokeratin
19-del	Exon 19 Deletion Mutation
L858R	Leucine 858 Arginine Point Mutation
S768I	Serine 768 Isoleucine Mutation
T790M	Threonine 790 Methionine Mutation
cfDNA	Circulating Free DNA
EGFR-TKI	Epidermal Growth Factor Receptor Tyrosine Kinase Inhibitor

References

1. Mederos, N.; Friedlaender, A.; Peters, S.; Addeo, A. Gender-specific aspects of epidemiology, molecular genetics and outcome: Lung cancer. *ESMO Open* **2020**, *5*, e000796. [CrossRef]
2. Ishiguro, S.; Alhakamy, N.A.; Uppalapati, D.; Delzeit, J.; Berkland, C.J.; Tamura, M. Combined Local Pulmonary and Systemic Delivery of AT2R Gene by Modified TAT Peptide Nanoparticles Attenuates Both Murine and Human Lung Carcinoma Xenografts in Mice. *J. Pharm. Sci.* **2017**, *106*, 385–394. [CrossRef]
3. Cheng, M.; Shao, S.; Xu, W.; Liu, D. Novel causes and assessments of intrapulmonary metastasis. *Biochem. Biophys. Rep.* **2025**, *42*, 102004. [CrossRef]
4. Zhu, M.; Zhai, Z.; Wang, Y.; Chen, F.; Liu, R.; Yang, X.; Zhao, G. Advancements in the application of artificial intelligence in the field of colorectal cancer. *Front. Oncol.* **2025**, *15*, 1499223. [CrossRef]
5. Kresse, S.H.; Brandt-Winge, S.; Pharo, H.; Flatin, B.T.B.; Jeanmougin, M.; Vedeld, H.M.; Lind, G.E. Evaluation of commercial kits for isolation and bisulfite conversion of circulating cell-free tumor DNA from blood. *Clin. Epigenetics* **2023**, *15*, 151. [CrossRef] [PubMed]
6. Meng, X.; Sun, P.; Xu, H.; Wang, Z. Folic acid-functionalized magnetic nanoprobes via a PAMAM dendrimer/SA-biotin mediated cascade-amplifying system for the efficient enrichment of circulating tumor cells. *Biomater. Sci.* **2020**, *8*, 6395–6403. [CrossRef]
7. Abouleila, Y.; Onidani, K.; Ali, A.; Shoji, H.; Kawai, T.; Lim, C.T.; Kumar, V.; Okaya, S.; Kato, K.; Hiyama, E.; et al. Live single cell mass spectrometry reveals cancer-specific metabolic profiles of circulating tumor cells. *Cancer Sci.* **2019**, *110*, 697–706. [CrossRef]
8. Li, M.; Wu, S.; Zhuang, C.; Shi, C.; Gu, L.; Wang, P.; Guo, F.; Wang, Y.; Liu, Z. Metabolomic analysis of circulating tumor cells derived liver metastasis of colorectal cancer. *Heliyon* **2023**, *9*, e12515. [CrossRef] [PubMed]
9. Cheung, K.J.; Ewald, A.J. A collective route to metastasis: Seeding by tumor cell clusters. *Science* **2016**, *352*, 167–169. [CrossRef] [PubMed]
10. Han, J.; Lu, C.; Shen, M.; Sun, X.; Mo, X.; Yang, G. Fast, Reusable, Cell Uniformly Distributed Membrane Filtration Device for Separation of Circulating Tumor Cells. *ACS Omega* **2022**, *7*, 20761–20767. [CrossRef]
11. Deng, Z.; Wu, S.; Wang, Y.; Shi, D. Circulating tumor cell isolation for cancer diagnosis and prognosis. *EBioMedicine* **2022**, *83*, 104237. [CrossRef]
12. Li, W.; Liu, J.B.; Hou, L.K.; Yu, F.; Zhang, J.; Wu, W.; Tang, X.M.; Sun, F.; Lu, H.M.; Deng, J.; et al. Liquid biopsy in lung cancer: Significance in diagnostics, prediction, and treatment monitoring. *Mol. Cancer* **2022**, *21*, 25. [CrossRef]
13. Chen, H.; Li, Q.; Hu, Q.; Jiao, X.; Ren, W.; Wang, S.; Peng, G. Double spiral chip-embedded micro-trapezoid filters (SMT filters) for the sensitive isolation of CTCs of prostate cancer by spectral detection. *Nanoscale Adv.* **2022**, *4*, 5392–5403. [CrossRef]
14. Xue, M.; Xiang, A.; Guo, Y.; Wang, L.; Wang, R.; Wang, W.; Ji, G.; Lu, Z. Dynamic Halbach array magnet integrated microfluidic system for the continuous-flow separation of rare tumor cells. *RSC Adv.* **2019**, *9*, 38496–38504. [CrossRef]
15. Peng, T.; Qiang, J.; Yuan, S. Sheathless inertial particle focusing methods within microfluidic devices: A review. *Front. Bioeng. Biotechnol.* **2023**, *11*, 1331968. [CrossRef] [PubMed]
16. Wu, M.; Huang, P.H.; Zhang, R.; Mao, Z.; Chen, C.; Kemeny, G.; Li, P.; Lee, A.V.; Gyanchandani, R.; Armstrong, A.J.; et al. Circulating Tumor Cell Phenotyping via High-Throughput Acoustic Separation. *Small* **2020**, *16*, e2004438. [CrossRef] [PubMed]
17. Yoon, H.J.; Kozminsky, M.; Nagrath, S. Emerging role of nanomaterials in circulating tumor cell isolation and analysis. *ACS Nano* **2014**, *8*, 1995–2017. [CrossRef] [PubMed]

18. Khan, T.; Becker, T.M.; Po, J.W.; Chua, W.; Ma, Y. Single-Circulating Tumor Cell Whole Genome Amplification to Unravel Cancer Heterogeneity and Actionable Biomarkers. *Int. J. Mol. Sci.* **2022**, *23*, 8386. [CrossRef]
19. Moon, G.Y.; Dalkiran, B.; Park, H.S.; Shin, D.; Son, C.; Choi, J.H.; Bang, S.; Lee, H.; Doh, I.; Kim, D.H.; et al. Dual Biomarker Strategies for Liquid Biopsy: Integrating Circulating Tumor Cells and Circulating Tumor DNA for Enhanced Tumor Monitoring. *Biosensors* **2025**, *15*, 74. [CrossRef]
20. Eliahoo, P.; Setayesh, H.; Hoffman, T.; Wu, Y.; Li, S.; Treweek, J.B. Viscoelasticity in 3D Cell Culture and Regenerative Medicine: Measurement Techniques and Biological Relevance. *ACS Mater. Au* **2024**, *4*, 354–384. [CrossRef]
21. Zhao, Y.; Shi, W.; Tang, Q. An eleven-gene risk model associated with lymph node metastasis predicts overall survival in lung adenocarcinoma. *Sci. Rep.* **2023**, *13*, 6852. [CrossRef] [PubMed]
22. Dong, W.; Wang, C.; Wang, C.; Zhao, K.; Ma, Z.; Hu, S. Inconsistent clinical outcomes following afatinib treatment in NSCLC patients harboring uncommon epidermal growth factor receptor mutation. *Front. Oncol.* **2022**, *12*, 999606. [CrossRef]
23. He, C.; Wei, C.; Wen, J.; Chen, S.; Chen, L.; Wu, Y.; Shen, Y.; Bai, H.; Zhang, Y.; Chen, X.; et al. Comprehensive analysis of NGS and ARMS-PCR for detecting EGFR mutations based on 4467 cases of NSCLC patients. *J. Cancer Res. Clin. Oncol.* **2022**, *148*, 321–330. [CrossRef] [PubMed]
24. Li, C.; He, Q.; Liang, H.; Cheng, B.; Li, J.; Xiong, S.; Zhao, Y.; Guo, M.; Liu, Z.; He, J.; et al. Diagnostic Accuracy of Droplet Digital PCR and Amplification Refractory Mutation System PCR for Detecting EGFR Mutation in Cell-Free DNA of Lung Cancer: A Meta-Analysis. *Front. Oncol.* **2020**, *10*, 290. [CrossRef]
25. Liang, C.; Wu, Z.; Gan, X.; Liu, Y.; You, Y.; Liu, C.; Zhou, C.; Liang, Y.; Mo, H.; Chen, A.M.; et al. Detection of Rare Mutations in EGFR-ARMS-PCR-Negative Lung Adenocarcinoma by Sanger Sequencing. *Yonsei Med. J.* **2018**, *59*, 13–19. [CrossRef]
26. Bai, X.; Song, B.; Chen, Z.; Zhang, W.; Chen, D.; Dai, Y.; Liang, S.; Zhang, D.; Zhao, Z.; Feng, L. Postoperative evaluation of tumours based on label-free acoustic separation of circulating tumour cells by microstreaming. *Lab Chip* **2021**, *21*, 2721–2729. [CrossRef]
27. Hashemiesfahan, M.; Christiaens, J.W.; Maisto, A.; Gelin, P.; Gardeniers, H.; De Malsche, W. Characterizing Acoustic Behavior of Silicon Microchannels Separated by a Porous Wall. *Micromachines* **2024**, *15*, 868. [CrossRef]
28. Gautam, G.P.; Gurung, R.; Fencl, F.A.; Piyasena, M.E. Separation of sub-micron particles from micron particles using acoustic fluid relocation combined with acoustophoresis. *Anal. Bioanal. Chem.* **2018**, *410*, 6561–6571. [CrossRef] [PubMed]
29. Uddin, M.R.; Sarowar, M.T.; Chen, X. Separation of CTCs from WBCs using DEP-assisted inertial manipulation: A numerical study. *Electrophoresis* **2023**, *44*, 1781–1794. [CrossRef]
30. Sonn, C.H.; Cho, J.H.; Kim, J.W.; Kang, M.S.; Lee, J.; Kim, J. Detection of circulating tumor cells in patients with non-small cell lung cancer using a size-based platform. *Oncol. Lett.* **2017**, *13*, 2717–2722. [CrossRef]
31. Cai, S.; Deng, Y.; Wang, Z.; Zhu, J.; Huang, C.; Du, L.; Wang, C.; Yu, X.; Liu, W.; Yang, C.; et al. Development and clinical validation of a microfluidic-based platform for CTC enrichment and downstream molecular analysis. *Front. Oncol.* **2023**, *13*, 1238332. [CrossRef]
32. Lin, A.Y.; Wang, D.D.; Li, L.; Lin, P.P. Identification and Comprehensive Co-Detection of Necrotic and Viable Aneuploid Cancer Cells in Peripheral Blood. *Cancers* **2021**, *13*, 5108. [CrossRef] [PubMed]
33. Niu, M.; Yi, M.; Li, N.; Luo, S.; Wu, K. Predictive biomarkers of anti-PD-1/PD-L1 therapy in NSCLC. *Exp. Hematol. Oncol.* **2021**, *10*, 18. [CrossRef] [PubMed]
34. Liang, N.; Liu, L.; Li, P.; Xu, Y.; Hou, Y.; Peng, J.; Song, Y.; Bing, Z.; Wang, Y.; Wang, Y.; et al. Efficient isolation and quantification of circulating tumor cells in non-small cell lung cancer patients using peptide-functionalized magnetic nanoparticles. *J. Thorac. Dis.* **2020**, *12*, 4262–4273. [CrossRef]
35. You, H.S.; Gao, C.X.; Wang, H.B.; Luo, S.S.; Chen, S.Y.; Dong, Y.L.; Lyu, J.; Tian, T. Concordance of Treatment Recommendations for Metastatic Non-Small-Cell Lung Cancer Between Watson for Oncology System and Medical Team. *Cancer Manag. Res.* **2020**, *12*, 1947–1958. [CrossRef] [PubMed]
36. Zhou, C.; Pan, S.; Liu, P.; Feng, N.; Lu, P.; Wang, Z.; Huang, C.; Wu, L.; Chen, Y. Polystyrene microsphere-mediated optical sensing strategy for ultrasensitive determination of aflatoxin M(1) in milk. *Talanta* **2023**, *258*, 124357. [CrossRef]
37. Sun, W.; Jia, C.; Huang, T.; Sheng, W.; Li, G.; Zhang, H.; Jing, F.; Jin, Q.; Zhao, J.; Li, G.; et al. High-performance size-based microdevice for the detection of circulating tumor cells from peripheral blood in rectal cancer patients. *PLoS ONE* **2013**, *8*, e75865. [CrossRef]
38. Fu, Q.; Zhang, Y.; Huang, T.; Liu, Y. Measurement of the Compressibility of Cell and Nucleus Based on Acoustofluidic Microdevice. *J. Vis. Exp. JoVE* **2022**, *14*, 185. [CrossRef]
39. Ding, X.; Peng, Z.; Lin, S.C.; Geri, M.; Li, S.; Li, P.; Chen, Y.; Dao, M.; Suresh, S.; Huang, T.J. Cell separation using tilted-angle standing surface acoustic waves. *Proc. Natl. Acad. Sci. USA* **2014**, *111*, 12992–12997. [CrossRef]
40. Ding, X.; Li, P.; Lin, S.C.; Stratton, Z.S.; Nama, N.; Guo, F.; Slotcavage, D.; Mao, X.; Shi, J.; Costanzo, F.; et al. Surface acoustic wave microfluidics. *Lab Chip* **2013**, *13*, 3626–3649. [CrossRef]

41. Lu, C.; Han, J.; Sun, X.; Yang, G. Electrochemical Detection and Point-of-Care Testing for Circulating Tumor Cells: Current Techniques and Future Potentials. *Sensors* **2020**, *20*, 6073. [CrossRef]
42. Topa, J.; Grešner, P.; Żaczek, A.J.; Markiewicz, A. Breast cancer circulating tumor cells with mesenchymal features-an unreachable target? *Cell. Mol. Life Sci. CMLS* **2022**, *79*, 81. [CrossRef]
43. Li, P.; Mao, Z.; Peng, Z.; Zhou, L.; Chen, Y.; Huang, P.H.; Truica, C.I.; Drabick, J.J.; El-Deiry, W.S.; Dao, M.; et al. Acoustic separation of circulating tumor cells. *Proc. Natl. Acad. Sci. USA* **2015**, *112*, 4970–4975. [CrossRef]
44. Liu, J.; Zhao, R.; Zhang, J.; Zhang, J. ARMS for EGFR mutation analysis of cytologic and corresponding lung adenocarcinoma histologic specimens. *J. Cancer Res. Clin. Oncol.* **2015**, *141*, 221–227. [CrossRef] [PubMed]
45. Li, J.; Xu, M.; Peng, J.; Wang, J.; Zhao, Y.; Wu, W.; Lan, X. Novel technologies in cfDNA analysis and potential utility in clinic. *Chin. J. Cancer Res. Chung-Kuo Yen Cheng Yen Chiu* **2021**, *33*, 708–718. [CrossRef] [PubMed]
46. Qi, J.; Hong, B.; Wang, S.; Wang, J.; Fang, J.; Sun, R.; Nie, J.; Wang, H. Plasma cell-free DNA methylome-based liquid biopsy for accurate gastric cancer detection. *Cancer Sci.* **2024**, *115*, 3426–3438. [CrossRef] [PubMed]
47. Lin, J.; Li, M.; Chen, S.; Weng, L.; He, Z. Efficacy and Safety of First-Generation EGFR-TKIs Combined with Chemotherapy for Treatment-Naïve Advanced Non-Small-Cell Lung Cancer Patients Harboring Sensitive EGFR Mutations: A Single-Center, Open-Label, Single-Arm, Phase II Clinical Trial. *J. Inflamm. Res.* **2021**, *14*, 2557–2567. [CrossRef]
48. Fu, K.; Xie, F.; Wang, F.; Fu, L. Therapeutic strategies for EGFR-mutated non-small cell lung cancer patients with osimertinib resistance. *J. Hematol. Oncol.* **2022**, *15*, 173. [CrossRef]
49. You, Q.; Wang, J.; Yu, Y.; Li, F.; Meng, L.; Chen, M.; Yang, Q.; Xu, Z.; Sun, J.; Zhuo, W.; et al. The histone deacetylase SIRT6 promotes glycolysis through the HIF-1α/HK2 signaling axis and induces erlotinib resistance in non-small cell lung cancer. *Apoptosis Int. J. Program. Cell Death* **2022**, *27*, 883–898. [CrossRef]
50. He, J.; Huang, Z.; Han, L.; Gong, Y.; Xie, C. Mechanisms and management of 3rd-generation EGFR-TKI resistance in advanced non-small cell lung cancer (Review). *Int. J. Oncol.* **2021**, *59*, 90. [CrossRef]
51. Oya, Y.; Yoshida, T.; Kuroda, H.; Shimizu, J.; Horio, Y.; Sakao, Y.; Inaba, Y.; Hida, T.; Yatabe, Y. Association Between EGFR T790M Status and Progression Patterns During Initial EGFR-TKI Treatment in Patients Harboring EGFR Mutation. *Clin. Lung Cancer* **2017**, *18*, 698–705.e2. [CrossRef] [PubMed]
52. Li, X.; Chen, J.; Zhang, C.; Han, Z.; Zheng, X.; Cao, D. Application value of CT radiomic nomogram in predicting T790M mutation of lung adenocarcinoma. *BMC Pulm. Med.* **2023**, *23*, 339. [CrossRef] [PubMed]
53. Paliogiannis, P.; Colombino, M.; Sini, M.C.; Manca, A.; Casula, M.; Palomba, G.; Pisano, M.; Doneddu, V.; Zinellu, A.; Santeufemia, D.; et al. Global prognostic impact of driver genetic alterations in patients with lung adenocarcinoma: A real-life study. *BMC Pulm. Med.* **2022**, *22*, 32. [CrossRef]
54. Cui, Q.; Hu, Y.; Cui, Q.; Wu, D.; Mao, Y.; Ma, D.; Liu, H. Osimertinib Rechallenge with Bevacizumab vs. Chemotherapy Plus Bevacizumab in EGFR-Mutant NSCLC Patients with Osimertinib Resistance. *Front. Pharmacol.* **2021**, *12*, 746707. [CrossRef] [PubMed]
55. Trachu, N.; Reungwetwattana, T.; Meanwatthana, J.; Sukasem, C.; Majam, T.; Saengsiwaritt, W.; Jittikoon, J.; Udomsinprasert, W. Leukocytes telomere length as a biomarker of adverse drug reactions induced by Osimertinib in advanced non-small cell lung cancer. *Sci. Rep.* **2024**, *14*, 26543. [CrossRef]
56. Li, Y.; Mao, T.; Wang, J.; Zheng, H.; Hu, Z.; Cao, P.; Yang, S.; Zhu, L.; Guo, S.; Zhao, X.; et al. Toward the next generation EGFR inhibitors: An overview of osimertinib resistance mediated by EGFR mutations in non-small cell lung cancer. *Cell Commun. Signal. CCS* **2023**, *21*, 71. [CrossRef]
57. Wu, S.G.; Yu, C.J.; Yang, J.C.; Shih, J.Y. The effectiveness of afatinib in patients with lung adenocarcinoma harboring complex epidermal growth factor receptor mutation. *Ther. Adv. Med. Oncol.* **2020**, *12*, 1758835920946156. [CrossRef]

micromachines

MDPI

Article

Pico-Scale Digital PCR on a Super-Hydrophilic Microarray Chip for Multi-Target Detection

Qingyue Xian [1,2], Jie Zhang [2], Yu Ching Wong [3], Yibo Gao [4,5], Qi Song [4,5], Na Xu [4,5] and Weijia Wen [1,2,3,*]

1 Academy of Interdisciplinary Studies, The Hong Kong University of Science and Technology, Clear Water Bay, Kowloon 999077, Hong Kong; qxianaa@connect.ust.hk
2 Thrust of Advanced Materials, The Hong Kong University of Science and Technology (Guangzhou), Nansha, Guangzhou 511400, China; jzhang151@connect.hkust-gz.edu.cn
3 Department of Physics, The Hong Kong University of Science and Technology, Clear Water Bay, Kowloon 999077, Hong Kong; ycwongak@connect.ust.hk
4 Shenzhen Shineway Biotech Co., Ltd., Futian, Shenzhen 518048, China; gyb@swtech.me (Y.G.); sq@swtech.me (Q.S.); xn@swtech.me (N.X.)
5 Zhuhai Shineway Biotech Co., Ltd., Zhuhai 519080, China
* Correspondence: phwen@ust.hk

Abstract: The technology of digital polymerase chain reaction (dPCR) is rapidly evolving, yet current devices often suffer from bulkiness and cumbersome sample-loading procedures. Moreover, challenges such as droplet merging and partition size limitations impede efficiency. In this study, we present a super-hydrophilic microarray chip specifically designed for dPCR, featuring streamlined loading methods compatible with micro-electro-mechanical systems (MEMS) technology. Utilizing hydrodynamic principles, our platform enables the formation of a uniform array of 120-pL independent reaction units within a closed channel. The setup allows for rapid reactions facilitated by an efficient thermal cycler and real-time imaging. We achieved absolute quantitative detection of hepatitis B virus (HBV) plasmids at varying concentrations, alongside multiple targets, including cancer mutation gene fragments and reference genes. This work highlights the chip's versatility and potential applications in point-of-care testing (POCT) for cancer diagnostics.

Keywords: dPCR; microarray; hydrophilic; POCT

1. Introduction

Digital polymerase chain reaction (dPCR) is a method for the absolute quantification of a nucleotide sequence [1], playing a crucial role in biomedical researches and advancing molecular diagnostics. First introduced by K.W. Kinzler and B. Vogelstein in 1999 [2], dPCR did not gain widespread development until the advent of the first commercial instruments [3]. While dPCR is based on the same biochemical principles as quantitative real-time polymerase chain reaction (qPCR), it differs in its quantification approach. In dPCR, samples, reagents, and probes are partitioned into numerous isolated reaction units, such as microwells or droplets. Ideally, each unit contains either 0 or 1 target template, allowing for absolute quantification based on the accumulated signal during PCR amplification, regardless of the amplification efficiency and background. Compared to qPCR, the binary nature of dPCR minimizes the influence of reaction inhibitors and impurities, leading to more reliable quantitative assessments [4], particularly for detecting biomarkers in diseases such as cancer and infectious diseases. Moreover, it is an endpoint detection method that typically does not require real-time data. The partitioning in dPCR heavily relies on microfluidic technology, which includes designs for droplet generation and lithography-based methods utilizing micro-electro-mechanical systems (MEMS) to create microfluidic chips with various compartments [5]. Recent advancements in microfluidics have enabled dPCR to be performed in nanoliter or picoliter reaction units, achieving high-throughput analysis with minimal reagent consumption and a broader dynamic range of Poisson distributions [1,4,6]. Currently, dPCR technology is widely applied in single cell analysis [6,7], cancer diagnostics [8–10], bacteria or virus detection [11–14] as well as quantitation and preamplification of next-generation sequencing libraries [1,15,16].

Existing dPCR platforms can be categorized into two primary types: chamber-based dPCR system (cdPCR) and droplet-based dPCR (ddPCR) system, based on their partitioning methods [1]. cdPCR utilizes physical structures as partitions, while ddPCR employs a continuous oil phase to separate water, forming droplets through mechanisms like T junctions, flow-focusing or coflowing structures [6]. For instance, the Stilla Naica System for Crystal Digital PCR generates monodisperse droplets through the injection nozzles with pressurization [17]. Other notable ddPCR systems include Bio-Rad's QX100/200 [18,19] and DPBio's Nebula Auto Digital PCR System [20]. Although these platforms can produce tens of thousands of droplets for high-throughput analysis, droplet merging is a common issue, and ddPCR systems require constant flow and surfactants for droplet stabilization [21]. In addition, the droplet generator is separated from the reaction or detection device, adding complexity in operation.

In contrast, cdPCR systems avoid these challenges due to their physical walls, which provide mechanical and thermal stability as well as chemical resistance [6]. Meanwhile, MEMS techniques facilitate the efficient fabrication of nano/micro chambers with high replication accuracy. A typical example is Thermo Fisher's Quantstudio 3D [22,23], which features an "open-array plate" with 20,000 etched through-holes with hydrophilic coating for capillary loading of 0.8 nL samples, followed by sealing with immiscible oil to prevent evaporation and cross-contamination. However, this kind of cdPCR systems based on open-array [21,24] suffer from sample evaporation during loading, affecting result accuracy and limiting partition unit size. Heyries, K. et al [25] presented a PDMS-based digital PCR device that employs surface tension-based sample partitioning and dehydration control to achieve high-fidelity amplification of single DNA molecules in 1,000,000 picoliter-volume reactors. Fluidigm's BioMark system [26] features microfluidic channels and chambers equipped with flexible valves that direct the sample into 770 partitions of 0.85-nL volume. Precise distribution is achieved with high throughput and low consumption, but the numerous pipelines and nanoflex valves increase the complexity of the design. Additionally, many systems often rely on chemical modifications, such as coatings with Sigmacote [4], fluoropolymers [21], or silanes [27], to achieve desired wettability, which may not withstand high temperatures, pressures, and voltages, thereby limiting their compatibility with MEMS processes. Therefore, there is an urgent need to develop a cdPCR platform that addresses these limitations with minimal complexity.

In this study, we present a super-hydrophilic microarray chip for dPCR analysis that overcome the above limitations. We reduced the volume of reaction units for cdPCR to 120 pL, which was found to be uniform and the size can be controlled. The loading and distribution of the reaction units can be achieved at atmospheric pressure solely through hydrodynamics based on the chip surface's wettability, eliminating the need for scraping, injection or pressurization. The super-hydrophilic surface was created by a silica layer formed inside the microwells, instead of chemical grafting, allowing it to withstand high temperatures, pressures and voltages, thus maintaining performance during subsequent anodic bonding with processed glass. Our protocol is fully compatible with MEMS technology, enhancing its potential for high-integration systems. A platform was established for dPCR reaction and detection, enabling precise quantification down to as few as 10^0 copies/µL of plasmid DNA containing hepatitis B virus (HBV). We further investigated the detection of B-Raf proto-oncogene serine/threonine kinase (BARF) mutation gene fragments, epidermal growth factor receptor (EGFR) mutation gene fragments, and human glyceraldehyde-3-phosphate dehydrogenase (GAPDH) endogenous reference genes. A performance comparison with the commercial Naica 6-color digital PCR system demonstrated comparable stability and accuracy in detection while reducing time and operational complexity. By eliminating the need for bulky pumping systems, our proposed dPCR platform offers a user-friendly, durable and cost-effective solution for nucleic acid quantification.

2. Materials and Methods

2.1. Fabrication of the dPCR Chip

dPCR chips were designed and fabricated using low-pressure chemical vapor deposition (LPCVD) and deep reactive ion etching (DRIE) technologies. The chip contains two channels, each of which has 11,500 independent uniform microwells formed on a silicon substrate. Each microwell measures 50 µm on each side. Detailed information on the fabrication process of the super-hydrophilic microarray chip is illustrated in Figure 1A.

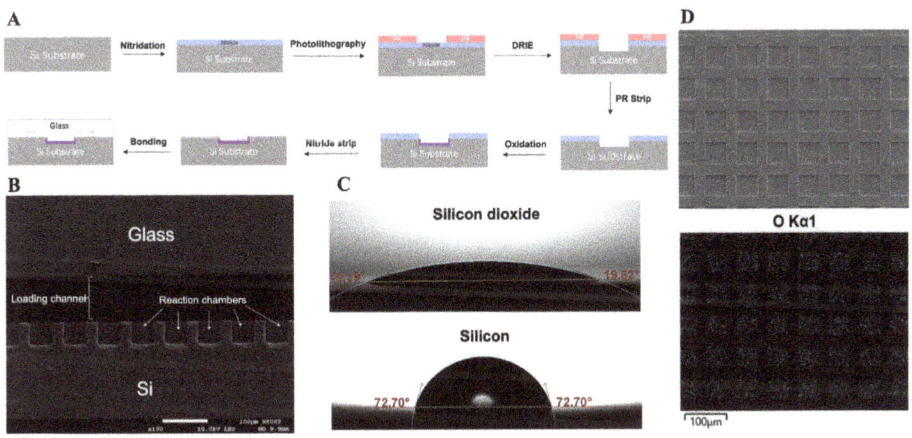

Figure 1. (**A**) Fabrication protocol of the super-hydrophilic microarray chip. (**B**) SEM image of the cross-section of the chip. (**C**) Contact angles of the silicon dioxide and silicon surface. (**D**) SEM image of the microarray surface and the corresponding EDX mapping result of O element.

Initially, a layer of silicon nitride was deposited on the silicon wafer, followed by DRIE to create the picoliter wells. The DRIE process facilitates the formation of microwells with a highly controlled vertical profile and uniform size. Subsequently, a layer of silicon dioxide was formed on the surfaces (walls and bottoms) of the microwells that were not covered by silicon nitride, employing a wet oxidation method. The silicon nitride was then removed using 85 wt % phosphoric acid at 160 °C. Finally, the silicon substrate was anodic bonded to a borosilicate glass layer that was processed to include wide loading channels and injection holes, resulting in a super-hydrophilic microwell array within a closed channel. Figure 1B presents a scanning electron microscope (SEM) image of the chip's cross-section. The microwells have a depth and width of approximately 50 µm. Energy-dispersive X-ray spectroscopy (EDX) analysis (Figure 1D) confirms the presence of an oxide layer on the inner surfaces of the microwells. The lower image shows the mapping of the O element, where the blue dots indicate the position of O element detected. Additionally, Figure 1C illustrates the contact angle of the silicon dioxide surface, which is significantly smaller than that of the bare silicon surface, providing a super-hydrophilic surface inside the microwells. The aqueous PCR reaction solution can more easily distribute within the microwells and be separated by the oil phase due to differences in surface tension, ultimately forming independent static reaction units. Figure 2A displays the layout of a chip with two layers and Figure 2B shows the overlook of a fabricated chip.

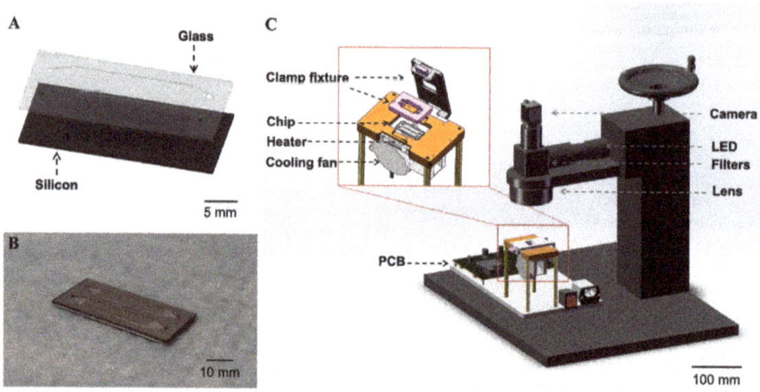

Figure 2. (**A**) Design of the dPCR chip. (**B**) Image of the dPCR chip. (**C**) Set up of the dPCR reaction and detection system.

2.2. Set Up of the dPCR Reaction and Detection System

The reaction solution was loaded onto the microarray, followed by the introduction of mineral oil at a flow rate of 10 µL/s to partition the reaction solution in the microwells and form independent reaction units. The inlets and outlets of the chip were then sealed with a PCR sealing film (Guangzhou Jet Bio-Filtration Co., Ltd., Guangzhou, China). Figure 2C reveals the reaction and real-time detection platform of the dPCR chip. The chip was placed on a microheater equipped with a cooling fan connected to a programmed temperature control PCB. The microheater integrated with the thermal sensor was constructed with the method presented in our previous works [24,28]. A clamp fixture was positioned atop the microheater to tightly seal the inlets and outlets of the chip, preventing liquid evaporation during thermal cycling. An optical inspection system, comprising a 470 nm LED light source and a monochrome camera (a2A1920-160umPRO, Basler, Ahrensburg, Germany), was set up for real-time observation of the reactions. Currently, the chip is for single-use only to avoid cross-contamination.

2.3. On-Chip Loading and Distribution

The aqueous solution for dPCR reactions was loaded to fill the entire chip under atmospheric pressure. The subsequent distribution of the aqueous solution into independent reaction units by oil is achieved through hydrodynamics, relying on the differences of surface tensions between the two phases on the hydrophilic and super-hydrophilic surfaces. The contact angles and surface tensions of water and mineral oil (Sigma-Aldrich, Inc., St. Louis, MO, USA) were measured on the surfaces of bare silicon, silicon dioxide, and glass, following standard cleaning procedures using a Theta Flow (Biolin Scientific AB, Gothenburg, Sweden). The interfacial tension between water and mineral oil was also quantified. The surface free energies of these three surfaces were calculated using contact angle data obtained with water and diiodomethane via the OWRK/Fowkes model. Using Equation (1), the interfacial energies between the liquids and solids were calculated to define the contact angles (θ_w) of the two-phase fluid on the wetted walls (Table 1), as shown in Figure 3A. Here, γ_{s1} and γ_{s2} refer to the interfacial energies between the two fluids and the solid, respectively, while σ represents the interfacial tension between the fluids.

$$cos(\theta_w) = \frac{\gamma_{s1} - \gamma_{s2}}{\sigma}, \quad (1)$$

Table 1. Parameters of the surface porperties.

Parameters	Values		
	Silicon	Silicon Dioxide	Glass
σ [1] (mN/m)		36.3636	
σ_s [2] (mN/m)	50.66	73.77	60.94
$\gamma_{water-solid}$ [3] (mN/m)	24.620	31.361	14.119
$\gamma_{oil-solid}$ [3] (mN/m)	29.056	52.818	39.375
θ [4] (rad)	1.4485	0.9396	0.8030

[1] σ: The interfacial tension coefficient between water and mineral oil. [2] σ_s: The surface free energy of the solid. [3] γ: The interfacial energies between fluid and solid. [4] θ: The contact angles of the two-phase fluid on the wetted walls.

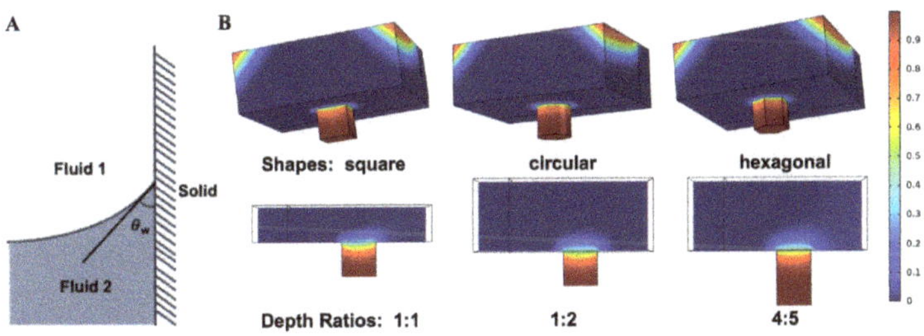

Figure 3. Numerical Simulation: (**A**) Schematic diagram of θ_w of the two-phase fluid on the wetted wall. (**B**) Volume fraction of water in different designs.

The parameters were incorporated into numerical simulations using COMSOL Multiphysics 5.6 to model the distribution of the aqueous phase as the oil phase was introduced. The simulation included modules for laminar flow and phase field. A time-dependent study was conducted, beginning with the loading of mineral oil at specific flow rates, with the initial state defined as completely filled with water. The surface tension coefficient of two fluids and wetted walls were defined based on the calculated results in Table 1. To optimize chip design, models with varying shapes (square, circular, hexagonal) and depth ratios of the microwells to the wide channel were tested to compare loading and distribution effects. Oil phase loading rates were varied equally to the loading of the entire microarray chip between 5 µL/s and 15 µL/s.

To assess the filling uniformity of the aqueous solution in the microwells, FAM solution was loaded and distributed on the chip. Fluorescence images of the microarray were captured, and cross-sectional images of the microarray chip with the fluorescent solution partitioned in the microwells were obtained using a confocal microscope (LSM980, Carl Zeiss AG, Jena, Germany).

2.4. On-Chip dPCR Analysis

2.4.1. dPCR of HBV

HBV plasmid served as the target DNA to validate the quantitative detection capabilities of the designed dPCR chip. Each 10 µL PCR solution comprised 5 µL of 2× TaqMan gene expression master mix (Thermo Fisher Scientific, Waltham, MA, USA), 1 µM probe, 0.25 µM each of forward and reverse primers (GenScript Biotech Corporation, Nanjing, China), 2 µL of template solution, and nuclease-free water. HBV plasmids were prepared with X_{dil}

of 1×10^{-1}, 5×10^{-2}, 2.5×10^{-2}, 1×10^{-2}, 5×10^{-3}, 1×10^{-3}, 5×10^{-4}, and 2.5×10^{-4}. Amplification involved denaturation at 95 °C for 30 s, followed by 45 cycles of 95 °C for 10 s and 60 °C for 30 s. After amplification, the number of positive wells was calculated using algorithms. The sample with $X_{dil} = 1 \times 10^{-2}$ was also tested on the Naica 6-color digital PCR system (Stilla Technologies, Paris, France) for comparison.

2.4.2. dPCR of Multiple Targets

Plasmids containing BRAF V600E and EGFR T790M mutations, along with GAPDH extracted from human sources, were detected in multiple assays using the fluorescence channels of FAM, TAMRA, and Cy5, respectively. The primers and probes (Sangon Biotech Co., Ltd., Shanghai, China) for the target DNAs were prepared (Table S1). Each 10 µL PCR reaction unit consisted of 1 µL of 10× Taq HS Buffer (Mg^{2+}), 0.2 µL of dNTP mix (10 mM each), 0.2 µL of Taq HS DNA polymerase (5 U/µL, Vazyme Biotech Co., Ltd., Nanjing, China), 0.4 µM of forward and reverse primers, and 0.2 µM of probes for each target, along with a specific amount of templates and nuclease-free water. The reaction was initiated at 95 °C for 30 s, followed by 45 thermocycles at 95 °C for 15 s and 60 °C for 45 s. Fluorescence images of the dPCR results were captured using a Nikon Eclipse Ni system for each fluorescence channel.

In the multiple dPCR experiment, the 4.4 µL template solution included 1.6 µL of BRAF plasmid, 0.8 µL of EGFR plasmid, and 2 µL of human genes. Samples containing only a single template (ST) and no template control (NTC) were also tested in parallel to establish positive thresholds and investigate nonspecific amplification. The same samples were tested on the Naica 6-color digital PCR system for reference.

3. Results

3.1. On-Chip Loading and Distribution

The volume fraction of water after distribution in different designs was determined through numerical simulations of the two-phase flow (Figure 3B). Table 2 presents the calculated filling rates of water based on the collected data. Among the geometrical designs tested, square microwells demonstrated superior performance compared to circular and hexagonal shapes, leading to their selection for subsequent experiments. Among the square designs, that with a 4:5 depth ratio (80-µm microwells and 100-µm wide channels) had most water remained in the microwells after distribution. As indicated in Figure 3B, the two-phase interface is in a concave state equilibrium. Since θ_w is the same, theoretically, when the flow rate and wide channel depth are the same, the concave distance of the two-phase interface is also the same. Therefore, regardless of the depth of the microwell, the volume of water replaced by oil within the microwell is the same. Because the volume of a deep well is larger, the filling rate obtained in a deeper microwell increases. Considering the desired volume of reaction unit, we adopted the 1:2 depth ratio design, with microwells having a depth of 50 µm and a width of 50 µm. Additionally, we observed that decreasing the loading velocity of the mineral oil during distribution (equally to loading the entire microarray chip at 15 µL/s, 10 µL/s and 5 µL/s) resulted in more water being retained in the microwells.

Table 2. Filling rate of water in different designs.

Designs	Shapes			Depth Ratios			Velocities		
	Square	Circular	Hexagonal	1:1	1:2	4:5	High	Middle	Low
Filling rates (%)	96.7338	96.7120	96.0452	92.82	96.7338	98.6455	96.7338	96.7674	97.4326

To assess partition conditions and uniformity, we employed FAM as a fluorescent dye. The filling condition of the aqueous fluorescent solution in the microwells (Figure 4A) closely matched the simulation results. Based on both simulation and experimental findings, the partition volume was defined as 120 pL. If the reaction solution is evenly partitioned, the volume should remain consistent across all microwells. Figure 4B illustrates the solution distribution in the microarray chip following loading. Meanwhile, Figure 4C presents a histogram depicting the fluorescence intensity of each microwell. Some microwells showed higher or lower intensity, which is expected given the variation in light intensity between the center and edges of the optical field, particularly in large views. The average fluorescence intensity across all microwells was approximately 138, with a relative standard deviation of 3.37%. This indicates an acceptable uniformity of fluorescence intensity among the microwells, affirming that the partition volume in each microwell can be considered consistent.

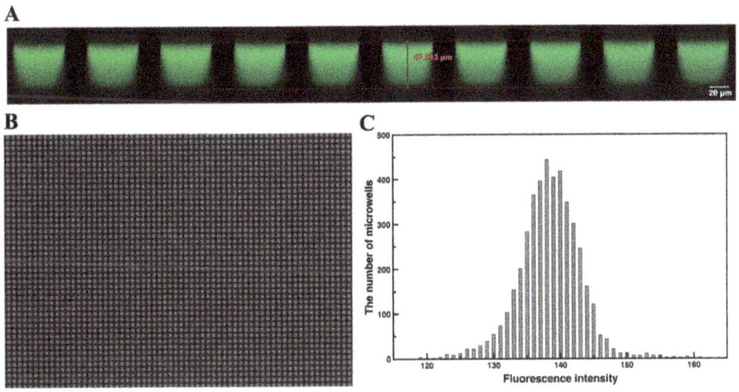

Figure 4. Loading and dirstribution of FAM solution: (**A**) Cross section of the microarray chip with the FAM solution patitioned in the microwells. (**B**) Fluorescence image of the FAM solution distributed on the microarray. (**C**) Histogram of the fluorescence intensity of each microwell.

3.2. On-Chip Thermal Performance

Precise temperature control and uniform thermal distribution are crucial for cdPCR. Since the temperature can be rapidly reached and maintained in the small reaction unit, the robust on-chip thermal performance avoids the reduction in detection performance caused by prolonged high or low temperatures due to slow temperature ramping. The on-chip heating and temperature sensing was achieved by thin-film resistors, a sputtering patterned layer of platinum on silicon substrate. The platinum sensor effectively reflected temperature changes, enabling control over the heating process. Temperature data were recorded during thermal cycling. As illustrated in the temperature profile in Figure 5A, the set temperatures were rapidly reached and maintained throughout the cycling process. The average heating rate exceeded 17.5 °C/s, while cooling rates were approximately 5.8 to 7 °C/s, comparable to other cdPCR designs and higher than many commercial rapid PCR devices [29]. The entire program for running the HBV dPCR was completed in 36 min, which was only 30% of the reaction time required by the Naica system. In addition, to validate temperature accuracy, an external temperature sensor was used to measure the surface temperature of the heater during a gradient heating from 55 °C to 95 °C. Deviations compared with the set temperature during the plateau phase were observed to be 0.1 °C at both 55 °C and 75 °C, and 0.2 °C at 95 °C. Additionally, the heater exhibited homogeneous thermal distribution, with only a 0.1 °C difference at 55 °C and 75 °C, 0.2 °C difference at 95 °C when tested across four different positions on the microheater (Figure 5B). Since the microheater is reusable and integrated with a clamp fixture, the operator only needs to replace the chip during use, making it user-friendly and a suitable choice for POCT.

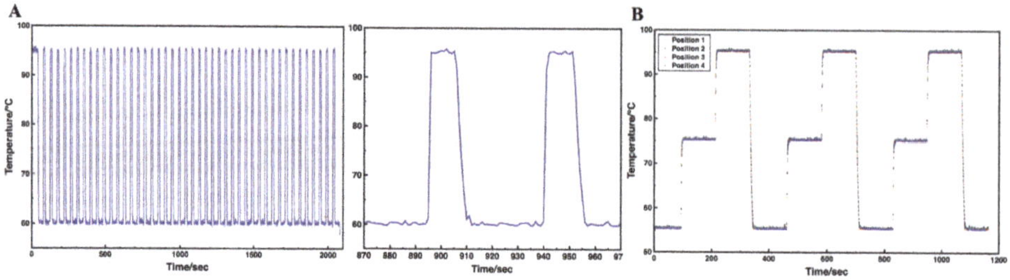

Figure 5. (**A**) PCR temperature profile of 45 cycles on the platform (left). Scatter diagram showing temperature pattern of two cycles (right). (**B**) Temperature uniformity of the microheater during gradient heating cycles.

3.3. On-Chip dPCR Analysis

3.3.1. dPCR of HBV

To validate the stability of the static reaction units on the array throughout the PCR thermal cycle, fluorescence images of the microarray were captured at the end of each cycle (Figure S1). No movement or evaporation of the reaction units was observed. Some wells at the edge of the channel were excluded from calculations due to reflections from the glass wall, as well as those not filled with the reaction solution because of fabrication defects or unstable flow at the inlets. A total of up to 10,100 wells were analyzed and classified as positive or negative.

Fluorescence images of the dPCR results for detecting different concentrations of HBV plasmid are shown in Figure 6A–H. As the dilution factor (X_{dil}) of the sample decreased, fewer positive wells were observed. The numbers of positive and negative microwells were calculated using algorithms in MATLAB (https://ww2.mathworks.cn/products/matlab.html, accessed on 4 March 2025). Ultimately, the copy number of target DNA in each sample (C_s) was deduced using Poisson's distribution, based on the equation below [1]:

$$-ln(1-f_0) = \frac{M}{N} = CV_p, \tag{2}$$

where f_0 was the fraction of positive units over the number of partitions, M was the copy number of target DNA, N was the total number of partitions, C was the copy number concentration in the reaction, and V_p was the partition volume. In the experiment, V_p was 120 pL, allowing for the calculation of C. The relationship is given by $C_s = 5C$ based on the reaction components. The linear correlation between C_s and X_{dil} is depicted in Figure 6I, where $R^2 > 0.99$. The detection limit was 6 copies/µL, with a 95% statistical confidence intervals, [2.1022, 10.4154].

The measured C_s for the sample $X_{dil} = 1 \times 10^{-2}$ was 3708 copies/µL, closely aligning with the result of 3806 copies/µL obtained from the Naica 6-color digital PCR system, thus validating the system's reliability. Furthermore, the results exhibited a dynamic linear range of detection from 10^4 to 10^0 copies/µL.

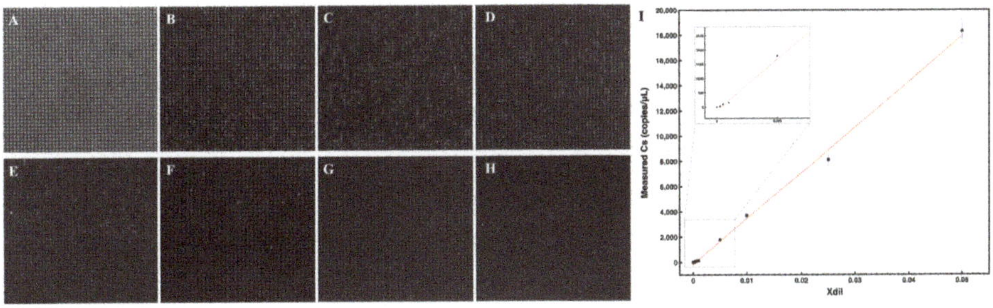

Figure 6. (**A–H**) Flourescence images of the results of the dPCR to detect different concentrations of HBV plasmids: the HBV plasmid dilution factors were (**A**) 1×10^{-1}; (**B**) 5×10^{-2}; (**C**) 2.5×10^{-2}; (**D**) 1×10^{-2}; (**E**) 5×10^{-3}; (**F**) 1×10^{-3}; (**G**) 5×10^{-4}; (**H**) no template control. (**I**) Linear correlation between the measured DNA concentration and the X_{dil}. Error bars are SD. for n = 3.

3.3.2. dPCR of Multiple Targets

Fluorescence images of the multiple dPCR results are presented in Figure 7, where the number of positive units can be identified. The copy number concentrations of the original samples were calculated using the previously described method, as shown in Table 3. The measured concentrations of the analytes on the microarray chip were closely aligned with those obtained from the Naica dPCR system. Minor differences may be attributed to sampling variance and variations in reaction components. Nonspecific amplification was not observed in the FAM and TAMRA channels, but small copy numbers were detected in the Cy5 channel even among its negative controls. This is acceptable, as the experimental environment was inevitably contaminated with human samples containing GAPDH endogenous reference genes. Overall, the microarray dPCR chip demonstrated reliability for multiple dPCR analyses.

Figure 7. Fluorescence images of the multiple dPCR results: (**A**) BRAF (FAM). (**B**) EGFR (TAMRA). (**C**) GAPDH (Cy5).

Table 3. Measured concentration (copies/μL) of the samples in three channels.

Sample Names	FAM	TAMRA	Cy5
Multiple templates	13,031	14,104	2273
BRAF ST	12,494	0	6
EGFR ST	0	14,495	15
GAPDH ST	0	0	2280
NTC	0	0	17
Multiple templates on Naica system	13,616	15,083	2003

4. Discussion

This study addresses the existing gaps in open-array cdPCR designs by presenting a closed-channel cdPCR microarray. Many current closed-channel cdPCR devices are based on polydimethylsiloxane (PDMS) [25]. PDMS PCR chips suffer from over 50% solution evaporation rates during thermal cycles because of its air permeability [30], leading to many works dedicated to address this issue, but certain evaporation rates still exist [31–33]. Our chip eliminated this problem by using glass as the cover slide, which is waterproof and provides high optical transmission, and high hydrophilicity. Moreover, the mechanical strength of glass wall and strong anodic bonding strength between glass and silicon provide robust physical support to stabilize the reaction units formed during the chip transfer and thermal cycling, which avoids cross-contamination among the reaction units.

Compared to existing chemical modification methods that improve hydrophilicity, our dPCR chip employed a method without chemical grafting, resulting in a stable super-hydrophilic layer that can sustain the subsequent anodic bonding process, and other processes where high temperature and voltage is applied. Therefore, the prepared super-hydrophilic array was fully compatible with MEMS technology, supporting the chip's potential for highly integrated applications.

This design simplifies sample loading and distribution, eliminating the need for manual scraping, robotic injection, or programmatic pressure sampling, thus streamlining system construction compared to its similar designs [4,24]. Since the inner surface of the entire chamber is hydrophilic, the aqueous solution for dPCR reactions can fill the entire chip including all microwells at atmospheric pressure via continuous pipetting. The subsequent distribution into independent reaction units by oil is achieved through hydrodynamics, relying on the chip's hydrophilicity and super-hydrophilicity, without the usage of bulk-pump.

Both simulation and experimental results confirm that the aqueous solution partitions uniformly within the microwells, exhibiting stable performance. The volume of the reaction units was approximately 65% less than most current cdPCR devices, normally 0.34–33 nL [1,4,34], which not only improves the detection sensitivity and accuracy, but also requires less reagents and samples. The rapid and uniform thermal cycles demonstrated reliable and efficient heating and cooling characteristics of the platform, which supports time-efficient reaction with the reaction time far less than commercial dPCR devices. Additionally, this platform enables real-time monitoring during the reaction, providing flexibility for various research applications.

The dPCR analyses of HBV across different dilution factors validated the method's reliability, demonstrated by the strong linear correlation between measured concentrations and dilution factors. The detection range (from 10^4 to 10^0 copies/μL) compares favorably to existing works, while requiring fewer partitions. The detection limit and sensitivity for quantifying DNA can be further enhanced by increasing the number of wells or reducing the chamber volume. Thus, the method presented here can adapt to various research and clinical needs with minor modifications.

Furthermore, BRAF and EGFR mutations are critical for optimizing targeted therapies in cancer patients [35,36]. Mutations in BRAF V600E and EGFR T790M significantly influence tumor progression and resistance to treatment, particularly in non-small cell lung cancer and colorectal cancer. Identifying these mutations can guide personalized treatment strategies and improve patient outcomes. By allowing for concurrent detection of these mutations alongside GAPDH reference genes, our platform highlights its clinical relevance and potential to improve personalized treatment strategies. The results obtained demonstrated no significant differences when compared to current commercial dPCR devices, affirming the compatibility of the super-hydrophilic microarray dPCR platform with diverse reaction systems and sample types, including plasmid and linear DNAs. Furthermore, this system can be easily upgraded to support 5/6-color detection with appropriate filters and analysis algorithms, further exemplifying its versatility for diverse needs.

In conclusion, our approach emphasizes the interdisciplinary nature of modern diagnostics by bridging microfluidic technology and molecular detection. The innovation of this work spans from manufacturing to the miniaturization of reaction units, and the simplification of operations. By facilitating accurate biomarker quantification and supporting multiplex detection capabilities, this system stands to enhance POCT applications, contributing to timely and effective disease monitoring and management.

5. Conclusions

In this study, we developed a dPCR platform utilizing a super-hydrophilic microarray chip equipped with a real-time monitoring system. Each channel can generate 10,100 independent reaction units, each with a volume of 120 pL and the reactions of two samples was allowed on a single chip. The sample loading and distribution require no specialized equipment and can be completed in just one minute. By using mineral oil to cover the reaction samples, we enabled the reactions to be conducted at atmospheric pressure, employing a clamp fixture to seal the inlets and outlets of the channels. This setup effectively prevented evaporation and cross-contamination among the reaction units. Meanwhile, the platform's reliable and efficient thermal characteristics significantly shortened the reaction time. We validated the dPCR system using gradient concentrations of HBV plasmid, X_{dil} ranging from 1×10^{-1} to 2.5×10^{-4} as DNA templates. The statistical results demonstrated that our dPCR system can achieve absolute quantitative detection of target DNA. Furthermore, this system is capable of detecting both single and multiple targets simultaneously, showing excellent performance in the combined detection of EGFR, BRAF, and GAPDH. This highlights the compatibility of our system with diverse reaction environments and template types, including circular and linear DNAs. Overall, this dPCR device, featuring MEMS-compatible fabrication and straightforward operation, has significant potential as a powerful POCT solution in pathogen detection and infectious disease monitoring.

Supplementary Materials: The following supporting information can be downloaded at: https://www.mdpi.com/article/10.3390/mi16040407/s1, Figure S1: Fluorescence images obtained during the thermal cycling testing with HBV plasmid $X_{dil} = 5 \times 10^{-3}$; Figure S2: Results of fluorescence images in each channel of the multiple dPCR testing with ST and NTC; Table S1: Sequences of primers and probes used in this study to test the dPCR chips; Table S2: Sequences of DNA synthesized in plasmids used in this study to test the dPCR chips.

Author Contributions: Conceptualization, Y.G. and W.W.; methodology, Q.X., Y.C.W. and N.X.; software, Q.X. and J.Z.; validation, Q.X.; formal analysis, Q.X.; investigation, Q.X., Q.S. and N.X.; resources, Y.G. and W.W.; data curation, Q.X.; writing—original draft preparation, Q.X.; writing—review and editing, Q.X., J.Z., Y.C.W., Q.S. and W.W.; visualization, Q.X.; supervision, W.W. and Q.S.; project administration, W.W. and Y.G.; funding acquisition, W.W. and Y.G. All authors have read and agreed to the published version of the manuscript.

Funding: This research was funded by Zhongshan-HKUST research program (ZSST20SC01), Guandong Province Program (2021JC02C22) and Zhuhai Industry-University-Research Cooperation Project (ZH22017002210019PWC).

Data Availability Statement: The data that support the findings of this study are available from the corresponding author upon reasonable request.

Acknowledgments: The authors would like to thank the support from the Nanosystem Fabrication Facility (CWB), Materials, Design and Manufacturing Facility (CWB), Materials Characterization and Preparation Facility (CWB), Novel IC Exploration Facility, Wave Functional Metamaterial Research Facility, Nanosystem Fabrication Facility (GZ) and Biosciences Central Research Facility (GZ) of the Hong Kong University of Science and Technology. The authors are also grateful for the help from Xiao Luo, Kai Lou, Menggui Li, Lihong Lin, Chenxuan Lou and Peng Hua.

Conflicts of Interest: Yibo Gao, Qi Song and Na Xu were employed by the Shenzhen Shineway Biotech Co., Ltd. and Zhuhai Shineway Biotech Co., Ltd.. The remaining authors declare that the research was conducted in the absence of any commercial or financial relationships that could be construed as a potential conflict of interest.

Abbreviations

The following abbreviations are used in this manuscript:

dPCR	Digital polymerase chain reaction
qPCR	Quantitative real-time polymerase chain reaction
MEMS	Micro-electro-mechanical systems
cdPCR	Chamber-based digital polymerase chain reaction
ddPCR	Droplet-based digital polymerase chain reaction
HBV	Hepatitis B virus
BRAF	B-Raf proto-oncogene, serine/threonine kinase
EGFR	Epidermal growth factor receptor
GAPDH	Glyceraldehyde-3-phosphate dehydrogenase
FAM	5-Carboxyfluorescein
TAMARA	5-Carboxytetramethylrhodamine
Cy5	Invitrogen Cyanine5
LPCVD	Low pressure chemical vapor deposition
DRIE	Deep reactive ion etching
SEM	Scanning electron microscope
EDX	Energy-dispersive X-ray spectroscopy
ST	Single template
NTC	No template control
PDMS	Polydimethylsiloxane
POCT	Point-of-care testing

References

1. Basu, A.S. Digital assays part I: Partitioning statistics and digital PCR. *Slas Technol. Transl. Life Sci. Innov.* **2017**, *22*, 369–386. [CrossRef] [PubMed]
2. Vogelstein, B.; Kinzler, K.W. Digital PCR. *Proc. Natl. Acad. Sci. USA* **1999**, *96*, 9236–9241. [CrossRef] [PubMed]
3. Morley, A.A. Digital PCR: A brief history. *Biomol. Detect. Quantif.* **2014**, *1*, 1–2. [CrossRef]
4. Peng, K.; Wu, Z.; Feng, Z.; Deng, R.; Ma, X.; Fan, B.; Liu, H.; Tang, Z.; Zhao, Z.; Li, Y. A highly integrated digital PCR system with on-chip heating for accurate DNA quantitative analysis. *Biosens. Bioelectron.* **2024**, *253*, 116167. [CrossRef]
5. Zhang, H.; Cao, L.; Brodsky, J.; Gablech, I.; Xu, F.; Li, Z.; Korabecna, M.; Neuzil, P. Quantitative or digital PCR? A comparative analysis for choosing the optimal one for biosensing applications. *TrAC Trends Anal. Chem.* **2024**, *174*, 117676. [CrossRef]
6. Xu, D.; Zhang, W.; Li, H.; Li, N.; Lin, J.M. Advances in droplet digital polymerase chain reaction on microfluidic chips. *Lab Chip* **2023**, *23*, 1258–1278. [CrossRef]
7. Wang, J.; Sun, N.; Lee, Y.T.; Ni, Y.; Koochekpour, R.; Zhu, Y.; Tseng, H.R.; Wang, S.; Jiang, L.; Zhu, H. A circulating tumor cell-based digital assay for the detection of EGFR T790M mutation in advanced non-small cell lung cancer. *J. Mater. Chem. B* **2020**, *8*, 5636–5644. [CrossRef] [PubMed]
8. Veyer, D.; Wack, M.; Mandavit, M.; Garrigou, S.; Hans, S.; Bonfils, P.; Tartour, E.; Bélec, L.; Wang-Renault, S.F.; Laurent-Puig, P.; et al. HPV circulating tumoral DNA quantification by droplet-based digital PCR: A promising predictive and prognostic biomarker for HPV-associated oropharyngeal cancers. *Int. J. Cancer* **2020**, *147*, 1222–1227. [CrossRef]
9. Ou, C.Y.; Vu, T.; Grunwald, J.T.; Toledano, M.; Zimak, J.; Toosky, M.; Shen, B.; Zell, J.A.; Gratton, E.; Abram, T.J.; et al. An ultrasensitive test for profiling circulating tumor DNA using integrated comprehensive droplet digital detection. *Lab Chip* **2019**, *19*, 993–1005.
10. Pekin, D.; Skhiri, Y.; Baret, J.C.; Le Corre, D.; Mazutis, L.; Salem, C.B.; Millot, F.; El Harrak, A.; Hutchison, J.B.; Larson, J.W.; et al. Quantitative and sensitive detection of rare mutations using droplet-based microfluidics. *Lab Chip* **2011**, *11*, 2156–2166. [CrossRef]
11. Basanisi, M.G.; La Bella, G.; Nobili, G.; Raele, D.A.; Cafiero, M.A.; Coppola, R.; Damato, A.M.; Fraccalvieri, R.; Sottili, R.; La Salandra, G. Detection of Coxiella burnetii DNA in sheep and goat milk and dairy products by droplet digital PCR in south Italy. *Int. J. Food Microbiol.* **2022**, *366*, 109583. [CrossRef] [PubMed]

12. Lv, X.; Wang, L.; Zhang, J.; He, X.; Shi, L.; Zhao, L. Quantitative detection of trace VBNC Cronobacter sakazakii by immunomagnetic separation in combination with PMAxx-ddPCR in dairy products. *Food Microbiol.* **2021**, *99*, 103831.
13. Alteri, C.; Cento, V.; Antonello, M.; Colagrossi, L.; Merli, M.; Ughi, N.; Renica, S.; Matarazzo, E.; Di Ruscio, F.; Tartaglione, L.; et al. Detection and quantification of SARS-CoV-2 by droplet digital PCR in real-time PCR negative nasopharyngeal swabs from suspected COVID-19 patients. *PLoS ONE* **2020**, *15*, e0236311. [CrossRef] [PubMed]
14. Sedlak, R.H.; Jerome, K.R. Viral diagnostics in the era of digital polymerase chain reaction. *Diagn. Microbiol. Infect. Dis.* **2013**, *75*, 1–4. [CrossRef]
15. Didelot, A.; Kotsopoulos, S.K.; Lupo, A.; Pekin, D.; Li, X.; Atochin, I.; Srinivasan, P.; Zhong, Q.; Olson, J.; Link, D.R.; et al. Multiplex picoliter-droplet digital PCR for quantitative assessment of DNA integrity in clinical samples. *Clin. Chem.* **2013**, *59*, 815–823. [CrossRef]
16. Tewhey, R.; Warner, J.B.; Nakano, M.; Libby, B.; Medkova, M.; David, P.H.; Kotsopoulos, S.K.; Samuels, M.L.; Hutchison, J.B.; Larson, J.W.; et al. Microdroplet-based PCR enrichment for large-scale targeted sequencing. *Nat. Biotechnol.* **2009**, *27*, 1025–1031.
17. Madic, J.; Zocevic, A.; Senlis, V.; Fradet, E.; Andre, B.; Muller, S.; Dangla, R.; Droniou, M. Three-color crystal digital PCR. *Biomol. Detect. Quantif.* **2016**, *10*, 34–46. [CrossRef]
18. Hindson, B.J.; Ness, K.D.; Masquelier, D.A.; Belgrader, P.; Heredia, N.J.; Makarewicz, A.J.; Bright, I.J.; Lucero, M.Y.; Hiddessen, A.L.; Legler, T.C.; et al. High-throughput droplet digital PCR system for absolute quantitation of DNA copy number. *Anal. Chem.* **2011**, *83*, 8604–8610. [CrossRef]
19. Bio-Rad. QX200 Droplet Digital Systems | Bio-Rad Droplet Digital™ PCR. Available online: https://www.bio-rad.com/life-science/droplet-digital-pcr/droplet-digital-pcr-budgeting?s_kwcid=AL!18120!3!709245417040!e!!g!!biorad%20qx200!21582536543!166846479060&WT.mc_id=241230044812&WT.srch=1&WT.knsh_id=_kenshoo_clickid_&gclid=CjwKCAiAnpy9BhAkEiwA-P8N4iqMn39GauYe6yT2ptRNOXMJwsILqF2gUYyIvAOxAiSTfNUaTG0vSBoCuWAQAvD_BwE&gad_source=1 (accessed on 8 February 2025).
20. DPBio. Nebula Digital PCR System. Available online: https://dp-bio.com/platform?id=4 (accessed on 8 February 2025).
21. Awashra, M.; Elomaa, P.; Ojalehto, T.; Saavalainen, P.; Jokinen, V. Superhydrophilic/superhydrophobic droplet microarrays of low surface tension biofluids for nucleic acid detection. *Adv. Mater. Interfaces* **2024**, *11*, 2300596. [CrossRef]
22. Thermo Fisher Scientific. OpenArray Technology Overview. Available online: https://www.thermofisher.cn/cn/zh/home/life-science/pcr/real-time-pcr/real-time-openarray/open-array-technology.html?adobe_mc=MCMID%7C90839306120932686505880960015367277842%7CMCAID%3D30984116708B72AE-40000FB356BA8C90%7CMCORGID%3D5B135A0C5370E6B40A490D44%40AdobeOrg%7CTS=1614293705 (accessed on 8 February 2025).
23. Broccanello, C.; Gerace, L.; Stevanato, P. QuantStudio™ 12K Flex OpenArray® System as a tool for high-throughput genotyping and gene expression analysis. *Quant. Real-Time PCR Methods Protoc.* **2020**, *2065*, 199–208.
24. Chen, X.; Song, Q.; Zhang, B.; Gao, Y.; Lou, K.; Liu, Y.; Wen, W. A rapid digital pcr system with a pressurized thermal cycler. *Micromachines* **2021**, *12*, 1562. [CrossRef] [PubMed]
25. Heyries, K.A.; Tropini, C.; VanInsberghe, M.; Doolin, C.; Petriv, O.I.; Singhal, A.; Leung, K.; Hughesman, C.B.; Hansen, C.L. Megapixel digital PCR. *Nat. Methods* **2011**, *8*, 649–651. [CrossRef] [PubMed]
26. Fluidigm. High-Definition Biology; Span Research Through Validation on Biomark HD. Available online: https://fluidigm.my.salesforce.com/sfc/p/#700000009DAw/a/4u0000019cyo/7fyPz7QqSX_v76cIgM9qZxesAbtX5Z3JrJs6Fsyqd_8 (accessed on 4 March 2025).
27. Arkles, B. Hydrophobicity, hydrophilicity and silane surface modification. *Gelest Inc.* **2011**, *215*, 547–1015.
28. Liu, Y.; Hu, Z.; Yang, S.; Xu, N.; Song, Q.; Gao, Y.; Wen, W. Fabrication of Cost-Effective Microchip-Based Device Using Sandblasting Technique for Real-Time Multiplex PCR Detection. *Micromachines* **2024**, *15*, 944. [CrossRef] [PubMed]
29. Madadelahi, M.; Agarwal, R.; Martinez-Chapa, S.O.; Madou, M.J. A roadmap to high-speed polymerase chain reaction (PCR): COVID-19 as a technology accelerator. *Biosens. Bioelectron.* **2024**, *246*, 115830. [CrossRef]
30. Zec, H.; O'Keefe, C.; Ma, P.; Wang, T.H. Ultra-thin, evaporation-resistent PDMS devices for absolute quantification of DNA using digital PCR. In Proceedings of the 2015 Transducers-2015 18th International Conference on Solid-State Sensors, Actuators and Microsystems (TRANSDUCERS), Anchorage, AK, USA, 21–25 June 2015; pp. 536–539.
31. Yang, S.; Xian, Q.; Liu, Y.; Zhang, Z.; Song, Q.; Gao, Y.; Wen, W. A silicon-based PDMS-PEG copolymer microfluidic chip for real-time polymerase chain reaction diagnosis. *J. Funct. Biomater.* **2023**, *14*, 208. [CrossRef]
32. Song, Q.; Gao, Y.; Zhu, Q.; Tian, Q.; Yu, B.; Song, B.; Xu, Y.; Yuan, M.; Ma, C.; Jin, W.; et al. A nanoliter self-priming compartmentalization chip for point-of-care digital PCR analysis. *Biomed. Microdevices* **2015**, *17*, 64. [CrossRef]
33. Gao, S.; Xu, T.; Wu, L.; Zhu, X.; Wang, X.; Jian, X.; Li, X. Overcoming bubble formation in polydimethylsiloxane-made PCR chips: mechanism and elimination with a high-pressure liquid seal. *Microsyst. Nanoeng.* **2024**, *10*, 136. [CrossRef]
34. Tan, L.L.; Loganathan, N.; Agarwalla, S.; Yang, C.; Yuan, W.; Zeng, J.; Wu, R.; Wang, W.; Duraiswamy, S. Current commercial dPCR platforms: Technology and market review. *Crit. Rev. Biotechnol.* **2023**, *43*, 433–464. [CrossRef]

35. Oxnard, G.R.; Hu, Y.; Mileham, K.F.; Husain, H.; Costa, D.B.; Tracy, P.; Feeney, N.; Sholl, L.M.; Dahlberg, S.E.; Redig, A.J.; et al. Assessment of Resistance Mechanisms and Clinical Implications in Patients with EGFR T790M–Positive Lung Cancer and Acquired Resistance to Osimertinib. *JAMA Oncol.* **2018**, *4*, 1527–1534. [CrossRef]
36. Kopetz, S.; Grothey, A.; Yaeger, R.; Van Cutsem, E.; Desai, J.; Yoshino, T.; Wasan, H.; Ciardiello, F.; Loupakis, F.; Hong, Y.S.; et al. Encorafenib, binimetinib, and cetuximab in BRAF V600E–mutated colorectal cancer. *N. Engl. J. Med.* **2019**, *381*, 1632–1643. [PubMed]

micromachines

MDPI

Review

Recent Advances in Microfluidic-Based Extracellular Vesicle Analysis

Jiming Chen [1,2,3,†], Meiyu Zheng [1,2,3,†], Qiaoling Xiao [1,2,3], Hui Wang [1,2,3], Caixing Chi [1,2,3], Tahui Lin [1,2,3], Yulin Wang [1,2,3], Xue Yi [1,2,3,*] and Lin Zhu [1,2,3,*]

1 Department of Basic Medicine, Xiamen Medical College, Xiamen 361023, China; jimmychen131@163.com (J.C.); zhengmeiyufjmu@163.com (M.Z.); 202200010531@xmmc.edu.cn (Q.X.); lyuxi8549@gmail.com (H.W.); ccx20221214@163.com (C.C.); tahui.lin@foxmail.com (T.L.); 202200010532@xmmc.edu.cn (Y.W.)

2 Key Laboratory of Functional and Clinical Translational Medicine, Fujian Province University, Xiamen 361023, China

3 Institute of Respiratory Diseases, Xiamen Medical College, Xiamen 361023, China

* Correspondence: yxue@xmmc.edu.cn (X.Y.); zhulinxmu@163.com (L.Z.)

† These authors contributed equally to this work.

Abstract: Extracellular vesicles (EVs) serve as vital messengers, facilitating communication between cells, and exhibit tremendous potential in the diagnosis and treatment of diseases. However, conventional EV isolation methods are labor-intensive, and they harvest EVs with low purity and compromised recovery. In addition, the drawbacks, such as the limited sensitivity and specificity of traditional EV analysis methods, hinder the application of EVs in clinical use. Therefore, it is urgent to develop effective and standardized methods for isolating and detecting EVs. Microfluidics technology is a powerful and rapidly developing technology that has been introduced as a potential solution for the above bottlenecks. It holds the advantages of high integration, short analysis time, and low consumption of samples and reagents. In this review, we summarize the traditional techniques alongside microfluidic-based methodologies for the isolation and detection of EVs. We emphasize the distinct advantages of microfluidic technology in enhancing the capture efficiency and precise targeting of extracellular vesicles (EVs). We also explore its analytical role in targeted detection. Furthermore, this review highlights the transformative impact of microfluidic technology on EV analysis, with the potential to achieve automated and high-throughput EV detection in clinical samples.

Keywords: extracellular vesicles; microfluidics; EV isolation; EV detection

1. Introduction

Extracellular vesicles (EVs) are membranous particles enclosed by lipid bilayers with diameters ranging from 50 nm to 1 μm, derived from almost all cells [1]. By inheriting abundant signaling molecules from donor cells, such as nucleic acids and proteins, EVs play a role in intercellular communication and are associated with disease progression [2–4]. EVs can be classified into several subgroups based on size and generation mechanism, including microvesicles, microparticles, and exosomes [3,5]. Among these subgroups, exosomes are microvesicles ranging in size from 50 to 150 nm, with an average of ~100 nm [3]. Multivesicular bodies (MVBs) contain multiple intraluminal vesicles (ILVs), which fuse with the plasma membrane to release ILVs as exosomes [6]. Therefore, exosomes contain proteins of the cell surface and soluble proteins linked to the extracellular environment [7,8]. Efficient isolation and reliable analysis of EVs are important prerequisites for the clinical application of EVs. Conventional techniques for isolating EVs, such as ultracentrifugation, ultrafiltration, particle size exclusion chromatography, and polymer precipitation, are time-consuming and inefficient [9]. Microfluidic technology achieves efficient EV isolation and detection with only a very small amount (10^{-3} to 10^{-12} microliters) of liquid specimens and reagents and has the advantages of integration and high throughput [10].

The high throughput capability of microfluidics-based capturing and detection technologies offers a significant advantage in the field of EV analysis. Through precise manipulation of fluids at the microscale level, these technologies enable rapid and efficient isolation and detection of EVs from complex biological samples. This not only expedites the process of EV analysis but also allows for the examination of a large number of samples in a relatively short period, facilitating high-throughput screening and analysis [11]. Furthermore, microfluidic-based EV capture and detection technologies support standardization of experimental protocols and assay conditions, promoting consistency and reproducibility across studies. Through precise control of fluidic parameters, such as flow rate, shear stress, and incubation time, microfluidic platforms enable reproducible isolation and characterization of EVs with minimal batch-to-batch variability. Furthermore, the integration of quality control features, such as on-chip calibration and reference standards, facilitates accurate quantification and comparison of EVs between different experiments and laboratories [12]. Therefore, the standardization of microfluidics-based platforms ensures reproducibility and reliability, making them valuable tools for EV research and clinical applications.

In this review, we will explore the latest developments in microfluidic-based EV isolation and analysis, as well as discuss the challenges and future directions of this rapidly evolving field.

2. Microfluidic-Based EV Isolation Strategies

2.1. *Label-Free Microfluidic Isolation*

Label-free isolation strategies depend on the physical characteristics of EVs, such as their size, density, and deformability, to separate them from other components present in bodily fluids. Conventional label-free methods, such as ultrafiltration (UF), ultracentrifugation (UC), and size-exclusion chromatography (SEC), usually require large sample volumes. UC differentiates EVs and other components based on size and density and has been recognized as one of the most commonly used EV isolation methods [13,14]. However, the long UC period might result in the coprecipitation of protein aggregates. In addition, the quality of EVs cannot be guaranteed due to repeated centrifugation processes and excessive centrifugal force [15]. UF adopts membrane filters with suitable pore sizes to purify EVs from other particles [11]. Regrettably, the accumulation of debris on the filter membranes will decrease the efficiency of EV isolation and shorten the lifespan of the membranes [16]. SEC isolates particles of different sizes based on their different flow rates in a column filled with porous beads. As a mild EV isolation method, SEC can obtain EVs with relatively complete structure and function [17,18]. However, the tedious procedure, low throughput, and recovery of SEC limit its wide application.

Holding the properties of easy integration, high-throughput, and low sample consumption, microfluidics technology provides a variety of strategies for efficient label-free EV isolation [19–21]. For example, Liang et al. proposed a double-filtration microfluidic device to isolate EVs with a size range of 30–200 nm [22]. Li et al. integrated filtration and microfluidics technology to design cascaded microfluidic circuits for preprogrammed, clog-free, and gentle isolation of EVs directly from blood within 30 min (Figure 1a) [23]. The problems of filter fouling and particle aggregation were solved by the pulsatile flows generated by the porous membrane, which lift particles away from the membrane. Researchers engineered an automated centrifugal microfluidic disc featuring membranes with specific functionalities (Exo-CMDS) for exosome isolation (Figure 1b) [24]. As a one-step method, Exo-CMDS enriched exosomes with an optimal exosomal concentration of 5.1×10^9 particles/mL from a small amount of blood samples (<300 μL) in 8 min. The above methods required tedious fabrication of microfluidic chips, which increased the uncertainty of trials. Viscoelastic microfluidic systems manipulate viscoelastic fluids based on their viscous and elastic characteristics under deformation without the complex design of microfluidic chips and have been used for micro-/nano-sized particles. Asghari et al.

developed sheathless oscillatory viscoelastic microfluidics for focusing and separating EVs [25]. Liu et al. proposed a viscoelastic-based microfluidic system for direct label-free isolation of exosomes (Figure 1c) [26]. A simple microfluidic device with two inlets, three outlets, and a straight microchannel was designed for the viscoelastic separation of exosomes. The presence of oxyethylene (PEO) causes elastic lift forces, which drive EVs toward the microchannel centerline according to their sizes. Under the size cutoff of 200 nm, exosomes were separated from large EVs. In addition, Yang et al. demonstrated a self-adaptive virtual microchannel for nanoparticle enrichment and separation in a continuous manner (Figure 1d) [27]. A gigahertz bulk acoustic resonator, in combination with microfluidics, triggers and stabilizes acoustic waves and streams. This process forms a virtual channel whose diameter can self-adjust, ranging from dozens to a few micrometers. Using a customized arc-shaped resonator, exosomes from patient plasma were purified. The system is stable and has high automation potential because of the self-adaptive and contactless continuous speration mode.

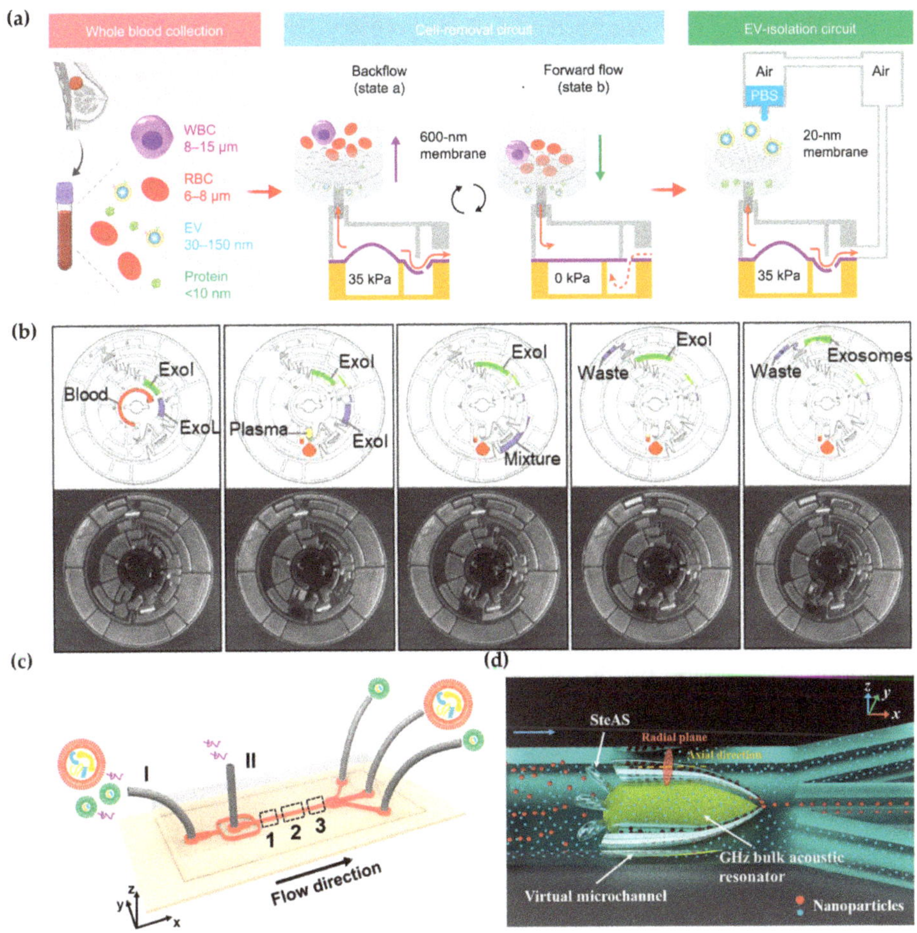

Figure 1. Label-free microfluidic-based EV isolation methods: (**a**) Cascaded microfluidic circuits for EV isolation. Reproduced from Ref. [23]. (**b**) Centrifugal microfluidic disc for EV enrichment. Reproduced from Ref. [24]. (**c**) Visual representation of the separation mechanism of viscoelastic microfluidics. I: inlet for sample fluids, II: inlet for sheath fluids. Reproduced from Ref. [26]. (**d**) Mechanism of the stereo acoustic stream (SteAS) platform. Reproduced from Ref. [27].

Deterministic lateral displacement (DLD)-patterned pillars separate particles based on the DLD critical diameter (D_c) and have been used to isolate circulating tumor cells (CTCs), stem cells, bacteria, and EVs [28]. Zeming et al. captured EVs with 1 μm polymer beads, which increased the size of the polymer beads. As a result, the DLD lateral displacement of beads can be translated to the amount of EVs [29]. To separate nanoscale exosomes, Wunsch et al. designed nanoscale DLD arrays with gap sizes from 25 to 235 nm to isolate human-urine-derived exosomes with a single-particle resolution [30]. Smith et al. designed 1024 nanoscale DLD (nanoDLD) arrays in a microfluidic chip to isolate EVs from serum and urine samples with a yield of ~50% [31]. Moreover, RNA sequencing was produced after separating EVs from prostate cancer (PCa) patient samples, which has the potential to indicate the aggression of PCa. However, the conventional DLD methods need input pressure to drive EVs or beads through pillars in the microfluidic chip, which limits the application of DLD technology in EV isolation. To avoid input pressure, Hattori et al. proposed an electroosmotic flow-driven DLD strategy that used electroosmotic flow to drive fluids for EV separation in a continuous manner [32]. The property of being easy-to-operate makes it a promising solution for clinical diagnostic applications.

Although label-free microfluidic isolation has the advantages of high throughput, rapidity, and cheapness, the low purity and inability to isolate EV subtypes limit its downstream applications.

2.2. Affinity-Based EV Isolation

Affinity-based isolation methods exploit the interaction between affinity ligands (antibodies, peptides, or aptamers) and receptors on EV membranes to isolate EVs specifically [33,34]. These ligands are typically modified on the surface of materials or interfaces, such as CIM® CDI disks [35], magnetic beads [36], carbon cloth [37], graphene [38], or Ti2CTx MXene membranes [39]. The strength and duration of the interaction between affinity ligands and receptors on surfaces determine the capture efficiency of EVs. However, the compromised interaction between EVs and traditional interfaces limits their capture efficiency. The micro-/nanoscale channels in microfluidic chips can enhance the contact frequency between recognition ligands on the chips and molecules on EV membranes, thus increasing EV capture efficiency. For example, the herringbone microfluidic chip, comprising patterned microgrooves, enhances fluid mixing efficiency by manipulating flow states and forming helical motions [40]. Therefore, collision between biological targets and affinity-trapping substrates is improved, resulting in improved EV capture efficiency [41].

However, the near-surface hydrodynamic resistance decreases mass transfer in the microchannel [42]. To overcome this near-surface hydrodynamic resistance, Li et al. developed a 3D porous sponge microfluidic chip made by salt crystallization, which provided a high surface-to-volume ratio [8]. Moreover, researchers also developed a fluid nanoporous microinterface (FluidporeFace) in a herringbone microfluidic chip for the efficient capture of tumor-derived EVs (Figure 2a) [41]. Supported lipid bilayers (SLBs) were encapsulated on the nanoporous herringbone microstructures, which not only improved the mass transfer but also enabled multivalent recognition of aptamers, thus achieving a multi-scale enhanced affinity reaction. As a result, the affinity increased by ~83-fold compared with the nonfluid interface. In addition, they designed a microfluidic chip to create a dynamic multivalent magnetic interface, enhancing the kinetics and thermodynamics of biomolecular recognition for the efficient isolation of EVs derived from tumors (T-EVs) [43]. Utilizing magnetic and flow fields, this engineered interface achieved a harmonious balance of affinity, selectivity, reversibility, and extendibility. As a result, they achieved a high-throughput recovery of T-EVs, facilitating comprehensive protein profiling. However, when utilizing a single ligand for EV capture and another ligand for EV detection, it is challenging to eliminate the interference of free proteins and obtain the requisite subtypes of EVs, thereby rendering it inadequate for clinical applications. To eliminate interference from free proteins, Zhang et al. designed a microfluidic differentiation method that accurately captured PD-L1$^+$ EVs [44]. PD-L1$^+$ EVs were labeled with biotin using DNA computation, incorporating dual inputs of lipid probes and PD-L1 aptamers. Subsequently, these labeled EVs were captured with streptavidin-modified microfluidic chips selectively.

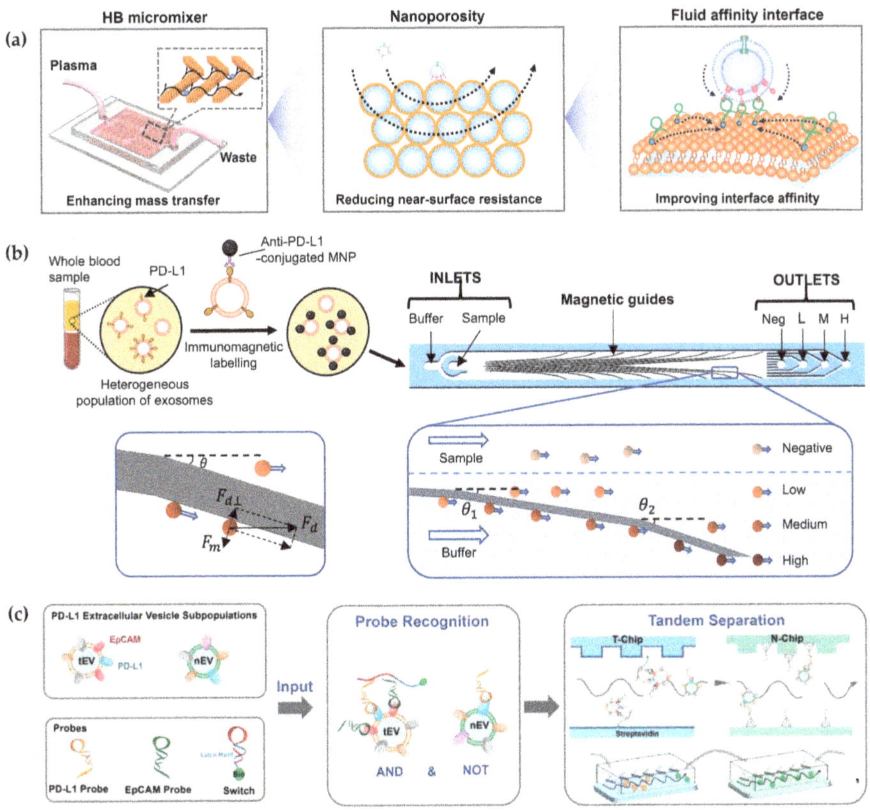

Figure 2. Affinity-based microfluidic EV isolation methods: (**a**) The schematic diagram of the FluidporeFace chip. Reproduced from Ref. [41]. (**b**) A magnetic deflection-based NanoEPIC system to achieve phenotypic profiling and nanoscale sorting of sEVs. Reproduced from Ref. [45]. (**c**) DNA computation-mediated microfluidic tandem separation for PD-L1$^+$ EV subpopulations. Reproduced from Ref. [46].

Different subtypes of EVs represent distinct sources and diverse biological functions. Therefore, the analysis of EV subtypes is crucial for studying the biological mechanisms of EVs. On the contrary, all antibody-based enrichment systems are limited to highly specific isolation protocols, which result in partial EV loss due to differences in the expression of EV membrane surface proteins. Chen et al. proposed a novel herringbone microfluidics device that not only possessed the advantages of herringbone microfluidics but also incorporated aptamer-functionalized core–shell bar codes (AFCSBs) [40]. Because their antiopal hydrogel shells have abundant interconnecting pores, barcodes can provide a rich surface area for the anchoring of multiple DNA aptamers, enabling the specific capture of multiple tumor-derived exosomes. However, its essence lies in using one type of aptamer for molecular capture and another type for molecular discrimination, making it still unable to distinguish different EVs. In order to isolate subtypes of EVs, MUN et al. developed a microfluidic chip-based magnetically labeled exosome isolation system (MEIS-chip) that involved magnetic nanoclusters (MNCs) conjugated with CD63 and HER2 with different degrees of magnetization (CD63 conjugated with low-saturation magnetized MNCs, CD63-LMC, and HER2 conjugated with high-saturation magnetized MNCs, HER2-HMC) [47]. Common exosomes were captured by CD63-LMC, while exosomes with HER overexpression bound to both CD63-LMC and HER2-HMC simultaneously. This allows for the acquisition of varying degrees of magnetic particles and magnetic separation in the MEIS chip via a magnetic field, ultimately resulting in the separation of different EV populations. Moreover, Chen et al. designed a nanoscale cytometry platform called NanoEPIC to enable the collection of small EVs (sEVs) bearing four different expression levels of PD-L1 by labeling them with antibody-functionalized magnetic nanoparticles (MNPs) (Figure 2b) [45]. EVs with higher PD-L1 expression levels had greater lateral deflection towards the edge of the device in the microfluidic flow channel due to their increased magnetic susceptibility resulting from binding to more MNPs. This facilitated the separation of sEVs based on four levels of PD-L1 expression: negative, low (exoL), medium (exo-M), and high (exo-H). Lu et al. achieved the isolation of tumor PD-L1$^+$ EVs and non-tumor PD-L1$^+$ EVs through DNA logic-mediated double aptamer recognition and a tandem chip for the first time (Figure 2c) [46]. Tumor-derived EVs were identified by EpCAM and PD-L1 nucleic acid ligands, induci ng the "AND" logic operation, whereas non-tumor-derived PD-L1$^+$ EVs only express PD-L1, thus invoking the "NOT" logic operation. These two independent outputs facilitated the separation of tumor- and non-tumor-origin PD-L1$^+$ EVs through tandem microfluidics, respectively. Consequently, utilizing a streptavidin-functionalized microfluidic chip (T-Chip), only tumor-derived PD-L1$^+$ EV populations can be isolated. After excluding tumor-derived PD-L1$^+$ EVs, the remaining PD-L1$^+$ EVs from normal cells can be captured through hybridization between the extension sequence on the PD-L1 probe and the corresponding cDNA modified on the second microfluidic chip (N-Chip).

Currently, microfluidic technology based on affinity separation has made tremendous advancements. There have been significant improvements in purification, capture efficiency, and subpopulation separation. Despite some progress in various studies, there are still certain limitations. For example, this method can suffer from non-specific binding to other entities present in the sample, such as proteins, lipoproteins, and cellular debris. This non-specific binding can lead to contamination and reduced purity in the isolated EV population [48]. Moreover, affinity-based EV isolation depends on the availability and specificity of surface markers for EV capture. However, not all EVs express the same surface markers, and the expression profile of EV surface markers can vary depending on cell type, physiological state, and environmental conditions. This limitation restricts the applicability of affinity-based methods, particularly when targeting specific subpopulations of EVs [14]. To effectively apply these technologies to clinical diagnosis and precision treatment, further innovation and improvement are still needed. In short, the efficient isolation of EVs is the premise for researching their biological function and clinical application. It is still necessary to develop new methods for the high-efficiency and high-purity isolation of EV subtypes.

In addition, here is a comparison presenting the effectiveness, efficiency, and practicality of traditional techniques versus microfluidic-based methodologies for EV analysis in Table 1 [49–51].

Table 1. Traditional techniques versus microfluidic-based methodologies for EV analysis [49–51].

Aspect	Traditional Techniques	Microfluidic-Based Methodologies
Effectiveness	Varied, depending on method (e.g., ultracentrifugation, precipitation)	High, with precise control over fluid manipulation and surface interactions
Efficiency	Time-consuming, labor-intensive	Rapid, automated processes with minimal sample and reagent consumption
Practicality	Limited scalability, manual operation	Scalable, integrated systems suitable for high-throughput analysis
Specificity	May lack specificity, leading to contamination and low yield	Enhanced specificity, with tailored devices for selective EV capture
Reproducibility	Variable due to manual handling and batch-to-batch variability	Improved reproducibility with standardized protocols and automated workflows

3. Microfluidic-Based EV Detection

3.1. Fluorescent Detection

Fluorescence technology combined with a microfluidic platform has been widely used in EV detection, which has the characteristics of fast response, good precision, and high sensitivity. After capturing EVs in microfluidic chips, membrane stains or fluorescent labeled antibodies are usually used to identify EVs [52]. For example, Kanwar et al. used a fluorescent carbanine dye (DiO) to stain the exosome membrane and counted the total number of exosomes captured on a microfluidic device (ExoChip) [12]. Antibodies against EV-specific biomarkers, such as CD63, CD9, and CD81, are used for EV identification and quantification [53]. Moreover, fluorescent-labeled antibodies against disease-related biomarkers on the EV membrane are usually used for quantitative and qualitative analysis of EV biomarkers, thus reflecting the progression of diseases. For example, Hisey et al. captured ovarian cancer exosomes in a herringbone groove microfluidic device and quantified EpCAM$^+$ exosomes [54]. The quantitative results showed that EpCAM$^+$ exosomes were related to HGSOC disease progression. After utilizing the integrated microfluidic exosome separation and detection system (EXID system), Lu et al. examined the abundance of exosomal PD-L1 [55]. Using the EXID system, a significant difference in fluorescence intensity was observed. The strategy had a limit of detection (LOD) of 10.76/μL, and the exosomal PD-L1 level reflected the sensibility for immune response. The conventional methods offer bulk information on proteins in EVs, which hardly enables absolute quantification. Liu et al. constructed "exosome-magnetic microbead-enzymatic reporter" complexes and encapsulated the complexes into droplets, which ensured a single complex was encapsulated in a droplet. As a result, cancer-specific exosomes were absolute counted with a low limit of detection of 10 exosomes μL^{-1} (Figure 3a) [56].

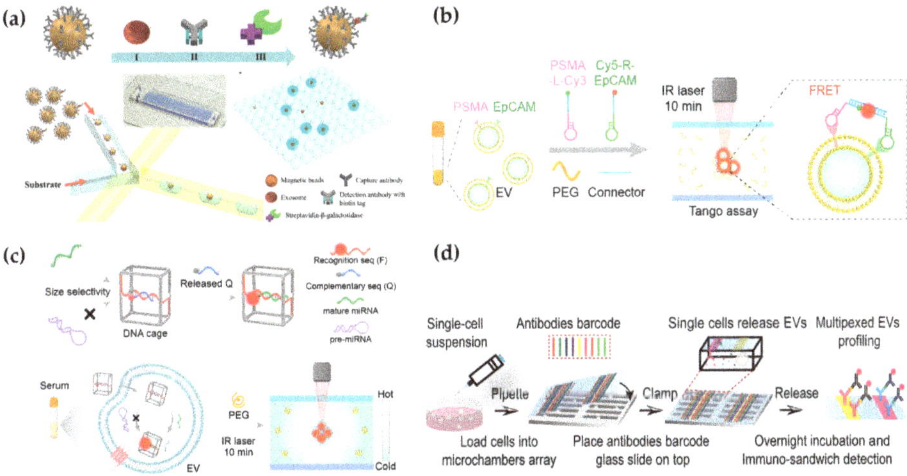

Figure 3. Microfluidic-based fluorescent detection: (**a**) The droplet digital ExoELISA method for quantifying EVs. Reproduced from Ref. [56]. (**b**) Schematic of the one-step thermophoretic AND gate operation (Tango) assay. Reproduced from Ref. [57]. (**c**) DTTA for detecting mRNA within EVs. Reproduced from Ref. [58]. (**d**) Schematic illustration depicting the workflow for the multiplexed profiling of single-cell EV secretion. Reproduced from Ref. [59].

The heterogeneity of EVs challenges the acknowledgement of their biological significance and clinical application. To explore the heterogeneity of EVs, Zhang's group developed a microfluidic chip featuring self-assembled 3D herringbone nanopatterns, enabling highly sensitive fluorescent detection of EV surface proteins [60]. The device was used to detect exosome subtypes expressing CD24, EpCAM, and FRalpha proteins in 2 μL plasma samples from 20 ovarian cancer patients and to suggest exosomal FRalpha as a promising biomarker in the early detection and monitoring of ovarian cancer progression. Furthermore, MMP14 on EVs holds potential for early detection and prognosis assessment in breast cancer metastasis [61]. The above microfluidic devices need cumbersome fabrication processes. Moreover, the interaction between proteins and antibodies may be limited due to the steric hindrance caused by the post-translational modification of proteins [62]. Sun's group developed a microfluidic thermophoresis device that accumulated particles in a size-dependent manner and amplified fluorescence signals based on the different diffusion rates of particles in a nonuniform temperature field [63]. Seven fluorescently labeled aptamers targeting different epitopes were employed for the subtyping analysis of EVs. This method demonstrated a sensitivity and specificity of 95% and 100%, respectively, in cancer detection, and it can also be utilized for cancer classification. Furthermore, the method is applicable for the proteomic analysis of EVs in breast cancer, enabling the identification of metastatic breast cancer, monitoring treatment responses, and predicting patients' progression-free survival rates [64].

In addition to phenotypic heterogeneity, the tracing of EV origins is also particularly important. It can accurately detect and monitor the progression of diseases. Sun's group utilized microfluidic thermophoresis devices for the specific detection of tumor-derived EVs, achieving an accuracy of 97% [65]. Similarly, a rapid and non-invasive diagnostic assay, named the one-step thermophoretic AND gate operation (Tango), has been developed for precise identification of prostate cancer (PCa)-derived EVs directly in serum samples within 15 min. This method demonstrated an impressive overall accuracy of 91% in discerning PCa from benign prostatic hyperplasia (BPH) using a streamlined single-step format. This innovative technique holds tremendous potential for the swift and non-invasive diagnosis of cancers (Figure 3b) [57].

In addition to proteins on the EV surface, EV nucleic acids are also important biomarkers and therapeutic targets for diseases [66]. Shao et al. developed a microfluidic chip to capture exosomes and analyze the mRNAs of MGMT and APNG in enriched tumor exosomes [67]. This strategy has the potential to predict the drug responses of GBM patients. Sun's group demonstrated a thermal swim sensor (TSN) employing nanoflares for the in situ detection of exosomal miRNA, eliminating the need for RNA extraction or target amplification. Through the thermophoretic accumulation of nanoflare-treated exosomes, a heightened fluorescent signal was produced upon binding with exosome miRNA, facilitating the direct quantitative assessment of exosome miRNA [68]. Afterwards, they devised a DNA tetrahedron-based thermophoretic analysis (DTTA) for the in situ detection of mRNA in EVs, achieving remarkable sensitivity and specificity [69]. Recently, they developed a DTTA for highly sensitive and selective in situ detection of mature miRNA in EVs. This assay achieved a detection limit of 2.05 fM for mature miRNA in EVs without interference from pre-miRNA and distinguished between breast cancer patients and healthy donors with an overall accuracy of 90% (Figure 3c) [58].

Moreover, cell culture supernatants, or EVs, in body fluids lose interaction information with other symbiotic cells in tissues, making them unable to accurately represent the role of EVs in intercellular communication [70]. Ji et al. employed spatially patterned antibody barcodes and achieved multiplexed profiling of single-cell EV secretion from over 1000 individual cells concurrently. This innovative approach enabled the comprehensive characterization of human oral squamous cell carcinoma, unraveling previously obscured single-cell heterogeneity in EV secretion dynamics (Figure 3d) [59]. This technology facilitates a thorough assessment of EV secretion diversity at the single-cell level, offering an invaluable tool to supplement existing single-cell analysis and EV research. Afterwards, they applied this platform to analyze the characteristic spectra of paired neuronal–microglial and neuronal–astrocyte single-cells in the human cell lineage. These results provide a basis for exploring how neurons and immune cells interact through complex secretion networks [71].

While fluorescence technology utilizing microfluidic techniques has been successfully employed in EV detection, offering commendable accuracy and high sensitivity, it is not without its drawbacks. Some issues still remain, including the intricate preparation of fluorescence labels and the occurrence of spectral interference. This interference encompasses background interference and spectral overlap issues, along with the phenomenon of photobleaching in fluorescent molecules. The resolution of these issues will contribute to enhancing the accuracy of fluorescence detection of EVs.

3.2. Visualization Detection

In recent years, the method of detecting EVs through microfluidic technology using the colorimetric method has undergone significant development and has simplified the equipment [34]. Visualization detection relies on the changes that can be observed by naked eyes, such as color changes within the detection system, which occur as a result of chemical or biochemical interactions between specific target analytes and colorimetric probes. One significant advantage of visualization assays is their independence from bulky off-chip detection systems. Consequently, visualization detection has garnered growing interest in biomedical research, particularly for disease diagnosis, owing to its distinct advantages in EV detection [70].

For instance, Chen et al. introduced a traditional colorimetric technique to detect EVs using a 3-D scaffold chip [72]. They proposed a ZnO nanowire-coated 3D scaffold chip device for effective immune capture and classical visual and colorimetric detection of EVs. In the work by Di et al., a rapid analysis method was introduced, utilizing nano-enzyme-assisted immunosorbent assays, eliminating the need for antibody detection [73]. The approach involved the immobilization of nanoparticles on the phospholipid membrane of exosomes, followed by the addition of chromogenic agents. For another, Jiang's group developed a sensor platform that can visually analyze EV surface proteins in minutes. The sensor consists of a gold nanoparticle (AuNP) and a set of aptamers [74]. In addition, Ko et al. engineered a photofluidic platform powered by smartphones to quantify brain-derived exosomes. This innovative chip enables rapid processing, delivering results within one hour, which is ten-fold quicker than conventional methods. The device boasts a detection limit of approximately 10^7 exosomes/mL [75]. Utilizing enzyme amplification, it can detect exosome biomarkers, with results easily read through a smartphone camera.

Although many researchers have made incredible progress in the field of visualization detection methods, which can rapidly detect a wide range of biomolecules, from infectious disease-related protein biomarkers to glucose and nucleic acids, the extensive application of visualization detection is limited. It is mainly used in underdeveloped regions, and low sensitivity is its major drawback.

3.3. Electrochemical Detection

Electrochemical methods transfer the signals of EV recognition to electrochemical signals, such as voltage, current, and resistance [76]. Microfluidic-based electrochemical techniques have attracted great attention in EV detection due to their broad detection range, high sensitivity, and specificity.

Electrochemical methods are usually used to profile EV surface proteins. For example, Akagi et al. presented an on-chip immunoelectrophoresis method for EV protein expression analysis based on the different positive charges on the EV surface caused by antibody binding [77]. Moreover, Akagi et al. found that exosomes from different cells had differential zeta potential. Thus, they developed an electrophoresis apparatus for tracking individual exosomes [78]. Moreover, Akagi et al. developed an on-chip microcapillary electrophoresis (μCE) system to detect the zeta potential distribution of exosomes from normal cells and prostate cancer cells [79]. They found that the huge negative charge of cancer exosomes was due to abundant sialic acids. Currently, affinity ligands are usually modified on microfluidic chips for EV capture. Then, the electrochemically responsive molecules were triggered to cause a change in electrochemical signals. To date, great efforts have been devoted to introducing various signal production and amplification strategies, such as metal nanoparticles, tetrahedral DNA nanostructures (TDNs), and nucleic acid-based amplification analysis, to electrochemical biosensors for EV detection. For example, Wang et al. have developed a new filter electrochemical microfluidic chip (FEMC) that integrates on-chip separation and in situ surface protein electrochemical analysis of exosomes in the whole blood of breast cancer patients [80]. In this system, zirconium-based metal–organic frameworks (Zr-MOFs) loaded with numerous electroactive methylene blue molecules (Zr-MOFMB@UiO-66) were attached to exosomes collected on electrode surfaces, leading to the amplification of electrical signals. The entire FEMC assay took 1 h to complete, enabling timely and more informed opportunities for the diagnosis of breast cancer. To highly sensitively detect colorectal cancer exosomes, a microfluidic electrochemical biosensing platform based on TDN-based signal amplification was constructed (Figure 4a) [81]. TDNs, including the EpCAM aptamer, were immobilized on Au nanoparticles (AuNPs) as a recognition element to harvest the exosomes. Then, the AuNPs had an obvious catalytic effect on the redox reaction of ferricyanide, enabling electrochemical detection. The platform had a broad measurement range (50–10^5 particles/μL) and a low limit of detection (42 particles/μL).

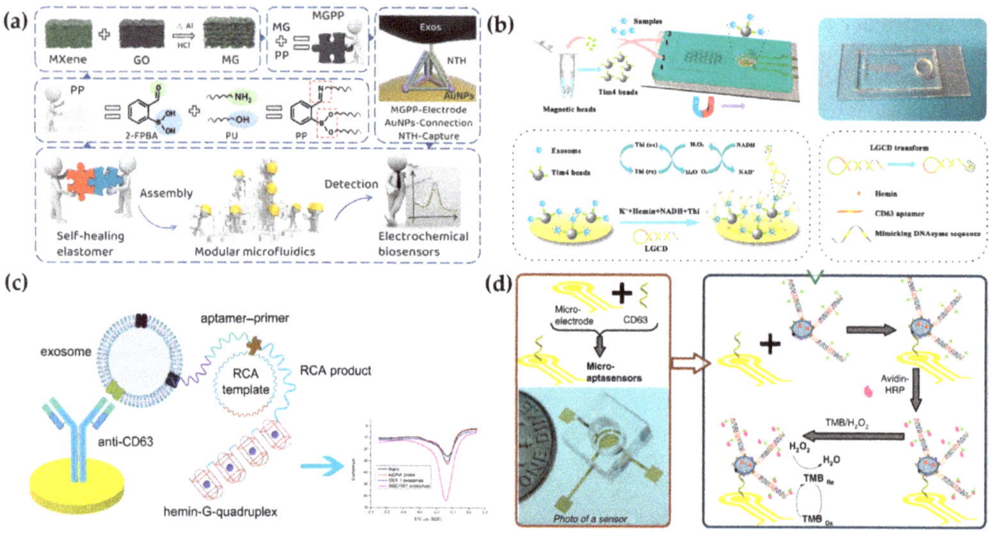

Figure 4. Microfluidic-based electrochemical detection: (**a**) TDN-based microfluidic electrochemical biosensing platform for exosome analysis. Reproduced from Ref. [81]. (**b**) Schematic diagram of the ExoPCD chip. Reproduced from Ref. [82]. (**c**) Hemin/G-quadruplex-assisted signal amplification. Reproduced from Ref. [83]. (**d**) Electrochemical micro-aptasensors based on HCR for exosome detection. Reproduced from Ref. [84].

Moreover, an increasing number of nucleic acid amplification methods have been employed in electrochemical biosensors. Xu et al. proposed a two-stage microfluidic platform (ExoPCD-chip) for the electrochemical analysis of hepatocellular exosomes in serum (Figure 4b) [82]. Particularly, exosomes captured by electrochemical aptasensors with a CD63 aptamer led to the accumulation of the hemin/G-quadruplex. This complex could function as a NADH oxidase and horseradish peroxidase (HRP)-mimicking DNAzyme simultaneously. Thus, the freshly formed H_2O_2 by NADH oxidation could be continuously catalyzed, accompanied by significant signal enhancement. Moreover, a staggered Y-shaped micropillar mixing pattern was introduced to create an anisotropic flow without any surface modification to improve exosome enrichment efficiency. Due to their flexible programmability, aptamers are easily engineered for signal amplification to improve EV detection sensitivity. For example, a hemin/G-quadruplex system and rolling circle amplification (RCA) were combined in an aptasensor for the selective and sensitive detection of gastric cancer exosomes (Figure 4c) [83]. RCA is recognized as a nucleic acid amplification analysis that can be performed at room temperature to preserve the integrity of exosomes. In addition, Zhang et al. designed a remarkably selective electrochemical micro-aptasensor with a detection limit of 5×10^2 exosomes/mL by integrating a micropatterned electrochemical aptasensor and a signal amplification strategy of hybridization chain reaction (HCR) (Figure 4d) [84]. Biotin-labeled HCR products were used to bind specifically to enriched exosomes, utilizing EpCAM aptamers as a bridge. This was followed by the attachment of multiple avidin-HRPs, producing a current signal through the enzyme reaction. Moreover, the proposed aptasensor was effective in discriminating serum samples from early-stage lung cancer patients and late-stage patients, indicating significant promise for early cancer diagnosis.

To sum up, electrochemistry proves highly suitable for EV analysis within an integrated microfluidic chip, offering a multitude of advantages. Furthermore, no requirements for optical transparency expand the choice of materials in electrochemical response. However, contamination and changes in pH, temperature, and ionic concentration often influence the lifetimes of electrodes, which needs to be addressed [85]. So far, only a limited number of microfluidic devices incorporated with electrochemical techniques have been developed for EV detection. We believe that there will be a growing number of microfluidic devices combined with electrochemical detection as a promising means for point-of-care diagnostics.

3.4. *Surface-Enhanced Raman Spectroscopy (SERS)*

Surface-enhanced Raman spectroscopy (SERS) effectively generates spectra on certain metal surfaces, providing vibrational and rotational energy information about molecules, which is reflected in spectral peaks used to specifically identify molecules [85]. However, SERS measurements present significant challenges to reproducibility and sensitivity [86]. In this aspect, microfluidic surface-enhanced Raman spectroscopy (MF-SERS) is making progress in resolving some significant and previously insurmountable issues and limitations of SERS detection to some extent, thus improving detection capability and extending its application [87]. Through the utilization of high-throughput nanosurface microfluidics control technology and unique fingerprint identification, precise testing of ultra-small populations of biochemical particles such as cancer EVs is made possible.

In efforts to amplify Raman signals from cancer-derived EVs, Mahsa et al. developed a nanosurface fluidic device for label-free, non-immunological SERS detection of EVs. This device effectively distinguished the SERS fingerprint of EVs from noncancerous glial cells (NHA) and two subpopulations of the GBM EVs (i.e., U87 and U373). The sample solution flowed from the input ports to the serpentine analysis channels ($50 \times 250\ \mu m^2$) to achieve a single-layer distribution of EVs on the nanosurface. At the same time, metal nanomaterials with SERS activity formed a hexagonal nanoscale triangular array, with each of the two triangles forming a bowtie structure with a suspended gap region area to amplify the EM field enhancement, with an electromagnetic field enhancement factor of 9×10^5 (Figure 5a) [88]. In addition to the above-mentioned label-free microfluidic Raman chip, Wang et al. developed a new one with immunoassays for quickly and sensitively detecting the exosomes (Figure 3b) [89]. Hybrid channels of triangular column arrays were used to enrich CD63-positive exosomes and were fixed in the Raman detection region. EpCAM-labeled Raman beads with high densities of nitrile were used as probes for detection, and the detection limit was 1.6×10^2 particles per mL with 20 µL samples.

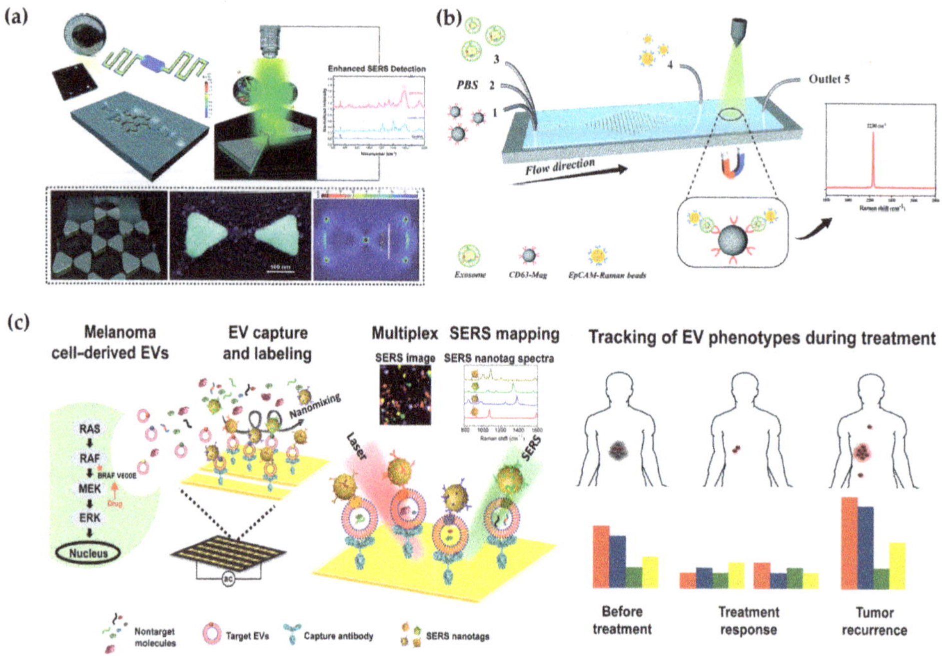

Figure 5. Microfluidic-based SERS detection. (**a**) Spatially suspended nanobowtie surface microfluidic device for SERS detection of EVs. Reproduced from Ref. [88]. (**b**) Microfluidic Raman chip for exosome capture and detection. Reproduced from Ref. [89]. (**c**) EV phenotyping by EPAC. Four colors represent four biomarkers: MCSP-MBA, red; MCAM-TFMBA, blue; ErbB3-DTNB, green; LNGFR-MPY, yellow. Reproduced from Ref. [90].

Various biomarkers can indicate diverse biological functions, making it significant to detect multiple exogenous biomarkers. Han et al. proposed a microfluidic-based SERS detection technique for profiling numerous exosomal biomarkers to diagnose osteosarcoma. Gold nanoparticles labeled with SERS tags can selectively bind to exosomes using specific antibodies in samples, forming exosome immunocomplexes. A microfluidic chip, comprising two symmetrical polydimethylsiloxane (PDMS) layers and a nanoporous polycarbonate track-etched (PCTE) membrane, was employed for exosome purification [91]. Microfluidic tangential flow filtration effectively eliminated plasma biomolecules and free SERS tags while enriching exosome immunocomplexes on the membrane for in situ SERS analysis [91,92]. Herein, Wang et al. also showcased a multiplex EV phenotype analyzer chip (EPAC). EPAC integrates a nanomixing-enhanced microchip and a multiplex surface-enhanced Raman scattering (SERS) nanotag system for direct EV phenotyping (Figure 5c) [90]. They observed the EV phenotypic heterogeneity and longitudinally monitored the EV phenotypic evolution, finding specific EV profiles involved in the development of drug resistance and the potential of EV phenotyping for monitoring treatment responses. Thus, the microfluidic-based SERS detection method offers great potential for the detection of external vesicles and cancer diagnosis.

4. Conclusions and Perspectives

EVs are intricately linked to numerous physiological processes as well as the onset and progression of diseases. Efficient EV isolation methods and sensitive EV detection methods will help to improve the understanding of the physiological and pathological effects of EVs and provide important support for the precision medicine of related diseases. At present, conventional EV isolation and detection technologies still have limitations. Microfluidic-based methods for isolation and detection of EVs have the obvious advantages of high integration, low consumption, fast speed, high separation efficiency, and high detection sensitivity, which opens up new ideas and directions for the research of EVs. With the help of microfluidic chips, the efficient isolation, enrichment, and multi-marker detection of EVs with different sizes can be integrated into a single chip, and a more diversified clinical detection instrument can be built.

The field of microfluidic-based EV isolation and detection is still in its infancy, and there are still a lot of theoretical and technical problems to be solved. Therefore, the means to achieve highly selective and accurate isolation of EVs in actual biological samples, sensitive and selective detection of these EVs, and even the biological information carried in them will be important topics in the study of EVs based on microfluidic chips. With the development of technology and in-depth research, the isolation and analysis of individual EVs can be realized by the microfluidic method, and commercial EV chips are also expected to be applied in clinical practice.

In this article, the existing methods of microfluidic-based EV isolation and analysis are reviewed. Compared with traditional ultracentrifugation, ultrafiltration, immunocapture, and co-precipitation, microfluidic chips are smaller and more flexible, and microfluidic immunoaffinity methods can isolate high-purity EVs with strong specificity. The isolation methods based on the physical characteristics of EVs do not need to add expensive reagents such as antigens and antibodies. Therefore, the cost is low, and the isolation process will not cause contamination, which is conducive to downstream analysis. Microfluidic-based EV analysis methods have the advantages of fast analysis speed, high throughput, and low reagent consumption, which can meet the needs of rapid detection of EVs in a large number of clinical samples. Therefore, microfluidic technology has significant advantages in the isolation of EVs in a small number of clinical samples and the rapid estimation of diseases.

Despite the significant advancements in microfluidics-based capturing and detection technologies for EV analysis, several challenges remain to be addressed. One of the primary challenges is the standardization of isolation and detection protocols to ensure consistency and reliability across different platforms and laboratories. Variability in sample preparation, device design, and operating conditions can lead to inconsistent results and hinder the reproducibility of findings. Therefore, efforts are needed to establish standardized protocols and quality control measures to facilitate the comparison and validation of results obtained from different microfluidic-based platforms.

Another challenge is the optimization of microfluidic devices for the analysis of specific EV subpopulations or cargo molecules. EVs exhibit heterogeneity in size, surface markers, and cargo content, which necessitates the development of tailored microfluidic devices capable of selectively capturing and analyzing desired EV subtypes. Furthermore, the integration of multiplexed detection modalities into microfluidic platforms would enable comprehensive profiling of EVs, facilitating the discovery of novel biomarkers and the elucidation of disease mechanisms.

In addition to technical challenges, the translation of microfluidic-based EV analysis from research laboratories to clinical settings requires overcoming regulatory and commercialization hurdles. Regulatory agencies, such as the Food and Drug Administration (FDA), require rigorous validation of diagnostic assays and devices to ensure their safety and efficacy for clinical use. Moreover, the scalability and cost-effectiveness of microfluidic-based platforms need to be optimized to enable widespread adoption in clinical diagnostics and personalized medicine.

In recent years, with the rapid development of micro/nano manufacturing, new materials, and information technology, the design of microfluidic chips and the performance of supporting devices have been further improved. It is mainly reflected in (1) the development of precision manufacturing technology, making it possible to integrate multiple EV isolation methods and realize the integration of EV isolation and detection on one chip; (2) by combining the chips and portable detection equipment, a miniaturized EV microfluidic isolation and analysis platform is constructed to realize the rapid detection of EVs and greatly expand its application space. With the miniaturization, integration, and automation of microfluidic EV isolation and analysis devices, microfluidic technology will play an increasingly important role in EV isolation, biochemical detection, and mechanism research.

In conclusion, microfluidics-based capturing and detection technologies offer powerful tools for the isolation, characterization, and analysis of extracellular vesicles. These technologies leverage precise fluid manipulation at the microscale level to enable rapid, efficient, and high-throughput analysis of EVs from complex biological samples. Despite remaining challenges in standardization, optimization, and translation to clinical applications, ongoing research efforts are poised to overcome these hurdles and unlock the full potential of microfluidic-based EV analysis in disease diagnosis, prognosis, and therapeutics. With continued innovation and collaboration across interdisciplinary fields, microfluidics-based EV analysis holds promise for revolutionizing personalized medicine and improving patient outcomes.

Author Contributions: Conceptualization, L.Z. and J.C.; writing—original draft preparation, J.C., M.Z., Q.X. and H.W.; writing—review and editing, L.Z., X.Y., J.C., C.C., T.L. and Y.W.; All authors have read and agreed to the published version of the manuscript.

Funding: This research was funded by the National Natural Science Foundation of China (22304145), the Xiamen Medical College Transverse Project (HX202302), the Natural Science Foundation of Fujian Province, China (2023J05285), the Medical and Health Guidance Project of Xiamen Science and Technology Bureau (3502Z20224ZD1299), the Project of Xiamen Medical College (K2023-12, K2023-45), the College Students' Innovative Entrepreneurial Training Plan Program of Fujian Province (202312631009), and the Education and Research Project of Young and Middle-aged Teachers of Fujian Province, China (JAT220406).

Conflicts of Interest: The authors declare no conflicts of interest.

References

1. Zhu, L.; Tian, W.; Yuan, L.; Chi, C.; Wang, Y.; Xiao, Q.; Zheng, M.; Yang, C.; Song, Y. Aptamer-Based Extracellular Vesicle Isolation, Analysis and Therapeutics. *Interdiscip. Med.* **2023**, *1*, e20220019. [CrossRef]
2. Robbins, P.; Morelli, A. Regulation of Immune Responses by Extracellular Vesicles. *Nat. Rev. Immunol.* **2014**, *14*, 195–208. [CrossRef] [PubMed]
3. Kalluri, R.; LeBleu, V. The Biology Function and Biomedical Applications of Exosomes. *Science* **2020**, *367*, 6977. [CrossRef]
4. Möller, A.; Lobb, R. The Evolving Translational Potential of Small Extracellular Vesicles in Cancer. *Nat. Rev. Can.* **2020**, *20*, 697–709. [CrossRef] [PubMed]
5. Mathieu, M.; Martin-Jaular, L.; Lavieu, G.; Théry, C. Specificities of Secretion and Uptake of Exosomes and Other Extracellular Vesicles for Cell-to-Cell Communication. *Nat. Cell Biol.* **2019**, *21*, 9–17. [CrossRef] [PubMed]
6. van Niel, G.; D'Angelo, G.; Raposo, G. Shedding Light on the Cell Biology of Extracellular Vesicles. *Nat. Rev. Mol. Cell Bio.* **2018**, *19*, 213–228. [CrossRef] [PubMed]
7. Han, Z.; Peng, C.; Yi, J.; Zhang, D.; Xiang, X.; Peng, X.; Su, B.; Liu, B.; Shen, Y.; Qiao, L. Highly Efficient Exosome Purification from Human Plasma by Tangential Flow Filtration Based Microfluidic Chip. *Sens. Actuators B Chem.* **2021**, *333*, 129563. [CrossRef]
8. Li, P.; Chen, J.; Chen, Y.; Song, S.; Huang, X.; Yang, Y.; Li, Y.; Tong, Y.; Xie, Y.; Li, J.; et al. Construction of Exosome Sorl1 Detection Platform Based on 3d Porous Microfluidic Chip and Its Application in Early Diagnosis of Colorectal Cancer. *Small* **2023**, *19*, e2207381. [CrossRef] [PubMed]
9. Coumans, F.; Brisson, A.; Buzas, E.; Dignat-George, F.; Drees, E.; El-Andaloussi, S.; Emanueli, C.; Gasecka, A.; Hendrix, A.; Hill, A.; et al. Methodological Guidelines to Study Extracellular Vesicles. *Circ. Res.* **2017**, *120*, 1632–1648. [CrossRef]
10. Whitesides, G. The Origins and the Future of Microfluidics. *Nature* **2006**, *442*, 368–373. [CrossRef]
11. Li, P.; Kaslan, M.; Lee, S.H.; Yao, J.; Gao, Z. Progress in Exosome Isolation Techniques. *Theranostics* **2017**, *7*, 789–804. [CrossRef] [PubMed]
12. Kanwar, S.; Dunlay, C.; Simeone, D.; Nagrath, S. Microfluidic Device (Exochip) for on-Chip Isolation, Quantification and Characterization of Circulating Exosomes. *Lab Chip* **2014**, *14*, 1891–1900. [CrossRef] [PubMed]
13. Royo, F.; Théry, C.; Falcón-Pérez, J.M.; Nieuwland, R.; Witwer, K.W. Methods for Separation and Characterization of Extracellular Vesicles: Results of a Worldwide Survey Performed by the Isev Rigor and Standardization Subcommittee. *Cells* **2020**, *9*, 1955. [CrossRef]
14. Witwer, K.W.; Buzás, E.I.; Bemis, L.T.; Bora, A.; Lässer, C.; Lötvall, J.; Nolte-t' Hoen, E.N.; Piper, M.G.; Sivaraman, S.; Skog, J.; et al. Standardization of Sample Collection, Isolation and Analysis Methods in Extracellular Vesicle Research. *J. Extracell. Vesicles.* **2013**, *2*, e1700716. [CrossRef]

15. Kalra, H.; Adda, C.G.; Liem, M.; Ang, C.S.; Mechler, A.; Simpson, R.J.; Hulett, M.D.; Mathivanan, S. Comparative Proteomics Evaluation of Plasma Exosome Isolation Techniques and Assessment of the Stability of Exosomes in Normal Human Blood Plasma. *Proteomics* **2013**, *13*, 3354–3364. [CrossRef]
16. Alvarez, M.; Khosroheidari, M.; Kanchi Ravi, R.; DiStefano, J. Comparison of Protein, Microrna, and Mrna Yields Using Different Methods of Urinary Exosome Isolation for the Discovery of Kidney Disease Biomarkers. *Kidney Int.* **2012**, *82*, 1024–1032. [CrossRef] [PubMed]
17. Nordin, J.Z.; Lee, Y.; Vader, P.; Mäger, I.; Johansson, H.J.; Heusermann, W.; Wiklander, O.P.; Hällbrink, M.; Seow, Y.; Bultema, J.J.; et al. Ultrafiltration with Size-Exclusion Liquid Chromatography for High Yield Isolation of Extracellular Vesicles Preserving Intact Biophysical and Functional Properties. *Nanomedicine* **2015**, *11*, 879–883. [CrossRef] [PubMed]
18. Sidhom, K.; Obi, P.O.; Saleem, A. A Review of Exosomal Isolation Methods: Is Size Exclusion Chromatography the Best Option? *Int. J. Mol. Sci.* **2020**, *21*, 6466. [CrossRef] [PubMed]
19. Poudineh, M.; Aldridge, P.; Ahmed, S.; Green, B.; Kermanshah, L.; Nguyen, V.; Tu, C.; Mohamadi, R.; Nam, R.; Hansen, A.; et al. Tracking the Dynamics of Circulating Tumour Cell Phenotypes Using Nanoparticle-Mediated Magnetic Ranking. *Nat. Nanotechnol.* **2017**, *12*, 274–281. [CrossRef]
20. Ibsen, S.; Wright, J.; Lewis, J.; Kim, S.; Ko, S.; Ong, J.; Manouchehri, S.; Vyas, A.; Akers, J.; Chen, C.; et al. Rapid Isolation and Detection of Exosomes and Associated Biomarkers from Plasma. *ACS Nano* **2017**, *11*, 6641–6651. [CrossRef]
21. Psaltis, D.; Quake, S.; Yang, C. Developing Optofluidic Technology through the Fusion of Microfluidics and Optics. *Nature* **2006**, *442*, 381–386. [CrossRef] [PubMed]
22. Li-Guo, L.; Meng-Qi, K.; Sherry, Z.; Ye-Feng, S.; Ping, W.; Tao, Y.; Fatih, I.; Winston Patrick, K.; Lan-Juan, L.; Utkan, D.; et al. An Integrated Double-Filtration Microfluidic Device for Isolation, Enrichment and Quantification of Urinary Extracellular Vesicles for Detection of Bladder Cancer. *Sci. Rep.* **2017**, *7*, 46224. [CrossRef] [PubMed]
23. Li, Z.; Liu, C.; Cheng, Y.; Li, Y.; Deng, J.; Bai, L.; Qin, L.; Mei, H.; Zeng, M.; Tian, F.; et al. Cascaded Microfluidic Circuits for Pulsatile Filtration of Extracellular Vesicles from Whole Blood for Early Cancer Diagnosis. *Sci. Adv.* **2023**, *9*, eade2819. [CrossRef] [PubMed]
24. Zhao, L.; Wang, H.; Fu, J.; Wu, X.; Liang, X.Y.; Liu, X.Y.; Wu, X.; Cao, L.L.; Xu, Z.Y.; Dong, M. Microfluidic-Based Exosome Isolation and Highly Sensitive Aptamer Exosome Membrane Protein Detection for Lung Cancer Diagnosis. *Biosens. Bioelectron.* **2022**, *214*, 114487. [CrossRef] [PubMed]
25. Mohammad, A.; Xiaobao, C.; Bogdan, M.; Daniel, V.L.; Mahmut Kamil, A.; Stavros, S.; Andrew, J.d. Oscillatory Viscoelastic Microfluidics for Efficient Focusing and Separation of Nanoscale Species. *ACS Nano* **2019**, *14*, 422–433. [CrossRef] [PubMed]
26. Liu, C.; Guo, J.; Tian, F.; Yang, N.; Yan, F.; Ding, Y.; Wei, J.; Hu, G.; Nie, G.; Sun, J. Field-Free Isolation of Exosomes from Extracellular Vesicles by Microfluidic Viscoelastic Flows. *ACS Nano* **2017**, *11*, 6968–6976. [CrossRef]
27. Yang, Y.; Zhang, L.; Jin, K.; He, M.; Wei, W.; Chen, X.; Yang, Q.; Wang, Y.; Pang, W.; Ren, X.; et al. Self-Adaptive Virtual Microchannel for Continuous Enrichment and Separation of Nanoparticles. *Sci. Adv.* **2022**, *8*, eabn8440. [CrossRef] [PubMed]
28. Salafi, T.; Zhang, Y.; Zhang, Y. A Review on Deterministic Lateral Displacement for Particle Separation and Detection. *Nano-Micro Lett.* **2019**, *11*, 77–109. [CrossRef] [PubMed]
29. Kerwin Kwek, Z.; Thoriq, S.; Swati, S.; Yong, Z. Fluorescent Label-Free Quantitative Detection of Nano-Sized Bioparticles Using a Pillar Array. *Nat. Commun.* **2018**, *9*, 1254. [CrossRef]
30. Benjamin, H.W.; Joshua, T.S.; Stacey, M.G.; Chao, W.; Markus, B.; Robert, L.B.; Robert, H.A.; Gustavo, S.; Yann, A. Nanoscale Lateral Displacement Arrays for the Separation of Exosomes and Colloids Down to 20 Nm. *Nat. Nanotechnol.* **2016**, *11*, 936–940. [CrossRef]
31. Joshua, T.S.; Benjamin, H.W.; Navneet, D.; Mehmet, E.A.; Kayla, L.; Kamlesh, K.Y.; Rachel, W.; Michael, A.P.; Jyotica, V.P.; Elizabeth, A.D.; et al. Integrated Nanoscale Deterministic Lateral Displacement Arrays for Separation of Extracellular Vesicles from Clinically-Relevant Volumes of Biological Samples. *Lab Chip* **2018**, *18*, 3913–3925. [CrossRef] [PubMed]
32. Yuya, H.; Taisuke, S.; Takao, Y.; Noritada, K.; Yoshinobu, B. Micro- and Nanopillar Chips for Continuous Separation of Extracellular Vesicles. *Anal. Chem.* **2019**, *91*, 6514–6521. [CrossRef] [PubMed]
33. Shao, H.; Im, H.; Castro, C.; Breakefield, X.; Weissleder, R.; Lee, H. New Technologies for Analysis of Extracellular Vesicles. *Chem. Rev.* **2018**, *118*, 1917–1950. [CrossRef] [PubMed]
34. Boriachek, K.; Masud, M.; Palma, C.; Phan, H.; Yamauchi, Y.; Hossain, M.; Nguyen, N.; Salomon, C.; Shiddiky, M. Avoiding Pre-Isolation Step in Exosome Analysis: Direct Isolation and Sensitive Detection of Exosomes Using Gold-Loaded Nanoporous Ferric Oxide Nanozymes. *Anal. Chem.* **2019**, *91*, 3827–3834. [CrossRef] [PubMed]
35. Evgen, M.; Crystal Jing Ying, T.; Mari, P.; Pia, S.; Marja-Liisa, R. Fast Isolation of Highly Specific Population of Platelet-Derived Extracellular Vesicles from Blood Plasma by Affinity Monolithic Column, Immobilized with Anti-Human Cd61 Antibody. *Anal. Chim. Acta* **2019**, *1091*, 160–168. [CrossRef] [PubMed]
36. Mondal, S.K.; Whiteside, T.L. Immunoaffinity-Based Isolation of Melanoma Cell-Derived and T Cell-Derived Exosomes from Plasma of Melanoma Patients. *Methods Mol. Biol.* **2021**, *2265*, 305–321. [CrossRef] [PubMed]
37. Ashraf, J.; Lau, S.; Akbarinejad, A.; Bryant, D.T.; Chamley, L.W.; Pilkington, L.I.; Barker, D.; Williams, D.E.; Evans, C.W.; Travas-Sejdic, J. Electrochemical Approach for Specific Capture and Rapid Release of Nanoscale Placental Extracellular Vesicles Using Aptamer-Modified Conducting Terpolymer-Coated Carbon Cloth. *ACS Appl. Nano Mater.* **2023**, *6*, 3981–3989. [CrossRef]

38. Huang, W.; Yu, Y.; Yang, C.; Zhang, X.; Shang, L.; Zu, Y.; Shi, K. Aptamer Decorated Magnetic Graphene Oxide Nanoparticles for Effective Capture of Exosomes. *Chem. Eng. J.* **2022**, *431*, 133849. [CrossRef]
39. Qiannan, Y.; Linlin, Z.; Zhimin, C.; Mingfeng, G.; Qian, M.; Li, Y.; Wen-Fei, D. Hierarchical Au Nanoarrays Functionalized 2d Ti(2)Ct(X) Mxene Membranes for the Detection of Exosomes Isolated from Human Lung Carcinoma Cells. *Biosens. Bioelectron.* **2022**, *216*, 14647. [CrossRef]
40. Chen, H.; Bian, F.; Guo, J.; Zhao, Y. Aptamer-Functionalized Barcodes in Herringbone Microfluidics for Multiple Detection of Exosomes. *Small Methods* **2022**, *6*, e2200236. [CrossRef]
41. Niu, Q.; Gao, J.; Zhao, K.; Chen, X.; Lin, X.; Huang, C.; An, Y.; Xiao, X.; Wu, Q.; Cui, L.; et al. Fluid Nanoporous Microinterface Enables Multiscale-Enhanced Affinity Interaction for Tumor-Derived Extracellular Vesicle Detection. *Proc. Natl. Acad. Sci. USA* **2022**, *119*, e2213236119. [CrossRef] [PubMed]
42. Yu, W.; Hurley, J.; Roberts, D.; Chakrabortty, S.K.; Enderle, D.; Noerholm, M.; Breakefield, X.O.; Skog, J.K. Exosome-Based Liquid Biopsies in Cancer: Opportunities and Challenges. *Ann. Oncol.* **2021**, *32*, 466–477. [CrossRef]
43. Niu, Q.; Shu, Y.; Chen, Y.; Huang, Z.; Yao, Z.; Chen, X.; Lin, F.; Feng, J.; Huang, C.; Wang, H.; et al. A Fluid Multivalent Magnetic Interface for High-Performance Isolation and Proteomic Profiling of Tumor-Derived Extracellular Vesicles. *Angew. Chem. Int. Ed.* **2023**, *62*, 202215337. [CrossRef] [PubMed]
44. Zhang, Y.; Wu, Q.; Huang, Y.; Wang, W.; Lu, Y.; Kang, S.; Yang, C.; Song, Y. Reliable Detection of Extracellular Pd-L1 by DNA Computation-Mediated Microfluidics. *Anal. Chem.* **2023**, *95*, 9373–9379. [CrossRef] [PubMed]
45. Chen, K.; Duong, B.; Ahmed, S.; Dhavarasa, P.; Wang, Z.; Labib, M.; Flynn, C.; Xu, J.; Zhang, Y.; Wang, H.; et al. A Magneto-Activated Nanoscale Cytometry Platform for Molecular Profiling of Small Extracellular Vesicles. *Nat. Commun.* **2023**, *14*, 5576. [CrossRef] [PubMed]
46. Yinzhu, L.; Bingqian, L.; Weizhi, L.; Jialu, Z.; Lin, Z.; Chaoyong, Y.; Yanling, S. Isolation of Pd-L1 Extracellular Vesicle Subpopulations Using DNA Computation Mediated Microfluidic Tandem Separation. *Small Methods* **2023**, *7*, e2300516. [CrossRef] [PubMed]
47. Mun, B.; Kim, R.; Jeong, H.; Kang, B.; Kim, J.; Son, H.Y.; Lim, J.; Rho, H.W.; Lim, E.K.; Haam, S. An Immuno-Magnetophoresis-Based Microfluidic Chip to Isolate and Detect Her2-Positive Cancer-Derived Exosomes Via Multiple Separation. *Biosens. Bioelectron.* **2023**, *239*, 115592. [CrossRef]
48. Théry, C.; Witwer, K.W.; Aikawa, E.; Alcaraz, M.J.; Anderson, J.D.; Andriantsitohaina, R.; Antoniou, A.; Arab, T.; Archer, F.; Atkin-Smith, G.K.; et al. Minimal Information for Studies of Extracellular Vesicles 2018 (Misev2018): A Position Statement of the International Society for Extracellular Vesicles and Update of the Misev2014 Guidelines. *J. Extracell. Vesicles.* **2018**, *7*, 535750. [CrossRef]
49. Cheng, S.; Li, Y.; Yan, H.; Wen, Y.; Zhou, X.; Friedman, L.; Zeng, Y. Advances in Microfluidic Extracellular Vesicle Analysis for Cancer Diagnostics. *Lab Chip* **2021**, *21*, 3219–3243. [CrossRef]
50. Greening, D.W.; Xu, R.; Ji, H.; Tauro, B.J.; Simpson, R.J. A Protocol for Exosome Isolation and Characterization: Evaluation of Ultracentrifugation, Density-Gradient Separation, and Immunoaffinity Capture Methods. *Methods Mol. Biol.* **2015**, *1295*, 179–209. [CrossRef]
51. Takov, K.; Yellon, D.M.; Davidson, S.M. Comparison of Small Extracellular Vesicles Isolated from Plasma by Ultracentrifugation or Size-Exclusion Chromatography: Yield, Purity and Functional Potential. *J. Extracell. Vesicles.* **2019**, *8*, 1560809. [CrossRef] [PubMed]
52. Cheng, N.; Du, D.; Wang, X.; Liu, D.; Xu, W.; Luo, Y.; Lin, Y. Recent Advances in Biosensors for Detecting Cancer-Derived Exosomes. *Trends. Biotech.* **2019**, *37*, 1236–1254. [CrossRef] [PubMed]
53. Dudani, J.; Gossett, D.; Tse, H.; Lamm, R.; Kulkarni, R.; Carlo, D. Rapid Inertial Solution Exchange for Enrichment and Flow Cytometric Detection of Microvesicles. *Biomicrofluidics* **2015**, *9*, 014112. [CrossRef] [PubMed]
54. Hisey, C.; Dorayappan, K.; Cohn, D.; Selvendiran, K.; Hansford, D. Microfluidic Affinity Separation Chip for Selective Capture and Release of Label-Free Ovarian Cancer Exosomes. *Lab Chip* **2018**, *18*, 3144–3153. [CrossRef] [PubMed]
55. Lu, Y.; Ye, L.; Jian, X.; Yang, D.; Zhang, H.; Tong, Z.; Wu, Z.; Shi, N.; Han, Y.; Mao, H. Integrated Microfluidic System for Isolating Exosome and Analyzing Protein Marker Pd-L1. *Biosens. Bioelectron.* **2022**, *204*, 113879. [CrossRef] [PubMed]
56. Liu, C.; Xu, X.; Li, B.; Situ, B.; Pan, W.; Hu, Y.; An, T.; Yao, S.; Zheng, L. Single-Exosome-Counting Immunoassays for Cancer Diagnostics. *Nano Lett.* **2018**, *18*, 4226–4232. [CrossRef] [PubMed]
57. Deng, J.Q.; Zhao, S.; Li, J.H.; Cheng, Y.C.; Liu, C.; Liu, Z.; Li, L.L.; Tian, F.; Dai, B.; Sun, J.S. One-Step Thermophoretic and Gate Operation on Extracellular Vesicles Improves Diagnosis of Prostate Cancer. *Angew. Chem. Int. Ed.* **2022**, *61*, e202207037. [CrossRef] [PubMed]
58. Zhao, S.; Zhang, S.H.; Hu, H.J.; Cheng, Y.C.; Zou, K.X.; Song, J.; Deng, J.Q.; Li, L.L.; Zhang, X.B.; Ke, G.L.; et al. Selective in Situ Analysis of Mature Micrornas in Extracellular Vesicles Using a DNA Cage-Based Thermophoretic Assay. *Angew. Chem. Int. Ed.* **2023**, *62*, 202303121. [CrossRef] [PubMed]
59. Ji, Y.H.; Qi, D.Y.; Li, L.M.; Su, H.R.; Li, X.J.; Luo, Y.; Sun, B.; Zhang, F.Y.; Lin, B.C.; Liu, T.J.; et al. Multiplexed Profiling of Single-Cell Extracellular Vesicles Secretion. *Proc. Natl. Acad. Sci. USA* **2019**, *116*, 5979–5984. [CrossRef]
60. Zhang, P.; Zhou, X.; He, M.; Shang, Y.; Tetlow, A.; Godwin, A.; Zeng, Y. Ultrasensitive Detection of Circulating Exosomes with a 3d-Nanopatterned Microfluidic Chip. *Nat. Biomed. Eng.* **2019**, *3*, 438–451. [CrossRef]

61. Zhang, P.; Wu, X.; Gardashova, G.; Yang, Y.; Zhang, Y.; Xu, L.; Zeng, Y. Molecular and Functional Extracellular Vesicle Analysis Using Nanopatterned Microchips Monitors Tumor Progression and Metastasis. *Sci. Transl. Med.* **2020**, *12*, eaaz2878. [CrossRef] [PubMed]
62. Lin, Z.; Yuanfeng, X.; Xinyu, W.; Haoting, L.; Mengjiao, H.; Bingqian, L.; Yanling, S.; Chaoyong, Y. Coupling Aptamer-Based Protein Tagging with Metabolic Glycan Labeling for in Situ Visualization and Biological Function Study of Exosomal Protein-Specific Glycosylation. *Angew. Chem. Int. Ed.* **2021**, *60*, 18111–18115. [CrossRef] [PubMed]
63. Liu, C.; Zhao, J.; Tian, F.; Cai, L.; Zhang, W.; Feng, Q.; Chang, J.; Wan, F.; Yang, Y.; Dai, B.; et al. Low-Cost Thermophoretic Profiling of Extracellular-Vesicle Surface Proteins for the Early Detection and Classification of Cancers. *Nat. Biomed. Eng.* **2019**, *3*, 183–193. [CrossRef] [PubMed]
64. Tian, F.; Zhang, S.H.; Liu, C.; Han, Z.W.; Liu, Y.; Deng, J.Q.; Li, Y.K.; Wu, X.; Cai, L.L.; Qin, L.L.; et al. Protein Analysis of Extracellular Vesicles to Monitor and Predict Therapeutic Response in Metastatic Breast Cancer. *Nat. Commun.* **2021**, *12*, 2536. [CrossRef] [PubMed]
65. Li, Y.; Deng, J.; Han, Z.; Liu, C.; Tian, F.; Xu, R.; Han, D.; Zhang, S.; Sun, J. Molecular Identification of Tumor-Derived Extracellular Vesicles Using Thermophoresis-Mediated DNA Computation. *J. Am. Chem. Soc.* **2021**, *143*, 1290–1295. [CrossRef] [PubMed]
66. Thind, A.; Wilson, C. Exosomal Mirnas as Cancer Biomarkers and Therapeutic Targets. *J. Extracell. Vesicles* **2016**, *5*, 31292. [CrossRef] [PubMed]
67. Huilin, S.; Jaehoon, C.; Kyungheon, L.; Leonora, B.; Changwook, M.; Bob, S.C.; Fred, H.H.; Xandra, O.B.; Hakho, L.; Ralph, W. Chip-Based Analysis of Exosomal Mrna Mediating Drug Resistance in Glioblastoma. *Nat. Commun.* **2015**, *6*, 6999. [CrossRef]
68. Zhao, J.; Liu, C.; Li, Y.; Ma, Y.; Deng, J.; Li, L.; Sun, J. Thermophoretic Detection of Exosomal Micrornas by Nanoflares. *J. Am. Chem. Soc.* **2020**, *142*, 4996–5001. [CrossRef] [PubMed]
69. Han, Z.; Wan, F.; Deng, J.; Zhao, J.; Li, Y.; Yang, Y.; Jiang, Q.; Ding, B.; Liu, C.; Dai, B.; et al. Ultrasensitive Detection of Mrna in Extracellular Vesicles Using DNA Tetrahedron-Based Thermophoretic Assay. *Nano Today* **2021**, *38*, 101203. [CrossRef]
70. Crescitelli, R.; Lässer, C.; Lötvall, J. Isolation and Characterization of Extracellular Vesicle Subpopulations from Tissues. *Nat. Protoc.* **2021**, *16*, 1548–1580. [CrossRef]
71. Deng, J.; Ji, Y.; Zhu, F.; Liu, L.; Li, L.; Bai, X.; Li, H.; Liu, X.; Luo, Y.; Lin, B.; et al. Mapping Secretome-Mediated Interaction between Paired Neuron–Macrophage Single Cells. *Proc. Natl. Acad. Sci. USA* **2022**, *119*, e2200944119. [CrossRef] [PubMed]
72. Chen, Z.; Cheng, S.; Cao, P.; Qiu, Q.; Chen, Y.; Xie, M.; Xu, Y.; Huang, W. Detection of Exosomes by Zno Nanowires Coated Three-Dimensional Scaffold Chip Device. *Biosens. Bioelectron.* **2018**, *122*, 211–216. [CrossRef] [PubMed]
73. Di, H.; Mi, Z.; Sun, Y.; Liu, X.; Liu, X.; Li, A.; Jiang, Y.; Gao, H.; Rong, P.; Liu, D. Nanozyme-Assisted Sensitive Profiling of Exosomal Proteins for Rapid Cancer Diagnosis. *Theranostics* **2020**, *10*, 9303–9314. [CrossRef] [PubMed]
74. Jiang, Y.; Shi, M.; Liu, Y.; Wan, S.; Cui, C.; Zhang, L.; Tan, W. Aptamer/Aunp Biosensor for Colorimetric Profiling of Exosomal Proteins. *Angew. Chem. Int. Ed.* **2017**, *56*, 11916–11920. [CrossRef] [PubMed]
75. Ko, J.; Hemphill, M.; Gabrieli, D.; Wu, L.; Yelleswarapu, V.; Lawrence, G.; Pennycooke, W.; Singh, A.; Meaney, D.; Issadore, D. Smartphone-Enabled Optofluidic Exosome Diagnostic for Concussion Recovery. *Sci. Rep.* **2016**, *6*, 31215. [CrossRef] [PubMed]
76. Park, J.; Park, J.S.; Huang, C.-H.; Jo, A.; Cook, K.; Wang, R.; Lin, H.-Y.; Van Deun, J.; Li, H.; Min, J.; et al. An Integrated Magneto-Electrochemical Device for the Rapid Profiling of Tumour Extracellular Vesicles from Blood Plasma. *Nat. Biomed. Eng.* **2021**, *5*, 678–689. [CrossRef] [PubMed]
77. Takanori, A.; Kei, K.; Masashi, K.; Nobuyoshi, K.; Takahiro, O.; Takanori, I. On-Chip Immunoelectrophoresis of Extracellular Vesicles Released from Human Breast Cancer Cells. *PLoS ONE* **2015**, *10*, e0123603. [CrossRef] [PubMed]
78. Kato, K.; Kobayashi, M.; Hanamura, N.; Akagi, T.; Kosaka, N.; Ochiya, T.; Ichiki, T. Electrokinetic Evaluation of Individual Exosomes by on-Chip Microcapillary Electrophoresis with Laser Dark-Field Microscopy. *Jpn. J. Appl. Phys.* **2013**, *52*, 06GK10. [CrossRef]
79. Akagi, T.; Kato, K.; Hanamura, N.; Kobayashi, M.; Ichiki, T. Evaluation of Desialylation Effect on Zeta Potential of Extracellular Vesicles Secreted from Human Prostate Cancer Cells by on-Chip Microcapillary Electrophoresis. *Jpn. J. Appl. Phys.* **2014**, *53*, 06JL01. [CrossRef]
80. Wang, Y.; Gao, W.; Sun, M.; Feng, B.; Shen, H.; Zhu, J.; Chen, X.; Yu, S. A Filter-Electrochemical Microfluidic Chip for Multiple Surface Protein Analysis of Exosomes to Detect and Classify Breast Cancer. *Biosens. Bioelectron.* **2023**, *239*, 115590. [CrossRef]
81. Wang, M.; Zhang, Z.; Li, G.; Jing, A. Room-Temperature Self-Healing Conductive Elastomers for Modular Assembly as a Microfluidic Electrochemical Biosensing Platform for the Detection of Colorectal Cancer Exosomes. *Micromachines* **2023**, *14*, 617. [CrossRef] [PubMed]
82. Xu, H.; Liao, C.; Zuo, P.; Liu, Z.; Ye, B. Magnetic-Based Microfluidic Device for on-Chip Isolation and Detection of Tumor-Derived Exosomes. *Anal. Chem.* **2018**, *90*, 13451–13458. [CrossRef]
83. Huang, R.; He, L.; Xia, Y.; Xu, H.; Liu, C.; Xie, H.; Wang, S.; Peng, L.; Liu, Y.; Liu, Y.; et al. A Sensitive Aptasensor Based on a Hemin/G-Quadruplex-Assisted Signal Amplification Strategy for Electrochemical Detection of Gastric Cancer Exosomes. *Small* **2019**, *15*, e1900735. [CrossRef]
84. Zhang, W.; Tian, Z.; Yang, S.; Rich, J.; Zhao, S.; Klingeborn, M.; Huang, P.; Li, Z.; Stout, A.; Murphy, Q.; et al. Electrochemical Micro-Aptasensors for Exosome Detection Based on Hybridization Chain Reaction Amplification. *Microsyst. Nanoeng.* **2021**, *7*, 63. [CrossRef]

85. Chen, M.; Lin, S.; Zhou, C.; Cui, D.; Haick, H.; Tang, N. From Conventional to Microfluidic: Progress in Extracellular Vesicle Separation and Individual Characterization. *Adv. Healthc. Mater.* **2023**, *12*, e2202437. [CrossRef]
86. Panneerselvam, R.; Sadat, H.; Höhn, E.; Das, A.; Noothalapati, H.; Belder, D. Microfluidics and Surface-Enhanced Raman Spectroscopy, a Win-Win Combination? *Lab Chip* **2022**, *22*, 665–682. [CrossRef]
87. Yue, S.; Fang, J.; Xu, Z. Advances in Droplet Microfluidics for Sers and Raman Analysis. *Biosens. Bioelectron.* **2022**, *198*, 113822. [CrossRef]
88. Jalali, M.; Isaac Hosseini, I.; AbdelFatah, T.; Montermini, L.; Wachsmann Hogiu, S.; Rak, J.; Mahshid, S. Plasmonic Nanobowtiefluidic Device for Sensitive Detection of Glioma Extracellular Vesicles by Raman Spectrometry. *Lab Chip* **2021**, *21*, 855–866. [CrossRef]
89. Wang, Y.; Li, Q.; Shi, H.; Tang, K.; Qiao, L.; Yu, G.; Ding, C.; Yu, S. Microfluidic Raman Biochip Detection of Exosomes: A Promising Tool for Prostate Cancer Diagnosis. *Lab Chip* **2020**, *20*, 4632–4637. [CrossRef]
90. Wang, J.; Wuethrich, A.; Sina, A.; Lane, R.; Lin, L.; Wang, Y.; Cebon, J.; Behren, A.; Trau, M. Tracking Extracellular Vesicle Phenotypic Changes Enables Treatment Monitoring in Melanoma. *Sci. Adv.* **2020**, *6*, eaax3223. [CrossRef]
91. Han, Z.; Peng, X.; Yang, Y.; Yi, J.; Zhao, D.; Bao, Q.; Long, S.; Yu, S.; Xu, X.; Liu, B.; et al. Integrated Microfluidic-Sers for Exosome Biomarker Profiling and Osteosarcoma Diagnosis. *Biosens. Bioelectron.* **2022**, *217*, 114709. [CrossRef] [PubMed]
92. Wang, Y.; Vaidyanathan, R.; Shiddiky, M.; Trau, M. Enabling Rapid and Specific Surface-Enhanced Raman Scattering Immunoassay Using Nanoscaled Surface Shear Forces. *ACS Nano* **2015**, *9*, 6354–6362. [CrossRef] [PubMed]

micromachines

Article

Rapid Construction of Liquid-like Surfaces via Single-Cycle Polymer Brush Grafting for Enhanced Antifouling in Microfluidic Systems

Feng Wu [1,*,†], Jing Xu [2,†], Yuanyuan Liu [3], Hua Sun [3], Lishang Zhang [1], Yixuan Liu [1], Weiwei Wang [1], Fali Chong [1], Dan Zou [4] and Shuli Wang [5,*]

1 School of Physics and New Energy, Xuzhou University of Technology, Xuzhou 221018, China; ztaiyang1@163.com (L.Z.); 20210802206@xzit.edu.cn (Y.L.); wwwang@xzit.edu.cn (W.W.); flchong2008@163.com (F.C.)
2 Medical Laboratory Department, The First People's Hospital of Xuzhou, Xuzhou 221116, China; xujingwlc@outlook.com
3 School of Material and Chemical Engineering, Xuzhou University of Technology, Xuzhou 221018, China; lyy8email@163.com (Y.L.); iamsunhua@xzit.edu.cn (H.S.)
4 School of Health Management, Xihua University, Chengdu 610039, China; zoudan@mail.xhu.edu.cn
5 Fujian Engineering Research Center for Solid-State Lighting, Department of Electronic Science, School of Electronic Science and Engineering, Xiamen University, Xiamen 361005, China
* Correspondence: wuf@xzit.edu.cn (F.W.); slwang@xmu.edu.cn (S.W.)
† These authors contributed equally to this work.

Abstract: Liquid-like surfaces have demonstrated immense potential in their ability to resist cell adhesion, a critical requirement for numerous applications across various domains. However, the conventional methodologies for preparing liquid-like surfaces often entail a complex multi-step polymer brush modification process, which is not only time-consuming but also presents significant challenges. In this work, we developed a single-cycle polymer brush modification strategy to build liquid-like surfaces by leveraging high-molecular-weight bis(3-aminopropyl)-terminated polydimethylsiloxane, which significantly simplifies the preparation process. The resultant liquid-like surface is endowed with exceptional slipperiness, effectively inhibiting bacterial colonization and diminishing the adherence of platelets. Moreover, it offers promising implications for reducing the dependency on anticoagulants in microfluidic systems constructed from PDMS, all while sustaining its antithrombotic attributes.

Keywords: liquid-like surface; slippery; antifouling; microfluidics

1. Introduction

Recent advancements in polydimethylsiloxane (PDMS)-based microfluidics have significantly transformed various fields, including biomedical engineering, chemical analysis, and materials science [1–6]. Central to the performance and functionality of these microfluidic systems is the interface, which plays a critical role in determining device efficiency [7,8]. A major challenge in microfluidics is fouling [9–11], which can severely impact system reliability and efficiency. This issue is particularly pronounced in in vitro blood testing, where the non-specific adhesion of non-target cells can compromise the subsequent cell analysis and identification [12]. Additionally, the use of soluble anticoagulant drugs, such as heparin, during detection processes can alter the biological characteristics of target cells, potentially affecting cell viability and protein expression [13,14]. Consequently, developing PDMS-based microfluidic chips with effective antifouling properties is of paramount importance.

To address the challenges of fouling in microfluidic systems, researchers have explored various antifouling strategies, including solid surface design [15,16] and liquid surface design [17,18]. Solid surface design involves techniques such as superhydrophobic or superoleophobic coatings, which create a physical barrier against cell adhesion [19,20]. This barrier reduces the likelihood of direct interactions between cells and the surface, thereby minimizing cellular attachment. However, the complex nanostructures required for these coatings can complicate their integration into microfluidic chips. In contrast, liquid surface design entails applying functional liquids to PDMS surfaces, creating extremely slippery interfaces that reduce cell adhesion [21,22]. This approach offers several advantages, including resistance to bacterial colonization, decreased macrophage penetration, and a reduced inflammatory response. Despite these benefits, liquid surface designs face challenges such as potential depletion and instability due to the inherent mobility of the liquid [23]. These limitations affect the practical application of both solid and liquid surface designs in achieving effective antifouling properties in PDMS-based microfluidic chips.

Recent research has underscored the exceptional properties of flexible polymer brushes, particularly bis (3-aminopropyl)-terminated PDMS (μPDMS), which exhibits an extraordinarily low glass transition temperature [1,24]. The chemical bonds within these brushes exhibit a notably low rotational conformation transition energy barrier, comparable to the energy present in thermal motion [25]. This characteristic endows μPDMS brushes with remarkable mobility at room temperature, resulting in a distinctly slippery "liquid-like" behavior. Such liquid-like surfaces have demonstrated considerable potential in various applications, including lossless or directional liquid transfer [26], resistance to fouling [25,27], and anti-icing [28] properties. In our recent work, we introduced a liquid-like surface functionalization strategy by grafting μPDMS in green solvent, which not only effectively resists cell adhesion but also exhibits superior biocompatibility, and realizes high-purity and high-efficiency circulating tumor cell (CTC) isolation in PDMS-based microfluidic chips [1]. However, the conventional preparation of this surface typically requires a multi-step modification process, which is both time-consuming and complex. Therefore, there is an urgent need to develop simplified methods to expedite the functionalization process, which is crucial for advancing the use of liquid-like surfaces in PDMS-based microfluidic systems.

In this paper, we established the rapid, liquid-like interfacial modification of PDMS surfaces by the single-cycle grafting of high-molecular-weight μPDMS brushes. The sliding performance of the modification process was better than the liquid-like surfaces with a four-cycle grafting of low-molecular-weight μPDMS brushes, demonstrating rapid and efficient functionalization. The flexibility of the modified μPDMS endow the liquid-like surfaces with an excellent antifouling performance for bacteria, platelets, and blood cells. Finally, we introduced the liquid-like interfacial design in the PDMS microfluidic channels and demonstrated its application prospects in anticoagulation.

2. Materials and Methods

The quartz crystals were obtained from Jiaxing Crystal Electronics Co. (Jiaxing, China). μPDMS with the molecular weights of 27,000 and 2500, N,N′-disuccinimidyl carbonate (DSC), and 3-aminopropyl triethoxysilane (APTES) were purchased from Sigma-Aldrich (St. Louis, MI, USA). Deionized water with a resistivity of 18.2 MΩ·cm was obtained from a Milli-Q system.

The ethanol solutions of DSC (1 μg mL^{-1}) and μPDMS (molecular weight of 27,000, 1 μg mL^{-1}) were prepared for the grafting of liquid-like polymer chains. The amino-functionalized substrates were immersed in the DSC ethanol solution for 60 min at room temperature, followed by rinsing with ethanol and rapid washing. Subsequently, the substrates were immersed in the μPDMS ethanol solution for an additional 60 min at room temperature, rinsed with ethanol, and then dried with compressed air. This process was repeated for several cycles when using the μPDMS with a molecular weight of 2500, and the concentration of the reactants and reaction conditions were without change.

For preparing the liquid-like surface-modified microfluidic inner surface, PDMS (Sylgard 184, Dow Corning Corporation (Midland, AL, USA), mixed at 10:1 base/cross-linker ratio) is first cast into single-channel molds and cured at 80 °C for 3 h. Subsequently, the PDMS was peeled off from the molds. The PDMS and the glass slide were treated with oxygen plasma to form silanol groups on the surface. Then, they were sequentially immersed in the ethanol solutions of APTES, DSC, and μPDMS and reacted at room temperature for 60 min. Finally, the PDMS and the glass slide were fixed together using dovetail clips to form the liquid-like surface-modified PDMS-based microfluidic chips.

The slippery property of the surface was measured by the tilted plate method. Briefly, a 10 μL blood droplet was placed on the surface. Subsequently, the surface was tilted relative to the horizontal plane until the blood droplet began to roll off the surface. The inclined angle can be considered the blood droplet sliding angle.

The density of amine groups on the samples was quantified using the acid orange II (AO II) colorimetric method [29]. In this procedure, the samples were first immersed in an aqueous solution of AO II at pH 4.0 for 4 h to enable reaction with the amine groups. Following this, the samples were thoroughly washed with a pH 4.0 aqueous hydrochloric acid solution to remove any unreacted AO II. Subsequently, the AO II bound to the samples was eluted using 200 μL of a sodium hydroxide solution at pH 11.0. The amount of eluted AO II was then measured by fluorescence spectroscopy, with excitation at 485 nm and emission at 520 nm, utilizing a Molecular Devices SpectraMax ID5 (San Jose, CA, USA). The amine group density was calculated by comparing the fluorescence data to a standard curve.

The amount of μPDMS grafted onto the surface was monitored using a Quartz Crystal Microbalance with Dissipation (QCM-D, Stockholm, Sweden). Initially, gold-coated quartz crystals were functionalized with DSC; subsequently, an ethanol solution of μPDMS was injected into the same cell. All the measurements were performed at room temperature to ensure consistent conditions throughout the experiment.

The chemical composition of the liquid-like polymer layer was analyzed using X-ray photoelectron spectroscopy (XPS). The measurements were performed with a PHI Quantum 2000 Scanning ESCA Microprobe (Physical Electronics, Eden Prairie, MN, USA), equipped with a monochromatic Al Kα X-ray source (1486.6 eV), operating at 15 kV and 35 W under a vacuum pressure of 5×10^{-7} Pa. The carbon peak at 284.4 eV was used as the reference for charge calibration.

The antibacterial activity of the liquid-like surface was assessed following the ISO 22196:2011 standard [30]. *Staphylococcus aureus* and *Escherichia coli* were precultured and diluted to 1/500 and 1/100, respectively, with nutrient broth to obtain a test inoculum with a concentration of approximately 6×10^5 cells mL^{-1}. On a sterile bench, sterilized PDMS plates were placed in a 24-well plate, and 100 μL of the diluted bacterial solution was added dropwise to the surface of each sample. Then, a clean PE film (Wuxi City, China) was used to cover the samples, which were then incubated in a 37 °C incubator for 24 h. Subsequently, a pipette was used to repeatedly pipette to completely detach the bacteria from the PDMS and the PE film. Finally, 100 μL of bacterial solution was drawn and evenly coated on the previously solidified solid medium and incubated at 37 °C for 24 h.

A dose of 10 U/mL heparin was added to fresh whole blood obtained from a volunteer. This heparinized blood was then applied to the sample surfaces and incubated for 1 h. Following incubation, the adherent blood cells were fixed with 2.5% glutaraldehyde solution at 4 °C for 2 h. The morphology of the adhered blood cells was subsequently examined using a scanning electron microscope (SEM, Hitachi S-2400, Tokyo, Japan).

Fresh whole blood, containing heparin at a concentration of 0.25 U/mL, was injected into the microfluidic chip at a flow rate of 0.1 mL/h. After allowing for 2 h, the PDMS and glass slide were removed, and the adherent blood cells on the chip were fixed with 2.5% glutaraldehyde solution at 4 °C for 2 h. The morphology of the adhered blood cells was then analyzed using a scanning electron microscope (SEM, Hitachi S-2400, Tokyo, Japan).

The data were expressed as mean ± standard deviation (SD), and all the experiments were repeated 3 times (mean ± SD, n = 3).

3. Results

The schematic of the grafting process of the μPDMS for preparing liquid-like surfaces is illustrated in Figure 1a. The utilization of high-molecular-weight μPDMS facilitates a rapid functionalization process through a single-cycle modification approach. In contrast, employing low-molecular-weight μPDMS necessitates a multi-cycle modification process involving several reaction steps, thereby extending the overall preparation time for the liquid-like surfaces.

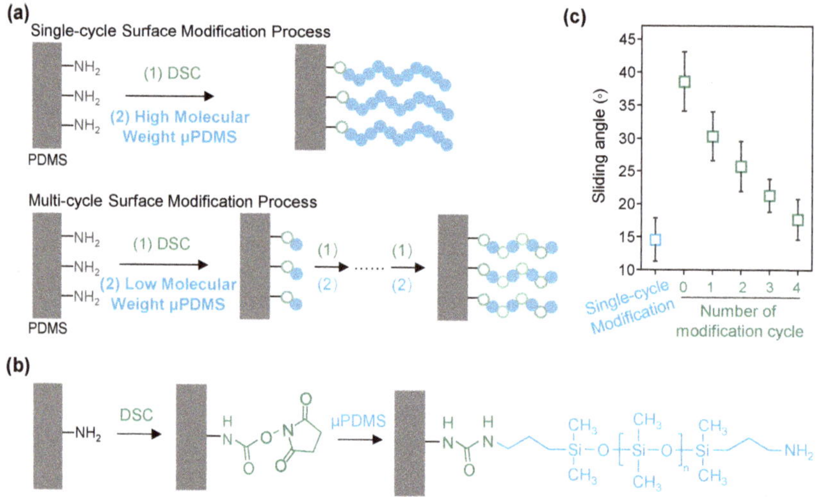

Figure 1. (**a**) The schematic of the preparation processes for liquid-like surfaces from the single-cycle grafting of high-molecular-weight μPDMS and multi-cycle grafting of low-molecular-weight μPDMS. The grass green hollow spheres denote DSC molecules, and the blue solid spheres represent μPDMS molecules. (**b**) The involved chemical reactions during the grafting of DSC and μPDMS on the APTES-modified PDMS surface. (**c**) The sliding angles of a 10 μL blood droplet on the liquid-like surfaces prepared from the single-cycle grafting of high-molecular-weight μPDMS, and 0–4 cycle grafting of low-molecular-weight μPDMS (mean ± SD, n = 3).

Figure 1b shows the chemical reaction that occurs in the single-cycle modification process for constructing the liquid-like surface. Initially, a monolayer of amino groups is grafted onto the PDMS surface via oxygen plasma treatment, followed by immersion in an ethanol solution containing APTES. Subsequently, the amino-modified surface is treated with an ethanol solution of DSC and μPDMS, facilitating the formation of liquid-like polymer chains. This process involves a ring-opening reaction between the succinimide groups and the amino groups on the PDMS surface [31]. This straightforward modification process enables the rapid functionalization of the liquid-like surface within a single processing cycle.

The μPDMS polymer brushes modified on the PDMS surface exhibit high mobility at room temperature due to their extremely low glass transition temperature, resulting in a highly melted state that imparts a liquid-like slippery property. To characterize this property, we measured the sliding angle of blood droplets (10 μL) on the liquid-like surfaces prepared from both the μPDMS with high and low molecular weight. As shown in Figure 1c, the blood droplet has a sliding angle of only 14.6° on liquid-like surfaces prepared through the single-cycle grafting of high-molecular-weight μPDMS. In contrast, the liquid-like surfaces modified from the single-cycle grafting of low-molecular-weight μPDMS exhibit a sliding angle of 30.3°. Increasing the grafting cycles of the low-molecular-weight μPDMS to four resulted in the reduction in the sliding angle to 17.7°. Thus, using the high-molecular-weight µPDMS allows for the rapid functionalization of the liquid-like surface, providing superior slipperiness and enhancing the antifouling performance.

The change in amino density on the PDMS surface offers a precise method for monitoring the reaction process during modification. As shown in Figure 2a, after the modification of DSC for 60 min, the surface amino density decreased from 6.72 nmol cm^{-2} to a minimum of 0.68 nmol cm^{-2}. Following the subsequent µPDMS modification for the same duration, the amino density nearly recovered to its original value (6.61 nmol cm^{-2}), indicating a relatively complete reaction and justifying the choice of 60 min as the optimal reaction time. To further quantify µPDMS grafting, we employed QCM-D for a single-cycle modification of 60 min. As shown in Figure 2b, the quantification results reveal that the high-molecular-weight µPDMS achieved a grafting amount of 433.6 ng cm^{-2}, whereas the low-molecular-weight µPDMS was only 45.3 ng cm^{-2}, approximately 10% of the former. This suggests successful µPDMS modification and demonstrates that the high-molecular-weight µPDMS yields longer polymer brushes within the same time frame. Long enough chains for the polymer brushes are one of the essential conditions for the construction of liquid-like surfaces. Therefore, the utilization of µPDMS with high molecular weight enables the rapid functionalization of the liquid-like surfaces. The surfaces' chemical properties after each modification step were confirmed using XPS. For the PDMS modified with APTES, the high-resolution C1s spectrum could be fit into the peaks of C-H at 284.9 eV and C-N at 285.3 eV. After being modified with DSC, the characteristic peak of C=O at 287.8 eV appeared, which provided evidence of successful modification with DSC. Finally, when µPDMS modified PDMS, the characteristic peak of C-Si at 282.4 eV appeared, which served as evidence of successful modification by µPDMS chains (Figure 2c).

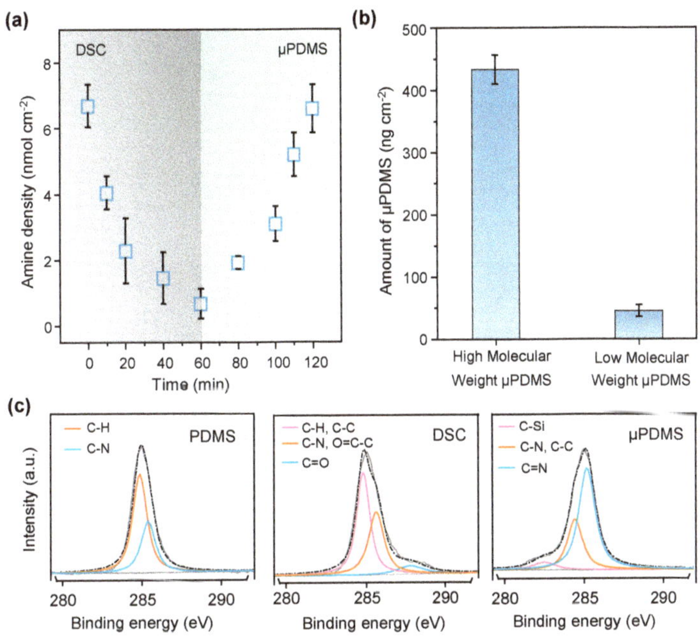

Figure 2. Physicochemical characterizations during the modification processes of liquid-like surfaces. (**a**) The amount of amine groups was determined by the acid orange method during the DSC and µPDMS grafting of 60 min. (**b**) The amount of µPDMS on the liquid-like surfaces prepared from the one-cycle grafting of high- and low-molecular-weight µPDMS (mean ± SD, n = 3). (**c**) The fitting curve of the peaks of C1s of XPS measurement before and after the modification of DSC and µPDMS.

Bacteria–material interactions are critical for applications requiring prolonged cellular proliferation in situ on the PDMS surface, as they may lead to bacterial adhesion and the formation of continuous bacterial biofilms [32], potentially compromising experimental accuracy [33]. The high flexibility of μPDMS imparts a liquid-like surface with reduced non-specific interaction forces for bacteria. This property renders the liquid-like surfaces highly promising for antibacterial applications. To assess these properties, we tested the interactions with *Staphylococcus aureus* and *Escherichia coli* as representative bacterial strains. As shown in Figure 3a, the solid surface is covered with a high density of bacterial colonies. In contrast, as depicted in Figure 3b, the colony density in the medium corresponding to the liquid-like surface is significantly lower than that corresponding to the solid surface. These findings indicate that the liquid-like surface possesses excellent antibacterial properties.

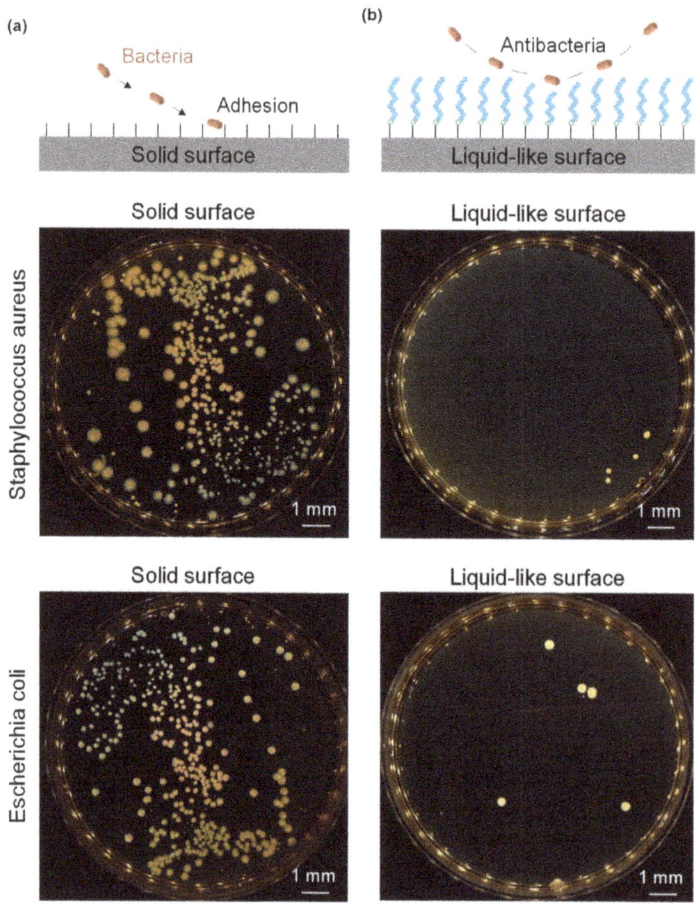

Figure 3. Colonization photographs of *Staphylococcus aureus* and *Escherichia coli* on (**a**) the solid surface and (**b**) the liquid-like surface.

Blood cell–material interactions are pivotal in applications involving blood contact with PDMS-based materials, as they can lead to platelet adhesion and the subsequent thrombus formation [34–36]. Notably, the liquid-like surface has shown reduced interactions with platelets due to the flexible nature of μPDMS, suggesting a potential for minimizing platelet adhesion. To evaluate this, we conducted whole blood adhesion experiments to compare platelet adhesion on solid and liquid-like surfaces. As depicted in the SEM images in Figure 4a, a significant number of highly activated platelets adhere to the solid surfaces. In contrast, only a minimal quantity of platelets was observed on the liquid-like surfaces, indicating a substantial reduction in the non-specific platelet adhesion. The liquid-like surfaces, resembling a smooth and slippery surface, effectively repel platelets, preventing their easy adherence. Statistical analysis further demonstrates that platelet adhesion on the liquid-like surface was only 3% of that observed on the solid surfaces (Figure 4b).

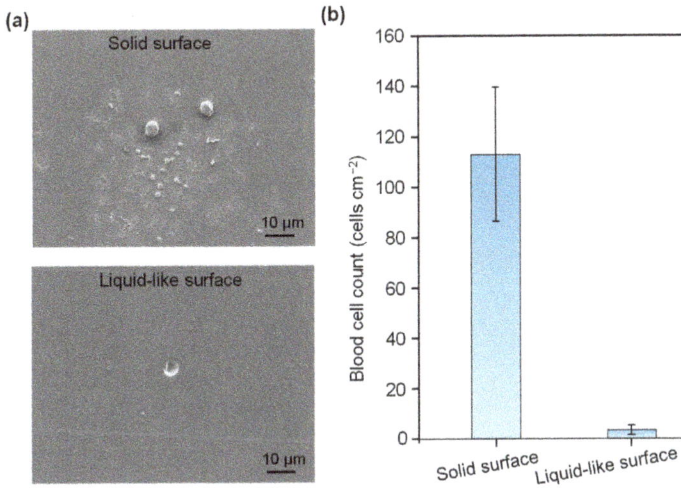

Figure 4. Adhesion of blood cells on solid and liquid-like surfaces. (**a**) SEM characterizations; (**b**) the number of adhered blood cells on both surfaces (mean ± SD, $n = 3$).

The liquid-like surface not only reduces blood cell adhesion but also has the potential to decrease the reliance on anticoagulants, such as heparin, in PDMS-based microfluidic devices. This capability helps maintain high cell viability for target cells. In this study, we used dovetail clips to fix the PDMS and the glass slide together to form the closed microchannels of the PDMS-based microfluidic chips (Figure 5a), reducing anticoagulant use while preventing blood cell adhesion. To simulate a complex physiological environment in vivo, fresh human blood with a low dose of heparin (0.25 U/mL) was introduced into the microfluidic chip to evaluate its anticoagulant performance. The SEM analysis revealed that the liquid-like surface significantly inhibited thrombus formation on its surface. In contrast, the solid surfaces exhibited severe clot formation, including activated platelets, red blood cells, and cross-linked fibrin (Figure 5b). The formation of thrombus may be attributed not only to the low dose of heparin but also to the low flow rate, as the low flow rate generates a small shear force that increases the adhesion of blood cells, thus contributing to thrombus formation [1]. On the liquid-like surface, only a few platelets in a resting, non-activated state were detected (Figure 5c). Thus, the liquid-like interfacial design in a PDMS-based microfluidic device can effectively reduce the need for anticoagulants while preserving antithrombotic properties.

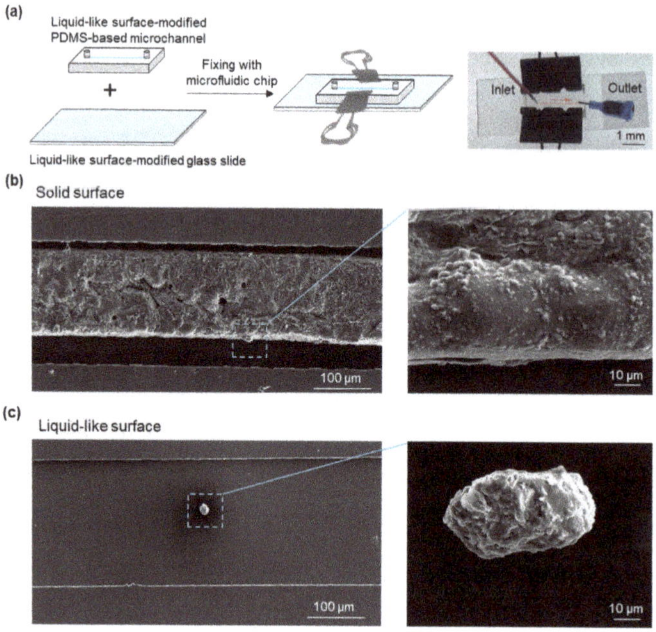

Figure 5. (**a**) Schematics of a closed microfluidic chip formed by fixing the PDMS and the glass slide together with dovetail clips and an image of the microfluidic chip after injection. Arrow shows the flow direction. The overall (left) and locally magnified views (right) of the SEM images of the microfluidic channels with (**b**) solid and (**c**) liquid-like interfacial design after the injection of fresh human blood.

4. Discussion

This work proposes a rapid construction method for liquid-like surfaces via single-cycle polymer brush grafting for enhanced antifouling in microfluidic systems. One advantage is that the modification strategy for rapidly constructing liquid-like surfaces is universal. The grafting process involves a ring-opening reaction between succinimide groups and amino groups on the glass surface. For materials, whether organic or inorganic, if there are amino groups on the surface, the liquid-like surface can be successfully modified. Another advantage is that ethanol is used as a green solvent to construct the liquid-like surface by grafting μPDMS. This strategy not only effectively resists cell adhesion but also exhibits superior biocompatibility.

This work not only broadens the versatility of liquid-like surfaces but also lays the foundation for future exploration in a wide range of biomedical applications. For instance, in biosensors, the liquid-like surfaces can improve sensitivity and accuracy by minimizing fouling. Future research could focus on optimizing the grafting process for different materials and applications, exploring the long-term stability and durability of the liquid-like surfaces, and investigating their performance in complex biological environments. This will further expand the applicability of this innovative approach in the biomedical field.

5. Conclusions

This study reports a rapid liquid-like surface functionalization strategy by the single-cycle grafting of high-molecular-weight μPDMS on the PDMS surfaces. The incorporation of this liquid-like surface imparts several advantageous properties to the PDMS, including a low sliding angle for blood droplets, excellent antibacterial characteristics, and reduced platelet adhesion. Additionally, the liquid-like surface improves PDMS-based microfluidics by not only diminishing blood cell adhesion and thrombus formation but also minimizing the requirement for heparin. Looking ahead, the liquid-like surfaces, which feature amino groups, offer the potential for further chemical modification or functionalization. For example, it allows for the grafting of specific ligands targeting the epithelial cell adhesion molecule, facilitating the capture of target cells while minimizing non-specific blood cell adhesion. This suggests that the liquid-like surface holds considerable promise for applications in cellular analysis and the assessment of gene or protein expression levels in target cells.

Author Contributions: F.W. and S.W. conceived the idea. F.W., D.Z. and S.W. designed the research. F.W., J.X., L.Z., Y.L. (Yixuan Liu), H.S., W.W. and D.Z. performed the experiments. F.W., J.X., L.Z., Y.L. (Yixuan Liu), H.S., W.W., F.C., D.Z. and S.W. analyzed and interpreted the results. F.W., J.X., D.Z. and S.W. drafted the manuscript, and all the authors contributed to the writing of the manuscript. F.W. and J.X. contributed equally to this work. All authors have read and agreed to the published version of the manuscript.

Funding: This work is supported by the National Natural Science Foundation of China (22005255), Fundamental Research Funds for the Central Universities (20720220085), and the Xiamen Natural Science Foundation Youth Project (3502Z202471002).

Data Availability Statement: The original contributions presented in the study are included in the article, further inquiries can be directed to the corresponding authors.

Acknowledgments: F.W. and F.C. acknowledge the Jiangsu Flexible Energy Storage Technology Engineering Center.

Conflicts of Interest: The authors declare no conflicts of interest.

References

1. Wu, F.; Chen, X.; Wang, S.; Zhou, R.; Wang, C.; Yu, L.; Zheng, J.; Yang, C.; Hou, X. Green synthesized liquid-like dynamic polymer chains with decreased nonspecific adhesivity for high-purity capture of circulating tumor cells. *CCS Chem.* **2024**, *6*, 507–517. [CrossRef]
2. Li, K.; Lu, X.; Liao, J.; Chen, H.; Lin, W.; Zhao, Y.; Tang, D.; Li, C.; Tian, Z.; Zhu, Z. DNA-DISK: Automated end-to-end data storage via enzymatic single-nucleotide DNA synthesis and sequencing on digital microfluidics. *Proc. Natl. Acad. Sci. USA* **2024**, *121*, e2410164121. [CrossRef] [PubMed]
3. Chen, X.; Ding, H.; Zhang, D.; Zhao, K.; Gao, J.; Lin, B.; Huang, C.; Song, Y.; Zhao, G.; Ma, Y. Reversible immunoaffinity interface enables dynamic manipulation of trapping force for accumulated capture and efficient release of circulating rare cells. *Adv. Sci.* **2021**, *8*, 2102070. [CrossRef] [PubMed]
4. Wu, F.; Kong, X.; Liu, Y.; Wang, S.; Chen, Z.; Hou, X. Microfluidic-based isolation of circulating tumor cells with high-efficiency and high-purity. *Chin. Chem. Lett.* **2024**, *35*, 109754. [CrossRef]
5. Xu, S.; Liu, Y.; Yang, Y.; Zhang, K.; Liang, W.; Xu, Z.; Wu, Y.; Luo, J.; Zhuang, C.; Cai, X. Recent progress and perspectives on neural chip platforms integrating PDMS-based microfluidic devices and microelectrode arrays. *Micromachines* **2023**, *14*, 709. [CrossRef]
6. Chen, Z.; Lv, Z.; Zhang, Z.; Weitz, D.; Zhang, H.; Zhang, Y.; Cui, W. Advanced microfluidic devices for fabricating multi-structural hydrogel microsphere. *Exploration* **2021**, *1*, 20210036. [CrossRef]
7. Wang, K.; Huang, K.; Wang, L.; Lin, X.; Tan, M.; Su, W. Microfluidic strategies for encapsulation, protection, and controlled delivery of probiotics. *J. Agric. Food Chem.* **2024**, *72*, 15092–15105. [CrossRef]
8. Lovegrove, J.T.; Kent, B.; Förster, S.; Garvey, C.J.; Stenzel, M.H. The flow of anisotropic nanoparticles in solution and in blood. *Exploration* **2023**, *3*, 20220075. [CrossRef]
9. Saxena, S.; Lu, Y.; Zhang, Z.; Li, Y.; Soleymani, L.; Hoare, T. Zwitter-repel: An anti-fouling coating promoting electrochemical biosensing in biological fluids. *Chem. Eng. J.* **2024**, *495*, 153522. [CrossRef]
10. Hou, X.; Li, J.; Tesler, A.B.; Yao, Y.; Wang, M.; Min, L.; Sheng, Z.; Aizenberg, J. Dynamic air/liquid pockets for guiding microscale flow. *Nat. Commun.* **2018**, *9*, 733. [CrossRef]
11. Cirillo, A.I.; Tomaiuolo, G.; Guido, S. Membrane fouling phenomena in microfluidic systems: From technical challenges to scientific opportunities. *Micromachines* **2021**, *12*, 820. [CrossRef] [PubMed]
12. Yao, S.; Yan, H.; Tian, S.; Luo, R.; Zhao, Y.; Wang, J. Anti-fouling coatings for blood-contacting devices. *Smart Mater. Med.* **2024**, *5*, 166–180. [CrossRef]
13. Mercader, A.; Ye, S.-H.; Kim, S.; Orizondo, R.A.; Cho, S.K.; Wagner, W.R. PDMS-zwitterionic hybrid for facile, antifouling microfluidic device fabrication. *Langmuir* **2022**, *38*, 3775–3784. [CrossRef]

14. Leslie, D.C.; Waterhouse, A.; Berthet, J.B.; Valentin, T.M.; Watters, A.L.; Jain, A.; Kim, P.; Hatton, B.D.; Nedder, A.; Donovan, K. A bioinspired omniphobic surface coating on medical devices prevents thrombosis and biofouling. *Nat. Biotechnol.* **2014**, *32*, 1134–1140. [CrossRef]
15. Tong, P.; Chen, L.; Sun, X.; Li, H.; Feng, Y.; Li, J.; Guan, S. Surface modification of biodegradable magnesium alloy with poly (L-lactic acid) and sulfonated hyaluronic acid nanoparticles for cardiovascular application. *Int. J. Biol. Macromol.* **2023**, *237*, 124191. [CrossRef] [PubMed]
16. Sahin, F.; Celik, N.; Ceylan, A.; Pekdemir, S.; Ruzi, M.; Onses, M.S. Antifouling superhydrophobic surfaces with bactericidal and SERS activity. *Chem. Eng. J.* **2022**, *431*, 133445. [CrossRef]
17. Wu, F.; Liu, Y.; Li, J.; Zhang, K.; Chong, F. Bioinspired strategies for functionalization of mg-based stents. *Crystals* **2022**, *12*, 1761. [CrossRef]
18. Wu, F.; Liu, Y.; Xu, J.; Pan, C. Bioinspired surface design for magnesium alloys with corrosion resistance. *Metals* **2022**, *12*, 1404. [CrossRef]
19. Rasitha, T.; Vanithakumari, S.; Krishna, D.N.G.; George, R.; Srinivasan, R.; Philip, J. Facile fabrication of robust superhydrophobic aluminum surfaces with enhanced corrosion protection and antifouling properties. *Prog. Org. Coat.* **2022**, *162*, 106560. [CrossRef]
20. Shi, R.; Shang, L.; Zhou, C.; Zhao, Y.; Zhang, T. Interfacial wettability and mass transfer characterizations for gas–liquid–solid triple-phase catalysis. *Exploration* **2022**, *2*, 20210046. [CrossRef]
21. Wu, F.; Xu, J.; Wang, Z.; Jiang, J.; Liu, Y.; Zhang, N.; Zhang, K.; Chong, F.; Li, J. Functional liquid-infused PDMS sponge-based catheter with antithrombosis, antibacteria, and anti-inflammatory properties. *Colloids Surf. B Biointerfaces* **2023**, *224*, 113208. [CrossRef] [PubMed]
22. Bai, M.; Tian, X.; Wang, Z.; Zhang, L.; Zhang, F.; Yang, Y.; Liu, L. Versatile dynamic bioactive lubricant-infused surface for effective isolation of circulating tumor cells. *Anal. Chem.* **2023**, *95*, 5307–5315. [CrossRef] [PubMed]
23. Chen, X.; Wen, G.; Guo, Z. What are the design principles, from the choice of lubricants and structures to the preparation method, for a stable slippery lubricant-infused porous surface? *Mater. Horiz.* **2020**, *7*, 1697–1726. [CrossRef]
24. Cheng, X.; Zhao, R.; Wang, S.; Meng, J. Liquid-like surfaces with enhanced de-wettability and durability: From structural designs to potential applications. *Adv. Mater.* **2024**, *36*, 2407315. [CrossRef]
25. Yang, C.; Wu, Q.; Zhong, L.; Lyu, C.; He, G.; Yang, C.; Li, X.; Huang, X.; Hu, N.; Chen, M. Liquid-like polymer-based self-cleaning coating for effective prevention of liquid foods contaminations. *J. Colloid Interface Sci.* **2021**, *589*, 327–335. [CrossRef]
26. Huang, S.; Li, J.; Liu, L.; Zhou, L.; Tian, X. Lossless fast drop self-transport on anisotropic omniphobic surfaces: Origin and elimination of microscopic liquid residue. *Adv. Mater.* **2019**, *31*, 1901417. [CrossRef]
27. Chen, Y.; Yu, X.; Chen, L.; Liu, S.; Xu, X.; Zhao, S.; Huang, S.; Tian, X. Dynamic poly(dimethylsiloxane) brush coating shows even better antiscaling capability than the low-surface-energy fluorocarbon counterpart. *Environ. Sci. Technol.* **2021**, *55*, 8839–8847. [CrossRef]
28. Zhang, L.; Guo, Z.; Sarma, J.; Dai, X. Passive removal of highly wetting liquids and ice on quasi-liquid surfaces. *ACS Appl. Mater. Interfaces* **2020**, *12*, 20084–20095. [CrossRef]
29. Kou, F.; Liu, C.; Wang, L.; Yasin, A.; Li, J.; Guan, S. Fabrication of citric acid/RGD multilayers on Mg-Zn-Y-Nd alloy via layer-by-layer self-assembly for promoting surface biocompatibility. *Adv. Mater. Interfaces* **2021**, *8*, 2002241. [CrossRef]
30. Hu, Q.; Du, Y.; Bai, Y.; Xing, D.; Wu, C.; Li, K.; Lang, S.; Liu, X.; Liu, G. Smart zwitterionic coatings with precise pH-responsive antibacterial functions for bone implants to combat bacterial infections. *Biomater. Sci.* **2024**, *12*, 4471–4482. [CrossRef]
31. Torma, V.; Gyenes, T.; Szakács, Z.; Noszál, B.; Némethy, Á.; Zrínyi, M. Novel amino acid-based polymers for pharmaceutical applications. *Polym. Bull.* **2007**, *59*, 311–318. [CrossRef]
32. Qin, X.-H.; Senturk, B.; Valentin, J.; Malheiro, V.; Fortunato, G.; Ren, Q.; Rottmar, M.; Maniura-Weber, K. Cell-membrane-inspired silicone interfaces that mitigate proinflammatory macrophage activation and bacterial adhesion. *Langmuir* **2018**, *35*, 1882–1894. [CrossRef] [PubMed]
33. Feng, C.; Tan, P.; Nie, G.; Zhu, M. Biomimetic and bioinspired nano-platforms for cancer vaccine development. *Exploration* **2023**, *3*, 20210263. [CrossRef] [PubMed]
34. Torres-Castro, K.; Acuña-Umaña, K.; Lesser-Rojas, L.; Reyes, D.R. Microfluidic Blood Separation: Key Technologies and Critical Figures of Merit. *Micromachines* **2023**, *14*, 2117. [CrossRef]
35. Corral-Nájera, K.; Chauhan, G.; Serna-Saldívar, S.O.; Martínez-Chapa, S.O.; Aeinehvand, M.M. Polymeric and biological membranes for organ-on-a-chip devices. *Microsyst. Nanoeng.* **2023**, *9*, 107. [CrossRef]
36. Shome, A.; Martinez, I.; Pinon, V.D.; Moses, J.C.; Garren, M.; Sapkota, A.; Crutchfield, N.; Francis, D.J.; Brisbois, E.J.; Handa, H. "Reactive" chemical strategy to attain substrate independent "liquid-like" omniphobic solid anti-biofouling coatings. *Adv. Funct. Mater.* **2024**, *34*, 2401387. [CrossRef]

micromachines

MDPI

Article

Effects of Fiber Arrangement on Flow Characteristics Along a Four-Fiber Element of Fiber Extractors

Oluwaseyi O. Ayeni [1], Holly A. Stretz [1,*] and Ahmad Vasel-Be-Hagh [2]

[1] Department of Chemical Engineering, Tennessee Technological University, Cookeville, TN 38501, USA; ayenioluseyio@gmail.com
[2] Department of Mechanical Engineering, University of South Florida, Tampa, FL 33620, USA; vaselbehagh@usf.edu
* Correspondence: hstretz@tntech.edu

Abstract: Fiber extractors, as process-intensified equipment, facilitate many applications, such as the purification of oils. The development of high-fidelity computational models is crucial to optimize the design. However, simulating microscale flows around tens of thousands of microfiber arrays is computationally unfeasible. Thus, it is necessary to identify smaller elements, consisting of only a few fibers, that can represent flow within massively arrayed fiber extractors. This study employed computational fluid dynamics to investigate different configurations of four-fiber elements to achieve this aim. Following previous modeling featuring flow around only one fiber, the goal was to understand how variations in inter-fiber distances affect the phase structures of a corn oil/water mixture, the steady-state interfacial surface area per unit of fluid volume, and the pressure drop along the flow direction. The study explored various total and relative flow rates and contact angles. The research characterized the flow as semi-restricted annular, noting the influence of neighboring fibers on phase complexity. The inter-fiber distance played a crucial role in generating high interfacial areas and reducing pressure. The chaotic nature of the slug interfaces facilitated intermixing between flows along different fibers. Interestingly, the specific interfacial area reached an optimum when the inter-fiber distance was between 10 and 50 μm.

Keywords: process intensification; advanced manufacturing; microfluidic; immiscible flow

1. Introduction

Bioprocessing relies significantly on purification processes, one of which is liquid/liquid extraction (one example: aqueous two-phase extraction for high-value pharmaceuticals) [1–8]. The early development of large interfacial surface areas affects the efficiency of such separation mechanisms. Conventional contactors such as mixer settlers, centrifugal contactors, agitated columns, and static mixers can facilitate relatively large interfacial areas. However, the complex hydrodynamics involved in typical contactors create a challenge in defining the interfacial surface areas [9,10].

Microfluidic devices, such as micro-structured reactors (MSRs) [11], have emerged as industrially viable alternatives to facilitate large interfacial surface areas, resulting in higher energy efficiency and product purity and lower waste, as desired by the biofeedstock process industry (example: purification of vegetable oils [7]). However, parallel processing of microfluidic tubes is challenging due to the complexities of scaling up to commercial throughput. Some approaches involve bundling parallel microfluidic tubes in order to increase throughput; however, process control of such bundles (when scales are at thousands or more of microfluidic channels operating in parallel) remains a challenge [11–13]. Bundled designs require the numbered microchannels to be separated, which makes it complex to control and maintain the uniformity of flow and heat transfer across the device. In addition, bundled designs demand the construction of special junctions [14,15].

Fiber extractors, FEs, have emerged as a novel and viable solution to these challenges as they preserve the sustainable benefits of microfluidics processing while working readily with current infrastructure for process control and scale-up [16–19]. (Note that a different type of processing, contactors with parallel hollow fibers where fluids flow through the fiber hollow, is being reported for areas such as carbon capture [20]. The fluid flows in these contactors are not the same as in the present work. Flows <u>inside</u> a hollow fiber experience "wall" constraints. Flows <u>around</u> a fiber do not experience "walls" in every direction radially outward, thus the need for modeling.) FEs can feature ~100,000 microfibers, even in a pipe as small as one inch in diameter. Due to massive arrays of microscale passages forming between these fibers for the fluids to flow through, FEs can achieve high production rates. Further, heat transfer and even electrification of a reaction become facile in process control by using metal fibers, such as steel fibers. Ayeni et al. have provided an expanded review of the state of the art of FE processing [21].

Designing an efficient FE requires a thorough understanding of its flow characteristics. Previous studies have primarily concentrated on varying process parameters to examine extraction efficiencies and overall mass transfer coefficients [22–25]. Consequently, there is a dearth of knowledge about flow dynamics through FEs, especially regarding the development of phase structures within these devices. Gaining this knowledge through experimental methods is quite challenging due to the large number of microfibers in FE devices and the difficulty in visualizing the microscale phase structures that form within the space between these fibers [21]. An alternative approach is using computational fluid dynamics (CFD). However, simulating multiphase flows through thousands of microscale fibers also poses significant challenges, explaining why, to our knowledge, no multi-physics model has ever been developed that effectively captures the unique flow characteristics of FEs.

Simulating an entire fiber extractor (FE) containing thousands of microfibers is computationally impractical. However, we can gain insights into the flow characteristics within such devices by simulating the flow around smaller components, referred to as a (potentially) representative volume element (RVE) hereafter. While these smaller simulations may not fully represent the entire system, they can still provide valuable information about flow along the microfibers. In a recent study, we reported a simulation of a core, one-fiber scenario where the RVE was conceptualized as a micron-scale rectangular area with a single 50 μm diameter fiber located at its center [21]. Two different materials, corn oil, and water, flowed along this fiber through the RVE, with the fiber treated as a wetted wall and the RVE sides functioning as symmetry planes. While the earlier one-fiber study shed light on how the phase structures formed and interacted along a single fiber, it lacked the influence of neighboring fibers.

Therefore, the current study builds upon that idealized one-fiber RVE [21] to include four fibers to understand how neighboring fibers potentially influence the flow and phase structures. Similar to the previously reported single-fiber case, this study also employs water and corn oil feeds as the working fluids, a situation commercially relevant to biodiesel production and vegetable oil refining. This study examines the effects of inter-fiber distance (i.e., packing density), contact angle, and flow rate on the evolution of phase structures, pressure drop, and specific interfacial areas. Specific interfacial surface area is a volume-normalized surface area that facilitates a fair comparison between interfacial surface areas produced within various representative volume elements. Interfacial surface area can be employed in mass transfer calculations to predict extraction efficiencies. The surface area is also useful for designing downstream coalescence mechanisms.

This study modeled the fluid environment via a finite element method (FEM), the details of which are presented in Section 2. The results of this computational analysis are then presented and discussed in Section 3, followed by a series of concluding remarks and recommendations for future work.

2. Materials and Methods, Numerical Analysis

2.1. The Configurations

Figure 1a shows a schematic of a 2D section of a fiber extractor's vessel when packed with arrayed fibers. The simulations reported in this study explored the flow and interactions of water and oil along representative volume elements (RVEs) that include four microfibers to take a step toward developing a better understanding of flow around solid fibers within fiber extractors. The study employed square RVEs comprised of four equally spaced fibers with a characteristic inter-fiber distance, "W" (Figure 1b). Another characteristic length is the distance between the fiber's edge and the RVE's edge, shown as "a" (Figure 1b). At the four outer edges of the RVE, symmetry conditions were applied. The inter-fiber distance was then varied across various cases using two approaches, i.e., constant dimensions, hereafter "CD", and constant voidage, hereafter "CV", resulting in inter-fiber distances ranging from 40 to 114 μm and fiber diameters between 5 and 59 μm. In all the cases studied, the fiber length was fixed at 1 cm, which was determined by the required computational resources. The high-fidelity simulations included in this research demanded several thousand CPU hours per case when utilizing a high-performance computing cluster. The term "voidage" is used to mean the volume around the fibers inside of the RVE available for fluid flow.

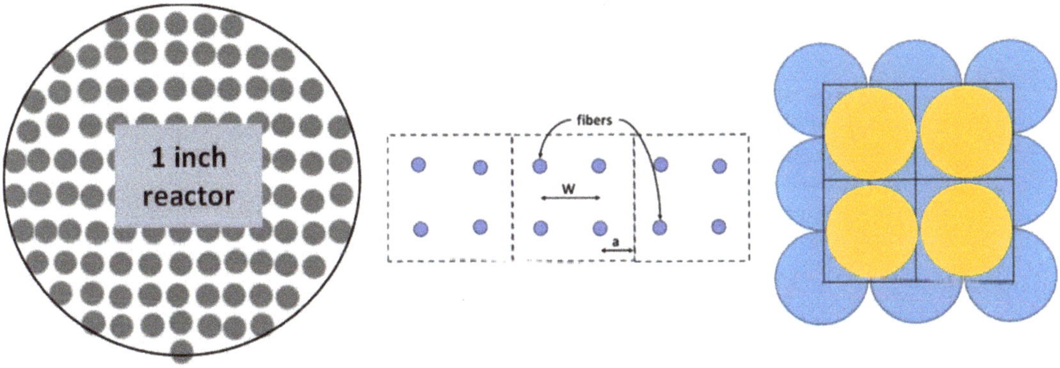

(**a**) (**b**) (**c**)

Figure 1. (**a**) A representation of a 2D cross-section of a hypothetical fiber extractor vessel. (**b**) Tessellating square-shaped RVEs containing four fibers, with W representing the inter-fiber distance and a being one-half of W. (**c**) Pictorial representation of channels or pathways formed by the fibers.

Using method "CD", the dimensions of the RVE were held constant while changing the diameter of the fiber from 40 to 70 μm (Table 1). Consequently, the inter-fiber distance and voidage also varied with these changes. If these variations were configured for a hypothetical extractor with a specific vessel diameter, all cases generated using the CD method would result in the same number of fibers (e.g., 87,700 fibers in a one-inch-diameter vessel, as shown in Table 1). Therefore, all scenarios created through the CD method can be considered as belonging to the same FE, sharing the same vessel diameter and the same number of fibers, but differing in fiber diameters and, consequently, in the dimensions of the flow pathways that form between the fibers (Figure 1c). Note that the fluids flowing in these pathways can intermix more with larger void space because these pathways are only semi-restricted, i.e., touching fiber surfaces but not on all sides. Also, note that for method "CD" to be a valid packing configuration, the four fibers had to fit inside the RVE without touching. Therefore, an 80 µm fiber diameter would be invalid. Table 1 summarizes the different RVE geometries modeled using method CD and presents how the changes in fiber diameter resulted in various packing densities, void percentages, fiber counts, and inter-fiber spacings.

Table 1. Fiber packing models.

Method CD: Varied Inter-Fiber Distance by Changing Voidage and Keeping the Number of Fibers Constant (Voidage Is the Space Between Fibers Where Fluid Flows)			
Fiber Diameter (µm)	**Voidage**	**Inter-Fiber Distance (µm)**	**Number of Fibers**
40	0.78	36	87,700
50	0.66	26	87,700
60	0.51	15	87,700
70	0.33	5	87,700
Method CV: Varied Inter-Fiber Distance by Keeping the Voidage Constant			
Fiber Diameter	**Voidage**	**Inter-Fiber Distance (µm)**	**Number of Fibers**
40	0.66	21	137,000
60	0.66	31	60,900
70	0.66	36	44,770
114	0.66	59	16,880

In method "CV", on the other hand, the voidage was held constant at 66%, identical to that of the previously studied single-fiber case [21]. Thus, the RVE's size changed when changing fiber diameters to maintain the same voidage. Also, keeping the voidage constant means that the fluid volume remains unchanged for various fiber diameters, although inter-fiber distance changes. Unlike the method "CD", this method could employ any fiber diameter without fiber surfaces coming into contact (Table 1).

The relationship between the voidage and the inter-fiber distance is presented graphically in Figure 2 for the two methods. As the voidage increased, so did the inter-fiber distance in method "CD", but in method "CV", voidage was constant. Figure 2 also shows the variation of the capillary number of the oil phase with inter-fiber distance. The capillary number changes in the "CD" method because it depends on the fluid velocity, which depends on the RVE area at a given constant flow rate. However, in the "CV" method, the flow area is always constant; hence, the capillary number remains unchanged. Note that since the fluid volume changes between some cases, we have reported all calculated interfacial surface areas as "specific surface area" by normalizing it via the fluid volume in the RVE. This also allowed comparing four-fiber model results to those previously reported for the single-fiber case. To study the effects of packing density on the phase structures and interfacial areas, models were run using the parameters provided in Table 1.

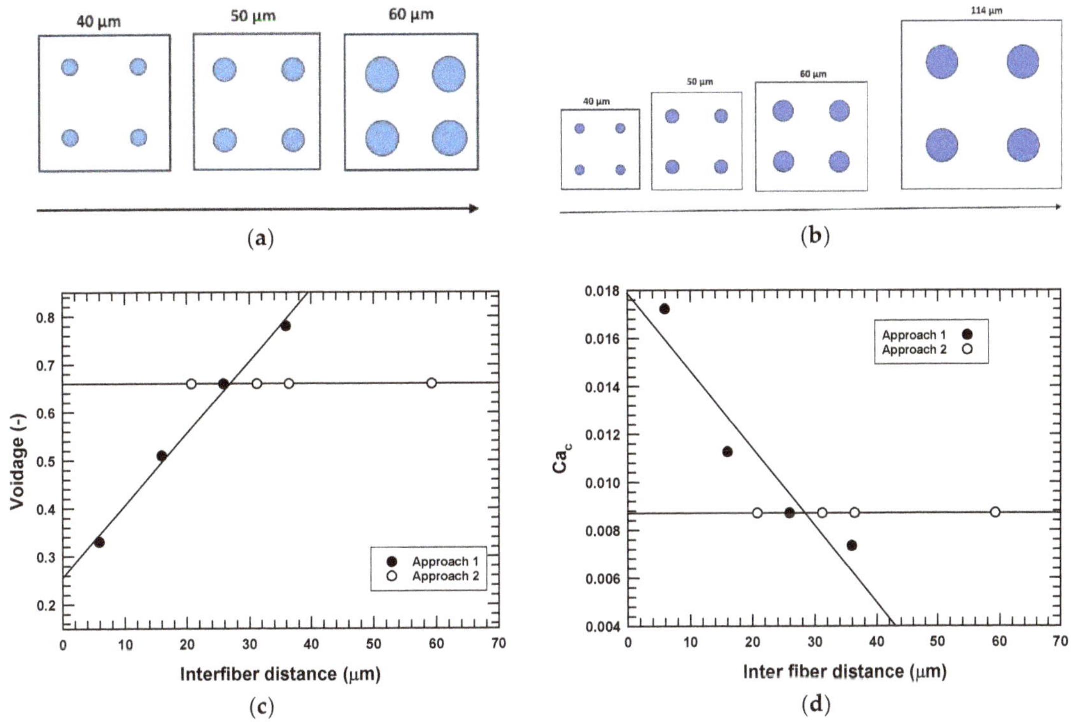

Figure 2. Comparison of approaches to packing the fibers into the fiber reactor and inter-fiber distances. (**a**) "CD" method, varying the fiber diameter but not the representative volume element dimensions, and (**b**) "CV" method varying the fiber diameter and the cell dimensions for constant voidage; (**c**) the voidage is plotted as a function of the inter-fiber distance; (**d**) effect of inter-fiber distance on capillary number.

2.2. Domain and Boundary Conditions

The inlet boundary and initial conditions are shown in Figure 3. The two phases were initialized by placing the fluid flows parallel to the fibers, with the oil phase in the inner and the water phase in the outer regions. The meshing sequence was created by meshing the inlet face with a triangular mesh and sweeping that inlet face down the entire domain.

A commercial fiber reactor for vegetable oil refining, using corn oil and water, was used as the target application. Water is represented by the subscript "w" and corn oil by the subscript "o". Table 2 presents the physical properties of the corn oil and water phases. For all studies except those for which the contact angle was the independent variable, a wetted wall boundary condition with a contact angle of 10° was imposed on the fiber walls. This choice was made based on Santos et al. [26], where corn oil made a contact angle of 10° on stainless steel in supercritical CO_2 applications. Symmetry boundary conditions were imposed on the outer boundaries of the RVE. The symmetry boundary condition was a semi-restricted flow condition discussed extensively in a previous study [21]. The velocities of the water and corn oil entering the RVE were selected to represent the experimental conditions for a tessellated 1-inch diameter fiber extractor vessel. Atmospheric pressure was specified as the outlet boundary condition at the outlet of the RVE.

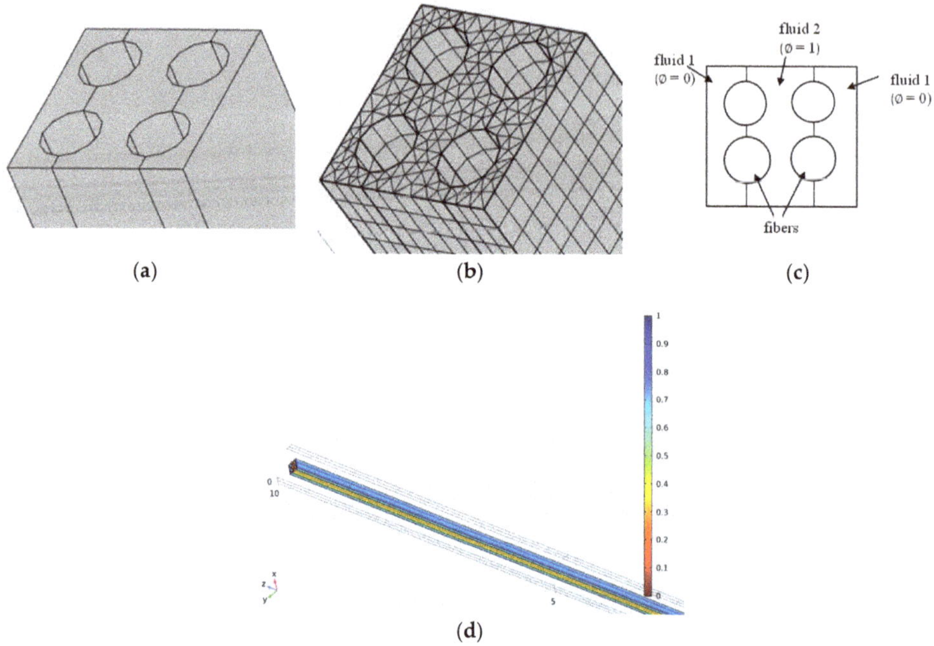

Figure 3. Schematic of (**a**) modeling domain at the inlet of the model showing the fluid volume and fibers, (**b**) meshed geometry, and (**c**) initial and inlet boundary conditions. Mid-planes drawn across the fibers in (**a**) and (**c**) are used to initialize the fluid phase boundaries (ϕ), (**d**): initialized phases of oil (red) and water (blue). The phases were initialized in a parallel manner adjacent to the fibers.

Table 2. Fluid properties and process conditions.

Density of water	1000 kg/m^3
Density of corn oil	920 kg/m^3
Surface tension @ temp = 35 °C	0.0316 N/m
Flow rate of water	45–150 mL/min
Flow rate of oil	75–150 mL/min
Viscosity of water	0.001 Pa.s
Viscosity of corn oil @ temp = 35 °C	0.0368 Pa.s

2.3. Governing Equations and Solution Method

The Coupled Level Set (LS)–Volume of Fluid (VOF) method has been proven suitable for tracking the interface between the materials and their associated phases, with LS capturing the interface and VOF modeling different phases or materials. LS-VOF also has a track record of successfully modeling microscale flows within both laminar and turbulent regimes [27–29]. Thus, this study simulated oil and water interactions along microfibers using a coupled LS-VOF model provided via COMSOL Microphysics, a commercially available CFD tool (version 5.6 (build 280), 6.0 (build 405), and 6.1 (build 357)). The level set model solved the transport equation for the level set function ϕ. The level set function (ϕ) was convected along with the flow using the same velocity field specified in the laminar flow interface. One set of the Navier–Stokes (N-S) equations was employed to solve the velocity, pressure, and viscous force fields for both phases. The fluid properties at the fluid–fluid interface were obtained using a weighted averaged function of the level set variable. The LS-VOF setup was identical to that used by the authors in their previous study of flow along a single microfiber [21], with the difference being that the present study explores water–oil interactions along four equidistant fibers, which would not affect the formulation of the governing equations and the numerical models applied to solve them. The specific equations are shown in detail in Supplemental Information.

2.4. Mesh Studies

The computational domain was discretized using a free tetrahedral mesh with a maximum element size of 10.9 μm, giving a total of 392,128 elements. Using the finite element method, the transient partial differential equations were converted to ordinary differential equations, and then from there, the temporal derivatives were discretized using the implicit Backward Differentiation Formula with a maximum order of 2 using a free time stepping scheme, resulting in Difference Algebraic Equations (DAEs). The average time step Δt used was about 100 μs. A linear iterative solver was employed to solve the DAEs iteratively until convergence was obtained. The convergence criterion was the relative tolerance and set at a value of 0.01. The model was run for 5 s with outputs printed every 0.1 s.

Meshing validation was performed to verify that the results were mesh-independent. Mesh convergence was performed for a study of the dilution of salt solution with pure water involving one fiber in an RVE. Table 3 and Figure S3 are summaries of the different meshing geometries modeled for mesh validation. The fine mesh and finer mesh were quite comparable. However, since the finer mesh took much more time to converge (up to one month), the fine mesh was employed.

Table 3. Results of meshing validation studies.

	Normal	Fine	Finer
Number of elements	133,532	294,438	949,482
Minimum element size	3.9×10^{-4} m	1.96×10^{-4} m	7.83×10^{-5} m
Maximum element size	1.3×10^{-3} m	1.04×10^{-3} m	7.24×10^{-4} m
Average quality (skewness)	65.99	66.36	66.2

3. Results and Discussion

In the following analysis, we first compare the one-fiber and four-fiber model results to understand their differences and whether a computationally simple one-fiber model is representative. We then present the inter-fiber distance, flow rate, and contact angle effects predicted when the neighboring fibers impact the various dependent variables such as phase structure, specific interfacial surface area, and pressure drop.

The channels formed by the fibers enclosed in a cm-scale pipe define a new type of pressure-driven microfluidic flow, here termed semi-restricted. Semi-restricted flow can be scaled to very high throughput. These boundary conditions, representing constraint only by the fiber(s) but without walls, contrast with a very new domain for microfluidics in the current literature termed "open channel microfluidics". Open-channel flows are recently driven by electrophoresis, and the volumetric throughput may be enhanced by fabricating dozens of channels in parallel [30–33]. In contrast, the semi-restricted pressure-driven flows in FEs easily have tens of thousands of parallel channels. Thus, semi-restricted microfluidics has some similarities with open-channel microfluidics but is scalable to the separations and reaction needs of commodity volume production, including vegetable oil purification, petroleum desulfurization, vaccine production, nanoparticle production and separations for nuclear species [17,18,25,34–36]. The latter three fields are reported in the literature at the lab scale but are not yet commercial, to the author's knowledge. Thus, the knowledge gap addressed here is using CFD to aid in the visualization of what types of phases develop in two-phase flows associated with microfluidic channels inside of FEs and how to control the surface area for extraction using inter-fiber distance, flow rate, and wetting or contact angle conditions. Note, however, that future work with fiber-packed modules can readily be modified to take advantage of the enhanced selectivity allowed by reactive fiber surfaces, such as is found in affinity chromatography (e.g., protein purification), or electrical fields, making the need for the base case of understanding pressure-driven flows even more compelling.

3.1. Single-Fiber Model and Four-Fiber Model Comparison

Including four fibers in the RVE instead of only one and then comparing results will illuminate how the nearest neighbor fibers might affect the flow and phase structures. A multi-fiber RVE also introduces the possibility of new independent variables: inter-fiber distance and fiber diameter. A base case was established, in common with the one-fiber model, using four 50 μm fibers with an inter-fiber distance of approximately 26 μm. To restate, four of the one-fiber models were tessellated into a larger unit cell, which was then termed the "four fiber" base case model, but with no boundary conditions inside of the new larger unit cell. Note that there was no difference in the initial conditions between the single- and four-fiber models. The flow rate of corn oil (Q_o) in the compared cases was equal to the flow rate of water (Q_W), i.e., $Q_o/Q_W = 1$. The contact angle was 10°. Interfacial tension was 0.0316 N/m, and the corn oil and water viscosities were 0.0368 Pa.s and 0.001 Pa.s, respectively.

The graphs in Figure 4a show the results of the interfacial areas plotted against the total flow rate for the single- versus four-fiber cases. The insets show that the phase structures at a select flow rate differed, indicating that neighboring fibers did affect phase structure. Thus, radial flows must have been present in the four-fiber RVE and flows between channels interacted and inter-mixed.

The four-fiber RVE showed a much more complex phase structure in all cases, which could not be readily described by slugs or slug lengths, the most typical phase structure found in the microfluidic literature. Also, the interfacial surface area increased in the presence of the near-neighbor fiber effects. The predicted increase was not in the same proportion as the increased fluid volume (which is theoretically four times the increase in volume from the one-fiber RVE and shown by the dashed line in Figure 4b). In fact, the nonlinear increase in normalized interfacial surface areas was at least 2.5 times more than theoretical. The greater interfacial area, even normalized, was attributed to the neighboring effect of the fibers in the four-fiber model. Thus, the channels numbering into hundreds or thousands in an actual fiber reactor would be expected to create more complex flows and greater specific interfacial areas than the idealized one-fiber or even four-fiber RVE. Massingill et al. [18] and Kim et al. [19] confirm these predictions. The complexity of attaining static equilibrium by spontaneous imbibition on a set of parallel fibers in two-phase flows has been previously noted [28]. Related studies by Duprat et al. [37] and Prortiere et al. reported that the spreading of a static droplet of fluid on two fibers depends on the inter-fiber distance. Duprat et al. reported that when a drop of liquid was deposited between two parallel fibers at a fixed distance apart, the evolution of the liquid interface between the fibers depended on the ratio of the distance between the fibers (d) and radii of the fibers (r), i.e., d/r. At least three regimes of phase or droplet evolution were observed. The existence of three regimes of phase development dependent on inter-fiber distance indicates, importantly, that there could be one more flow regime beyond microfluidic versus pipe flow.

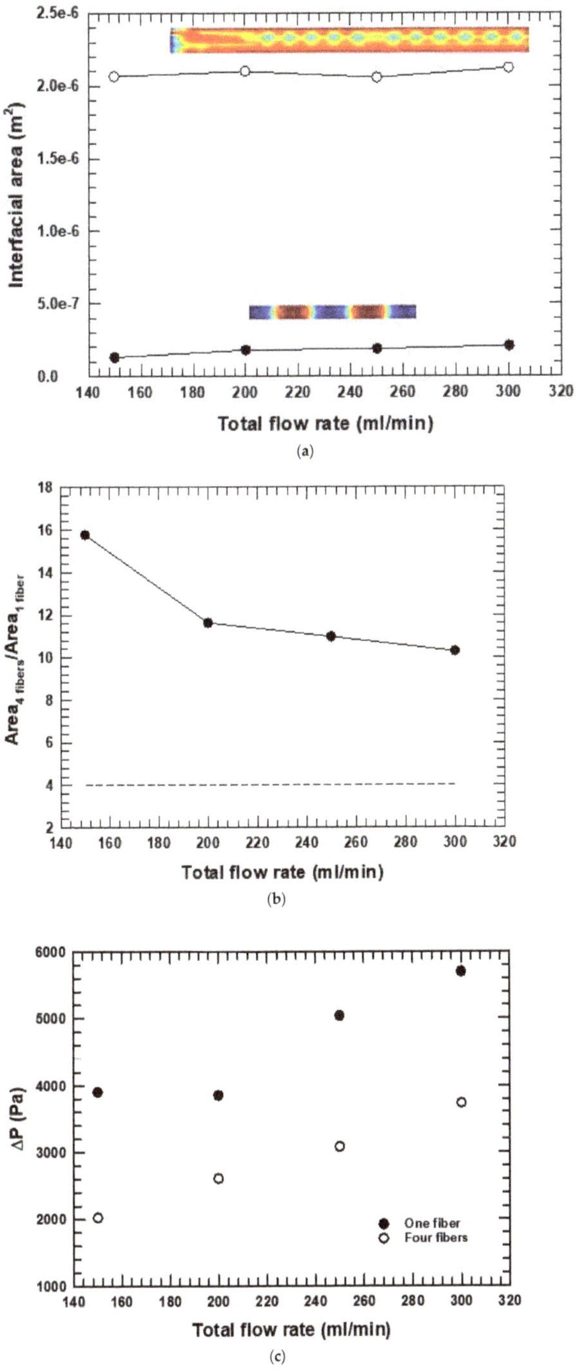

Figure 4. Comparison of single-fiber model and the four-fiber models using the total flow rate as a case study. (**a**) Interfacial areas vs. flow rate; (**b**) area scale factor from one fiber to four fibers; (**c**) pressure drop from one fiber to four fibers.

The Reynolds number (Re) in the present study was significantly small (Re << 1) and would not account for the phase complexity and mixing observed. Clearly, multi-dimensional flow paths were present in the fiber extractor. Hence, the fiber extractor might provide excellent guidance to one possible solution to the issues associated with microfluidic mixing [12]. Allowing creeping flows to interact may lead to localized mixing.

Figure 4c compares the pressure drops for single-fiber and four-fiber models as a function of the total flow rate. In both models, the pressure drop increased with the flow rate. However, at a given constant total flow rate, the pressure drop in the four-fiber model was at least 70% lower than in the single-fiber model, consistent with extra degrees of freedom for flow in the four-fiber case. Note that phase structures, previously reported as "slugs", and which changed with the total flow rate, were quite different between the single-fiber and four-fiber cases. These different phase structures, however, did not affect the predicted pressure drop as significantly as did the total flow rate.

3.2. *Effect of Inter-Fiber Distance on Phase Structures*

The consideration of multiple fibers in the model introduces new possibilities for independent variables. As Duprat et al. [37] and Protiere et al. [38] mentioned, fiber diameter and inter-fiber distance might affect droplet structure, consistent with changes in interfacial surface area and pressure drop. These authors provided three different "dispersed" phase structures as inter-fiber distance increased in their studies, which they termed bridge, barrel, and column, if one considers the "matrix" phase to be air and the "dispersed" phase to be water. Different inter-fiber distances or packing geometries were therefore compared in the present study. For all configurations of fiber diameter and channel space considered here, the flow conditions were as follows: Q_W = 75 mL/min, Q_o = 150 mL/min (relative flow rate ratio, Q_o/Q_W = 2), the interfacial tension, σ = 0.0316 N/m, the viscosities of the two phases, μ_o = 0.0368 Pa.s, μ_W = 0.001 Pa.s, respectively, and the contact angle, θ = 10°. One complication of varying packing density was that the fluid volumes were different across the cumulative data set; therefore, to compare the models, interfacial surface areas were divided by the fluid volumes to provide a specific interfacial area (m^2/m^3).

The steady-state phase structures predicted by the models are shown in Figure 5i–ix. The predictions correspond to t = 5 s to cover at least one full residence time. Note that Figure 5i,iii,iv belong to one approach to varying packing density, the CD method (constant dimensions). The others are for an alternative approach, the CV method (constant voidage). When the RVE dimensions are constant, the inter-fiber distance changes with fiber diameter, as does voidage (CD). When the voidage is constant, the inter-fiber distance changes with fiber diameter, simultaneously changing RVE dimensions (CV). The phase structures noted were quite complex, not always slugs, with marked qualitative differences. To enable discussion, we characterized the various phases as columnar with droplets (CWD), mixed slugs with droplets (MSWD), slugs (S), and core continuous with drops (CCWD). Note that Figure 5ii,vi depict predictions using the same base case study, i.e., the same RVE. Generally, periodic wave-like phase structures were more evident for the smaller RVEs (on the left), whereas more complex ragged-edged slug-like phase structures were seen as the void space increased (on the right). Since the microfluidic literature generally describes slugs, drops, or films, these two different phase structures shown here support the hypothesis that at least three different flow regimes can exist within creeping flows in microfluidics.

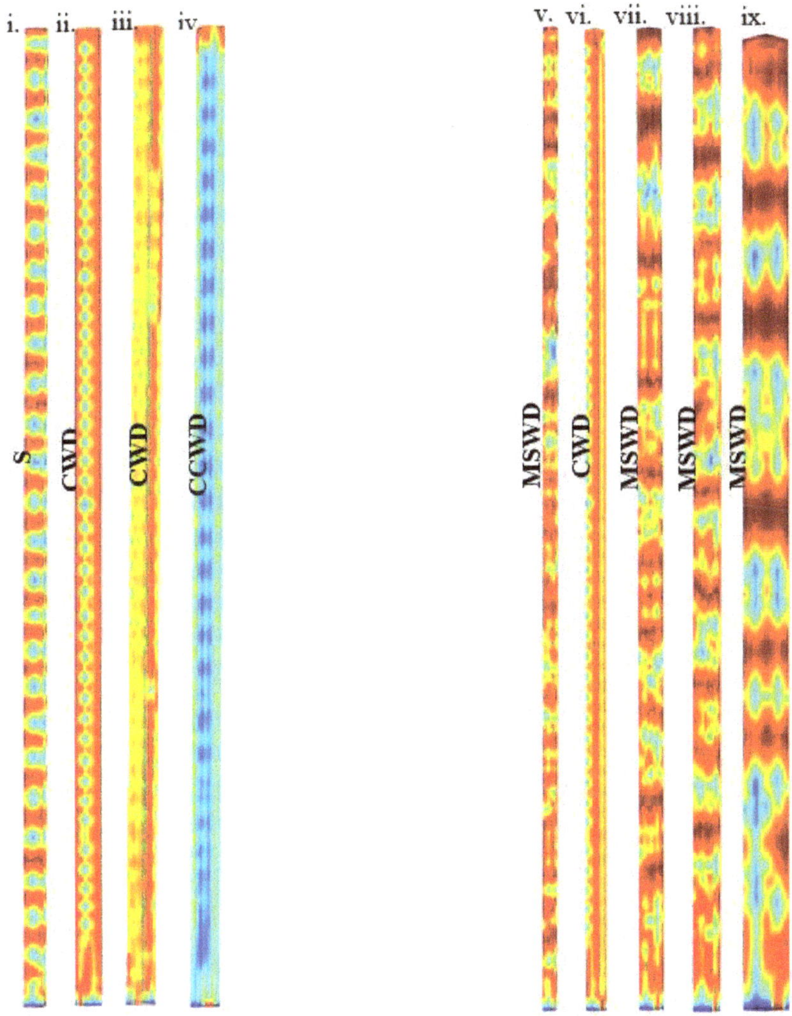

Figure 5. Scale is given by the diameter of the channel in microns, termed "W" below. Instantaneous phase structures generated from a four-fiber model using 40–114 µm diameter fibers as follows: (**i**) d_{fib} = 40 µm, W = 35.9 µm; (**ii**) d_{fib} = 50 µm, W= 25.9 µm; (**iii**) d_{fib} = 60 µm, W = 15.9 µm; (**iv**) d_{fib} = 70 µm, W = 5.9 µm; (**v**) d_{fib} = 40 µm, W = 20.8 µm; (**vi**) d_{fib} = 50 µm, W = 25.9 µm; (**vii**) d_{fib} = 60 µm, W = 31.2 µm; (**viii**) d_{fib} = 70 µm; W = 36.4 µm; (**ix**) d_{fib} = 114 µm; W= 59.6 µm. Figures ii and vi are the same, the base case. (**Left**) Models where RVE dimensions were constant, CD. (**Right**) Models where RVE void space was held constant, CV. Acronyms for phase structure: columnar with droplets (CWD), mixed slugs with droplets (MSWD), slugs (S), and core continuous with drops (CCWD).

Figure 6 shows the specific interfacial surface area against the inter-fiber distance for all the CD and CV models. The studies predicted an optimum inter-fiber distance for the generation of surface area, ranging between 10 and 50 µm. Since the surface area is essential for the overall mass transfer coefficients and the extraction efficiency of the vessel, this observation of an optimum is another key outcome of this CFD study. Figure 7 compares the effects of voidage or inter-fiber distance at a constant fiber diameter for three different fiber diameters. At constant fiber diameter, the more tightly packed the fibers, i.e., smaller void space, the greater the specific interfacial surface areas generated. An independent effect of fiber diameter may exist based on the trend for the 0.66 void space data as well.

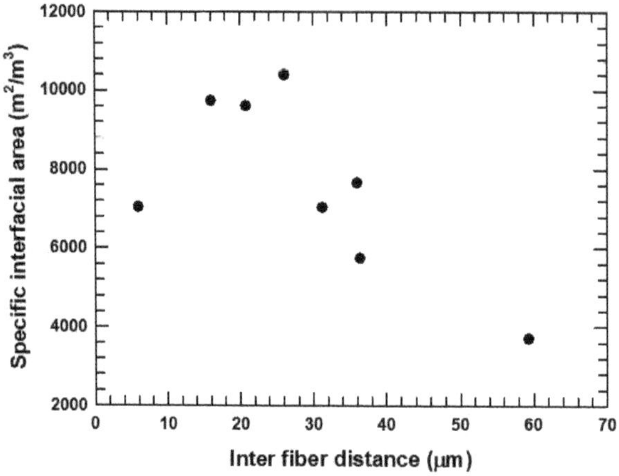

Figure 6. Effect of inter-fiber distance on the specific interfacial areas. Flow rate of corn oil Q_o = 150 mL/min, flow rate of water Q_W = 75 mL/min, relative flow rate Q_o/Q_W = 2, contact angle = 10°, corn oil viscosity = 0.0368 Pa.s, interfacial tension = 0.0316 N/m.

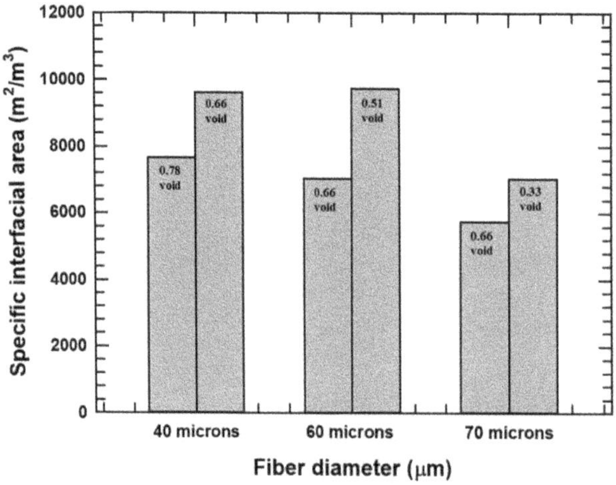

Figure 7. Specific area generated at constant fiber diameter and different voids. Voids left to right were 0.78, 0.66, 0.66, 0.51, 0.66, and 0.33.

Overall, the results show that the biggest predictor of interfacial surface area was the inter-fiber distance and that an optimal inter-fiber distance would exist (in terms of developing surface area) for a given set of feedstock, fluid properties, and process conditions.

Figure 8 shows the pressure drop versus inter-fiber distance. Regardless of the type of phase structure (which was quite different for all of these cases), pressure drop generally decreased with wider channels, thereby lowering energy (pumping) requirements. Again, the contrasting phase structures observed and interfacial surface areas formed in each of the cases did not appear to have a large effect on pressure drop, which appeared to be governed mostly by the area available for flow or the wetted surface resisting flow.

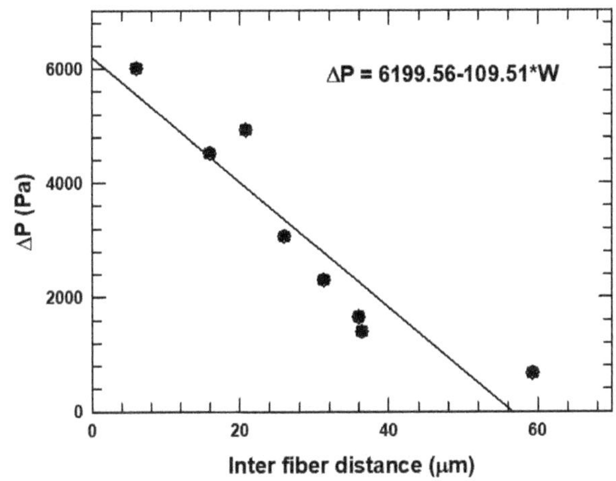

Figure 8. Effect of varying inter-fiber distance on the pressure drop. Inset is the effect of varying the hydraulic diameter on the pressure drop.

3.3. Flow Rate Effect

Overall flow rate affects the phase structure in these two-phase fluids, and phase structure in turn affects the interfacial surface area available for mass transfer/extraction efficiency. Thus, having explored process conditions that affect phase structure, the focus now is the "bottom line", which is how to connect to process efficiency. For a given overall volumetric flow rate, the relative flow rates of the two phases can also affect phase structure, again in turn yielding more or less surface area for extraction between phases. Thus, both overall and relative flow rates and their impact on specific interfacial areas were explored. Additionally, the overall flow rate was expected, of course, to affect pressure drop, one determinant of the energy efficiency of the process. CFD was used to predict pressure drop, but it can also uniquely predict whether, for a given overall flow rate, the phase structure alone affected the pressure drop. For example, in an extreme case, could formation of an emulsion with tiny droplets could increase the pressure drop.

The flow rate experiments were analyzed for 50 μm and 114 μm fiber diameters, the two extreme cases using the CV method. These two setups had a constant voidage of 0.66 but inter-fiber distances (W) of 26 μm and 59 μm, respectively. The number of fibers packed into a 1 inch reactor for the 50 μm diameter fibers would be approximately 87,000, and the number of fibers for a 114 μm diameter fiber would be approximately 17,000. Hence, there was a need to normalize the interfacial areas by fluid volume for comparison of output variables and ease of scale-up predictability. Total flow rates were varied from 100 mL/min to 300 mL/min. Relative flow rates were varied by keeping the flow rate of water constant at 75 mL/min and changing the flow rate of the corn oil from 45 mL/min to 125 mL/min. The contact angle was held at 10°, the interfacial tension, σ = 0.0316 N/m, and viscosity of corn oil and water, μ_o and μ_w, were equal to 0.0368 Pa.s and 0.001 Pa.s, respectively. Intuitively, one would hypothesize that the pressure drop would be lower given wider clearances, but predicting how flow rates impacted interfacial surface areas would be more complex.

Figure 9a compares the plots of specific interfacial surface area as a function of the total flow rate at a relative flow rate of 1 for the two different geometries (26 μm and 59 μm inter-fiber distance). Figure 9b compares the plots of relative flow rate against the specific interfacial surface area for the two different geometries. Figure 9c,d show the phase structures for the total flow rates study when the distances were 26 μm (c) and

59 μm (d). The phase structure observed was unique to the RVE geometry regardless of the flow rates. In the 26 μm distance case, we observed periodic phase structures resembling the surface of a wave (CWD), whereas slugs with a unique ragged meniscus appeared (MSWD) for the 59 μm distance. Consequently, due to the uniqueness of phase structures developed for the two geometries, specific interfacial surface areas were almost constant for a given packing despite varying flow rates. However, the difference in the specific interfacial area was 2.5 times greater with the smaller (26 μm) clearance between fibers versus the larger clearance (59 μm) for a constant flow rate (Figure 9a,b). Note that the smaller clearance with the larger surface area falls within the predicted range of optimal inter-fiber distance described previously. To summarize, smaller clearances led to greater surface areas for extraction.

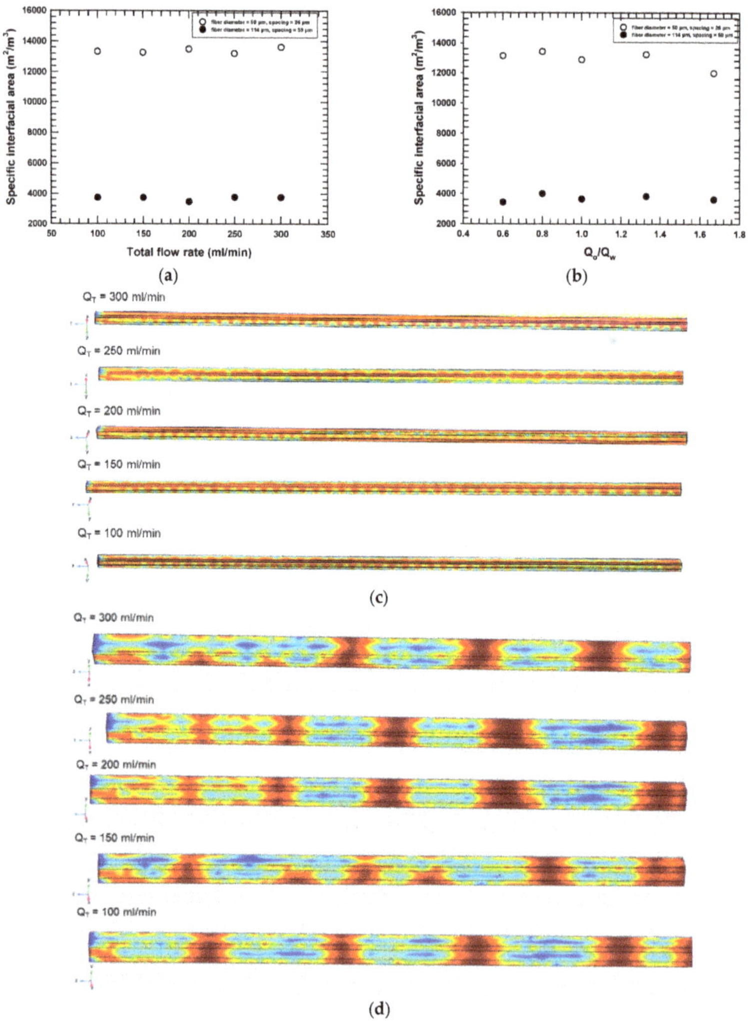

Figure 9. Effect of the flow rate on specific interfacial area in 50 μm fiber diameter with 26 μm inter-fiber distance and 114 μm fiber diameter with 59 μm inter-fiber distance models. (**a**) Total flow rate (relative phase ratio = 1). (**b**) Relative flow rate, $0.6 < Q_o/Q_W < 1.8$ where Q_W = 75 mL/min and contact angle = 10°. (**c**) Phase structures for distances of 26 μm and (**d**) phase structures for distances of 59 μm.

As the flow rate through constrained channels increases, it is well understood that pressure drop can be affected. Any increases in pressure drop can involve an energy cost when processing. In Figures 10 and 11, the pressure drop is plotted against the flow rate for the two cases (the main graph represents the 26 μm distance, and the inset represents the 59 μm distance). Both graphs show an increase in pressure drop with flow rate; however, the slope is 10–20% less in the 59 μm case as the total flow rate of oil was increased. Thus, the sensitivity of the pressure drop to flow rate was greater for smaller clearances.

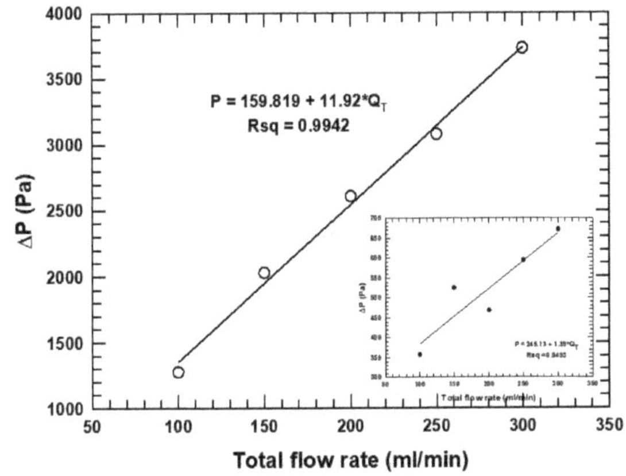

Figure 10. Effect of total flow rate on pressure drop for 50 μm with 26 μm. Inset shows 114 μm fiber diameter with 59 μm spacing.

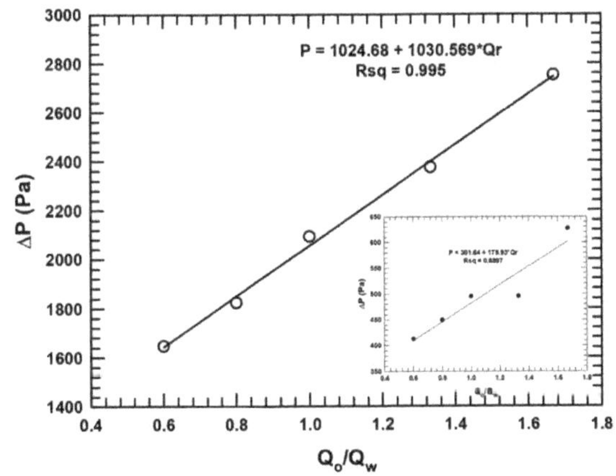

Figure 11. Effect of relative flow rate on pressure drop for 50 μm with 26 μm. Inset shows 114 μm fiber diameter with 59 μm spacing.

The overall flow rates also changed in the relative flow rate study. Hence, the effect of the overall flow rate might be a confounding variable. The sensitivity to the overall flow rate for the relative flow rate data set was plotted. Comparing two graphs, particularly their slopes, in Figure 10 to those in Figure S2 (see Supplementary Information) might tease out this potential artifact. The slopes were higher in both cases for the relative flow rate study (11.2 Pa.min/mL vs. 13.8 Pa.min/mL for the W = 26 μm case and 1.4 Pa.min/mL versus 2.4 Pa.min/mL for the W = 59 μm case). (Units of slope arise from calculating $\Delta y/\Delta x$ in the graph.) Thus, the relative flow rate did affect pressure drop beyond overall flow rate changes in the study.

These model predictions were compared to the experimental work of Massingill et al. [18]. In their work, they utilized different fiber reactors with different packing geometries for the transesterification of triglyceride to biodiesel. Their results showed that a 100% conversion was achievable in a tube 30.5 cm in length and 10.9 mm in diameter packed with 570,000 fibers with only 18 mL of fluid volume at a total flow rate of 3.5 mL/min. In contrast, the conversion dropped to 98.8% when the same reactor was packed with fewer fibers (540,000 fibers) and operated at a higher total flow rate of 7 mL/min. The pressure in the first reactor was, however, higher as a result of the tighter clearances. The periodicity of the 26 μm inter-fiber distance phase structure in the present CFD results indicates behavior like a microfluidic channel even though it has four interacting (semi-restricted, semi-open) channels. The distinctive phase structures that were observed are similar to the three regimes observed as a consequence of static wetting between two parallel fibers separated by a given distance, as reported by Duprat et al. [37,38].

In conclusion, the flow rate effect on the specific interfacial area is not pronounced for a given packing geometry. However, the increase in specific interfacial surface areas becomes significant when the clearances are tighter, with an optimal inter-fiber distance noted. The confinement arising from the small clearances, however, does come with an energy (pumping) expense.

3.4. Wettability (Contact Angle) Effect

Estimating the surface tension force on the fiber walls involved specifying the two-phase contact angle on the fiber walls. The contact angle was defined as the angle between the fiber, the wetting, and the non-wetting phases. Two geometries were compared: a model with a fiber diameter d = 50 μm and inter-fiber distance W = 26 μm and a model with a fiber diameter d = 114 μm and inter-fiber distance W = 59 μm. The voidage in both models was 0.66. The contact angle varied from 10° to 150°. The fluid properties are those provided in Table 2. The effects of wettability are useful to explore the effects of the capillary action (capillary number, $Ca_c = v_c \mu_c / \sigma$) within each individual microchannel because the capillary and surface tension values are more significant than inertial and viscous forces. The capillary number for the contact angle study ranged from 7×10^{-3} to 1.7×10^{-2}. As seen in Figure 2, the capillary number increased with decreasing inter-fiber distance. As the capillary number increased, the phase structures observed became more periodic and wave-like. Figure 12 compares the specific interfacial area for the two cases studied. Figure 12b,c indicate the representative phase structures observed. The phase structures changed significantly with the contact angle. For the W = 26 μm case (Figure 12b), the structures transitioned from regular periodic columnar with droplet structures at 10° and 20° to elongated double gyroid-like structures at 30° and to parallel type structures at 90°. The patterns transitioned further with increasing contact angle in reverse order from 120° all the way to 150°. We also note that the continuous phase underwent phase inversion when crossing the 90° angle. Thus, the oil phase acted as the wetting phase at lower angles, but after phase inversion, the water phase controlled spreading and was the wetting phase. For the W = 59 μm case (Figure 12c), the phase structures followed patterns with more irregular slug-like features with ragged edges. As the contact angle increased from 10°, the influence of fibers decreased. Hence, at angles greater than 90°, where the aqueous phase was the continuous phase, and the oil phase acted as the dispersed phase, the meniscus became rough-edged, and large slugs of both phases formed further downstream. Summarily, the contact angle effect revealed that more regular periodic phase structures were present in the case of the 26 μm distance between fibers, and more chaotic, random slug shapes were present at the larger distance, 59 μm. These different flow regimes were possibly a result of increasing clearance between fibers, resulting in smaller shear stresses and the added degree of freedom (DOF) for flows. The extra DOF suggests a third intermediate flow regime, resulting from conditions between microfluidic regimes and traditional pipe flow. Figure 12a, showing specific interfacial surface areas in the W = 26 µm case, confirms a transition relating to phase inversion. The system's instability around phase inversion can be clearly seen in the plot (open circles in Figure 13). In the W = 59 µm case, the specific interfacial surface area curve follows an S-shape, again indicating the effects of phase inversion, but more stable. However, the magnitude of the specific surface areas produced was not strongly governed by the contact angle.

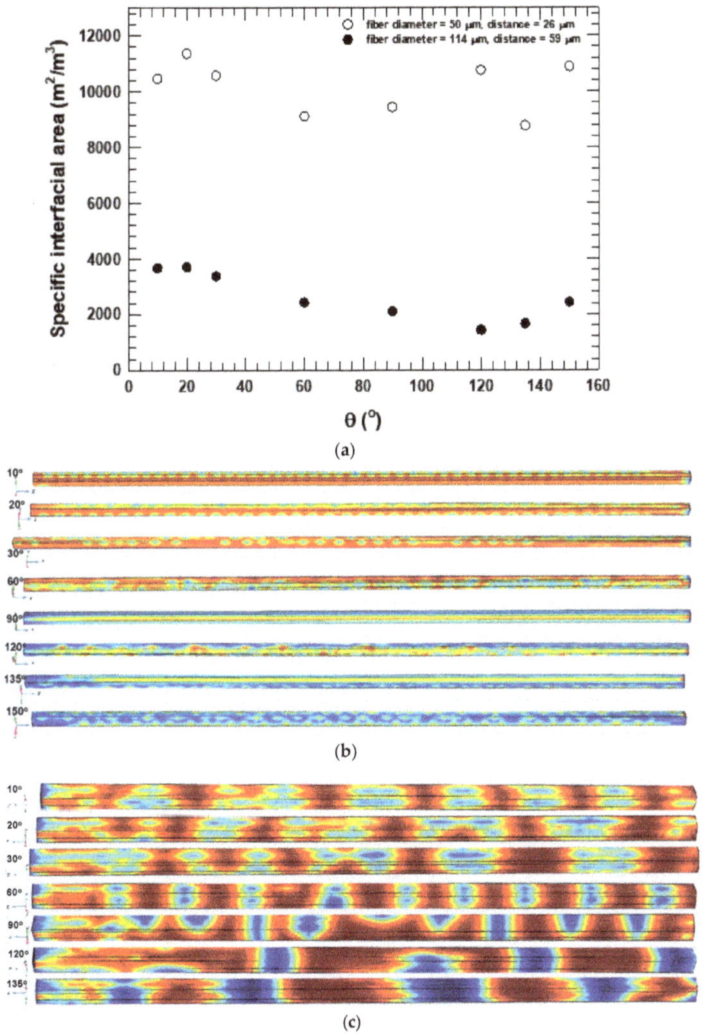

Figure 12. (**a**) Effect of contact angle on specific interfacial area for two cases: 50 µm diameter fiber with 26 µm distance between fibers (open circles) and 114 µm diameter fibers with 59 µm distance between fibers (filled circles). (**b**) Phase structures at different contact angles in the 50 µm diameter fiber geometry, W = 26 µm. (**c**) Phase structures at different contact angles in the 114 µm diameter fiber geometry, W = 59 µm. Note: both four-fiber RVEs have the same voidage.

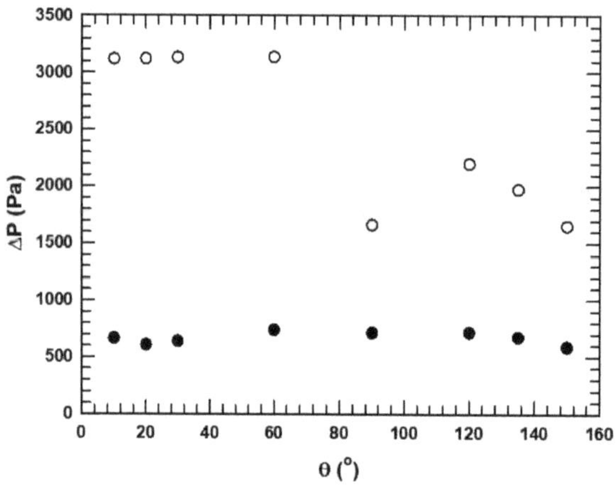

Figure 13. Effect of contact on pressure drop for two cases: 50 μm diameter fiber with 26 μm distance between fibers (open circles) and 114 μm diameter fibers with 59 μm distance between fibers (filled circles).

The effect of the contact angle on the pressure drop in both models and a graph of the pressure drop against the contact angle is shown in Figure 13. We observed two regimes for the 26 μm channels, shown by open circles. For the regime to the left of 90°, the corn oil phase was the wetting phase, and because corn oil is more viscous, the pressure drop was high. At angles beyond the inversion point, the pressure drop decreased because the low-viscosity aqueous phase was the wetting phase, and hence, there was less resistance against the flow. Again, the instability at phase inversion conditions can be noted. For the 59 μm distance case, the pressure drop was lower and not affected to any significant extent by phase inversion.

In summary, the oil phase was the wetting phase at smaller contact angles. As the contact angle increased, a phase inversion was observed. At a given contact angle, the specific interfacial surface area in the 26 μm clearances (d = 50 μm) was much larger than at the 59 μm clearances (d = 114 μm). Regardless of the wetting properties of the fiber, the most important factor governing the generation of interfacial surface area was the inter-fiber distance. Lower pressure drops were observed when water was in the wetting phase (at higher contact angles). Biofeedstock properties are, therefore, an important consideration for pressure drop and energy efficiency.

4. Conclusions

The fiber extractor (FE), a process vessel associated with process intensification for purification of biofeedstock, is a massively arrayed (i.e., numbered up) micro-structured extractor that provides several custom advantages of processing in a microfluidic environment. Such equipment offers a low footprint yet provides high throughput. The fluid flowed around the fibers, not through them. A CFD model of a four-fiber representative volume element was developed on a microfluidic scale. The effects of changing inter-fiber distance and process conditions on the phase structures, specific interfacial surface area, and pressure drop were presented. Predictions of the four-fiber CFD model include the following.

The presence of neighboring fibers clearly affected phase structure and indicated radial flows and mixing were present. Thus, the unique geometry of a fiber extractor has the potential for enhanced local mixing in microfluidic channels dependent on degrees of freedom for flow.

The interfacial surface area developed in the four-fiber case was more complex and proportionally much greater area than that of the idealized single-fiber annular flow, attributed to near-neighbor fiber effects.

The total flow rate, regardless of the type of phase structure, had the most significant effect on pressure drop.

The inter-fiber distance was the best predictor of the developed interfacial surface area. An optimum inter-fiber distance for generating surface area appeared to be between 10 and 50 μm.

The effect of contact angle (fiber wettability) on interfacial surface area and pressure drop was not pronounced. Contact angle contributed to the pressure drop by causing phase inversion. When the lower viscosity fluid (water) began to govern flow conditions, the pressure drop was also lower, particularly for the smaller clearances. However, at smaller inter-fiber distances, instabilities were noted during phase inversion, which could indicate processing instability.

Compared to the single-fiber case, the four-fiber model predicted new types of phase structures were noted such as standing waves and combinations of slugs and films.

At least three different flow regimes in creeping microfluidic flow were supported.

The phase structure in two-phase flows determines the interfacial surface area and, therefore, the potential for the extractor to enhance mass transfer and optimize extraction efficiency. These findings provide guidance for design and operation to optimize surface area and extraction efficiency by controlling the distances between fibers, the wetting or contact angle, and the flow rates. As the present work focused only on the unique fluidics possible in these microchannels, certainly CFD modeling of mass transfer, heat transfer, and interfacial reaction in two-phase flows with semi-restricted boundary conditions would extend the theoretical guidance available for these extractors. In addition, exciting work in open-channel microfluidics is recently reported with electrophoretic-driven flows, and extending fundamental modeling into other driving forces beyond pressure-driven flows could be relevant to some of the open-channel microfluidics work currently being reported.

Supplementary Materials: The following supporting information can be downloaded at https://www.mdpi.com/article/10.3390/mi16040425/s1. Table S1: Calculation of area for flows; Figure S1: Comparing the influence of the external wall on fluid flow around the fiber and the influence of fibers on the fluid flow; Figure S2: Pressure drop versus total flow rate (where Q_W = 75 mL/min; Q_O = 45–125 mL/min); Figure S3: Comparison of changes in the concentration profile for different mesh sizes.

Author Contributions: O.O.A. contributed conceptualization, methodology, software, validation, investigation, analysis, draft writing, editing and review, and visualization. H.A.S. contributed conceptualization, methodology, analysis, editing and review, visualization, supervision, administration, and funding acquisition. A.V.-B.-H. contributed conceptualization, methodology, software, validation, analysis, editing and review, visualization, and supervision. All authors have read and agreed to the published version of the manuscript.

Funding: This research was funded by Visionary Fibers Technologies and by the Tennessee Technological Center for Manufacturing Research.

Data Availability Statement: The original contributions presented in this study are included in the article/Supplementary Materials. Further inquiries can be directed to the corresponding author.

Acknowledgments: We gratefully acknowledge discussions and advice from Wade Gosset, Tessa Eskander, Asif Rony, and Adeel Hafeez. Funding was provided both by Visionary Fibers Technologies and the Tennessee Technological Center for Manufacturing Research. Computing help and access to the cluster were enabled by Mike Renfro at Tennessee Technological University.

Conflicts of Interest: The authors declare that they have no known competing financial interests or personal relationships that could have appeared to influence the work reported in this paper at the time of the work.

References

1. Bekavac, N.; Benković, M.; Jurina, T.; Valinger, D.; Gajdoš Kljusurić, J.; Jurinjak Tušek, A.; Šalić, A. Advancements in aqueous two-phase systems for enzyme extraction, purification, and biotransformation. *Molecules* **2024**, *29*, 3776. [CrossRef] [PubMed]
2. Venkataraman, S.; Rajendran, D.S.; Kumar, P.S.; Vo, D.-V.N.; Vaidyanathan, V.K. Extraction, purification and applications of biosurfactants based on microbial-derived glycolipids and lipopeptides. A review. *Environ. Chem. Lett.* **2022**, *20*, 949–970.
3. Molino, J.V.D.; Marques, D.D.A.V.; Júnior, A.P.; Mazzola, P.G.; Gatti, M.S.V. Different types of aqueous two-phase systems for biomolecule and bioparticle extraction and purification. *Biotechnol. Prog.* **2013**, *29*, 1343–1353. [CrossRef]
4. Ramaswamy, S.; Huang, H.-J.; Ramarao, B.V. *Separation and Purification Technologies in Biorefineries*; John Wiley & Sons: Chichester, UK, 2013.
5. Vicente, F.A.; Plazl, I.; Ventura, S.P.; Žnidaršič-Plazl, P. Separation and purification of biomacromolecules based on microfluidics. *Green Chem.* **2020**, *22*, 4391–4410.
6. Boodhoo, K.; Flickinger, M.; Woodley, J.; Emanuelsson, E. Bioprocess intensification: A route to efficient and sustainable biocatalytic transformations for the future. *Chem. Eng. Process. Process Intensif.* **2022**, *172*, 108793.
7. Sonawane, A.; Waghmode, S.R. A Review on Vegetable Oil Refining: Process, Advances and Value Addition to Refining by-Products. In *Bioremediation for Global Environmental Conservation*; IntechOpen: London, UK, 2023; p. 161.
8. Jafari, S.M.; Akhavan-Mahdavi, S. *Extraction Processes in the Food Industry: Unit Operations and Processing Equipment in the Food Industry*; Woodhead Publishing: Sawston, UK, 2023.
9. Kashid, M.; Harshe, Y.; Agar, D.W. Liquid−liquid slug flow in a capillary: An alternative to suspended drop or film contactors. *Ind. Eng. Chem. Res.* **2007**, *46*, 8420–8430. [CrossRef]
10. Mansourizadeh, A.; Rezaei, I.; Lau, W.J.; Seah, M.Q.; Ismail, A.F. A review on recent progress in environmental applications of membrane contactor technology. *J. Environ. Chem. Eng.* **2022**, *10*, 107631.
11. Kashid, M.N.; Gupta, A.; Renken, A.; Kiwi-Minsker, L. Numbering-up and mass transfer studies of liquid–liquid two-phase microstructured reactors. *Chem. Eng. J.* **2010**, *158*, 233–240.
12. Al-Azzawi, M.; Mjalli, F.S.; Husain, A.; Al-Dahhan, M. A Review on the hydrodynamics of the Liquid–Liquid two-phase flow in the microchannels. *Ind. Eng. Chem. Res.* **2021**, *60*, 5049–5075.
13. Singh, S.; Kumar, U.A. Hydrodynamics and mass transfer studies of liquid-liquid two-phase flow in parallel microchannels. *Int. J. Multiph. Flow* **2022**, *157*, 104248.
14. He, K.; Lin, Y.; Hu, Y.; Huang, S.-M. Phase separation features of oil–water parallel flow at hydrophobic and hydrophilic micro-T-junctions. *Chem. Eng. Sci.* **2022**, *253*, 117520. [CrossRef]
15. Chen, J.; Wang, S.; Cheng, S. Experimental investigation of two-phase distribution in parallel micro-T channels under adiabatic condition. *Chem. Eng. Sci.* **2012**, *84*, 706–717. [CrossRef]
16. Aghda, N.H.; Lara, E.J.; Patel, P.; Betancourt, T. High throughput preparation of poly (lactic-co-glycolic acid) nanoparticles using fiber fluidic reactor. *Materials* **2020**, *13*, 3075. [CrossRef] [PubMed]
17. Jamkhindikar, S.P.; Stretz, H.A.; Massingill, J.L., Jr.; Betancourt, T. High throughput fiber reactor process for organic nanoparticle production: Poly (N-isopropylacrylamide), polyacrylamide, and alginate. *J. Appl. Polym. Sci.* **2017**, *134*, 45524. [CrossRef]
18. Massingill, J.L., Jr.; Patel, P.N.; Guntupalli, M.; Garret, C.; Ji, C. High efficiency nondispersive reactor for two-phase reactions. *Org. Process Res. Dev.* **2008**, *12*, 771–777.
19. Kim, K.-W.; Song, K.-C.; Lee, E.-H.; Yoo, J.-H. A liquid-liquid static contactor using capillary phenomena induced by a highly packed fiber bundle. *J. Radioanal. Nucl. Chem.* **1999**, *241*, 461–468. [CrossRef]
20. Ramezani, R.; Di Felice, L.; Gallucci, F. A review on hollow fiber membrane contactors for carbon capture: Recent advances and future challenges. *Processes* **2022**, *10*, 2103. [CrossRef]
21. Ayeni, O.O.; Stretz, H.A.; Vasel-Be-Hagh, A. A Core Element CFD Model for a Two Phase Microfluidic Fiber Extractor. *Chem. Eng. Process. Process Intensif.* **2024**, *202*, 109842. [CrossRef]
22. Pan, S. A New Type of Continuous Countercurrent Extraction: An Extractor with Two Continuous Phases. *Sep. Sci.* **1974**, *9*, 227–248. [CrossRef]
23. Clonts, K.E. Liquid-Liquid Mass Transfer Apparatus. U.S. Patent 3,758,404, 11 September 1973.
24. Tallapudi-Kelley, S.; Ayeni, O.O.; Stretz, H.A. Formation of lithium carbonate nano-particles using a massively arrayed microfluidic fiber reactor. *Ceram. Int.* **2023**, *49*, 25649–25657. [CrossRef]
25. Zhang, F.; Shen, B.-X.; Sun, H.; Liu, J.-C. Removal of organosulfurs from liquefied petroleum gas in a fiber film contactor using a new formulated solvent. *Fuel Process. Technol.* **2015**, *140*, 76–81.

26. Santos, E.; Waghmare, P.R.; Temelli, F. Interfacial tension and equilibrium contact angle of corn oil on polished stainless steel in supercritical CO_2 and N_2. *J. Supercrit. Fluids* **2020**, *156*, 104665.
27. Venkateshwarlu, A.; Bharti, R.P. Effects of capillary number and flow rates on the hydrodynamics of droplet generation in two-phase cross-flow microfluidic systems. *J. Taiwan Inst. Chem. Eng.* **2021**, *129*, 64–79.
28. Charpentier, J.-B.; de Motta, J.B.; Ménard, T. Capillary phenomena in assemblies of parallel cylindrical fibers: From statics to dynamics. *Int. J. Multiph. Flow* **2020**, *129*, 103304.
29. Ho, C.-D.; Chang, H.; Chen, H.-J.; Chang, C.-L.; Li, H.-H.; Chang, Y.-Y. CFD simulation of the two-phase flow for a falling film microreactor. *Int. J. Heat Mass Transf.* **2011**, *54*, 3740–3748.
30. Berthier, J.; Theberge, A.B.; Berthier, E. *Open-Channel Microfluidics: Fundamentals and Applications*; IOP Publishing: Bristol, UK, 2024.
31. Glavan, A.C.; Martinez, R.V.; Maxwell, E.J.; Subramaniam, A.B.; Nunes, R.M.; Soh, S.; Whitesides, G.M. Rapid fabrication of pressure-driven open-channel microfluidic devices in omniphobic RF paper. *Lab A Chip* **2013**, *13*, 2922–2930.
32. Ertsgaard, C.T.; Yoo, D.; Christenson, P.R.; Klemme, D.J.; Oh, S.-H. Open-channel microfluidics via resonant wireless power transfer. *Nat. Commun.* **2022**, *13*, 1869. [CrossRef]
33. Berthier, E.; Dostie, A.M.; Lee, U.N.; Berthier, J.; Theberge, A.B. Open microfluidic capillary systems. *Anal. Chem.* **2019**, *91*, 8739–8750. [CrossRef]
34. Betancourt, T.; Byrne, J.D.; Sunaryo, N.; Crowder, S.W.; Kadapakkam, M.; Patel, S.; Casciato, S.; Brannon-Peppas, L. PEGylation strategies for active targeting of PLA/PLGA nanoparticles. *J. Biomed. Mater. Res. Part A* **2009**, *91*, 263–276.
35. Tallapudi, S. *Synthesis of High Throughput Lithium Carbonate Nanoparticles in a Scalable Microfluidic Reactor*; Tennessee Technological University: Cookeville, TN, USA, 2017.
36. Gupta, Y.; Saroha, A.K. Fibre film contactors. *Indian Chem. Eng.* **2012**, *54*, 113–124.
37. Duprat, C.; Protiere, S.; Beebe, A.; Stone, H.A. Wetting of flexible fibre arrays. *Nature* **2012**, *482*, 510–513. [PubMed]
38. Protiere, S.; Duprat, C.; Stone, H.A. Wetting on two parallel fibers: Drop to column transitions. *Soft Matter* **2013**, *9*, 271–276.

micromachines

Article

Model-Based Optimization of Solid-Supported Micro-Hotplates for Microfluidic Cryofixation

Daniel B. Thiem [1,*], Greta Szabo [1] and Thomas P. Burg [1,2,*]

[1] Integrated Micro-Nano-Systems Laboratory, Technische Universität Darmstadt, 64283 Darmstadt, Germany; greta.szabo@tu-darmstadt.de

[2] Centre for Synthetic Biology, Technische Universität Darmstadt, 64289 Darmstadt, Germany

* Correspondence: daniel.thiem@tu-darmstadt.de (D.B.T.); thomas.burg@tu-darmstadt.de (T.P.B.)

Abstract: Cryofixation by ultra-rapid freezing is widely regarded as the gold standard for preserving cell structure without artefacts for electron microscopy. However, conventional cryofixation technologies are not compatible with live imaging, making it difficult to capture dynamic cellular processes at a precise time. To overcome this limitation, we recently introduced a new technology, called microfluidic cryofixation. The principle is based on micro-hotplates counter-cooled with liquid nitrogen. While the power is on, the sample inside a foil-embedded microchannel on top of the micro-hotplate is kept warm. When the heater is turned off, the thermal energy is drained rapidly and the sample freezes. While this principle has been demonstrated experimentally with small samples ($<0.5\ mm^2$), there is an important trade-off between the attainable cooling rate, sample size, and heater power. Here, we elucidate these connections by theoretical modeling and by measurements. Our findings show that cooling rates of $10^6\ K\,s^{-1}$, which are required for the vitrification of pure water, can theoretically be attained in samples up to $\sim$1 mm wide and 5 μm thick by using diamond substrates. If a heat sink made of silicon or copper is used, the maximum thickness for the same cooling rate is reduced to $\sim$3 μm. Importantly, cooling rates of $10^4\ K\,s^{-1}$ to $10^5\ K\,s^{-1}$ can theoretically be attained for samples of arbitrary area. Such rates are sufficient for many real biological samples due to the natural cryoprotective effect of the cytosol. Thus, we expect that the vitrification of millimeter-scale specimens with thicknesses in the 10 μm range should be possible using micro-hotplate-based microfluidic cryofixation technology.

Keywords: cryofixation; vitrification; cooling rate; heat conduction model

1. Introduction

Micro-hotplates are widely used in the field of conductometric gas sensors [1–3]. These devices heat a metal oxide layer to specific temperatures, stimulating gas–surface interactions and changing the conductance, which is then measured to detect different gases. These hotplate designs are usually optimized for low power consumption [4–7]. This is achieved through a high isolation by suspending the hotplate with support beams, minimizing thermal contact with the substrate. Due to their low thermal mass, the transient response of these devices is typically very fast (up to <200 μs) [8–13]. Recent advances have further reduced the thermal mass by using nanowires to decrease power consumption while increasing the sensitivity of these gas sensors [14,15]. This power reduction is stated as a goal to improve the portability of such sensors [14]. While micro-hotplates for gas sensors can achieve fast transient responses [16,17] by optimizing the thermal isolation of the micro-heater; microfluidic heaters, in contrast, must heat the thermal mass of the sample itself, which inhibits a fast thermal response.

Fast temperature changes of a thermal mass are used in a variety of microfluidic and microbiological applications. For example, rapid heating and cooling are needed to improve cycle time for polymerase chain reactions (PCRs) [18–23], droplet actuation using thermocapillary effects [24], or microfluidic actuation with microbubbles [25,26] or thermally expandable polymers [27]. The use of rapid temperature changes for the operation of thermal micromixers in microfluidic systems has also been proposed in the literature [28]. As illustrated in Figure 1, different applications require different trade-offs between the size, power consumption, and response time or rate.

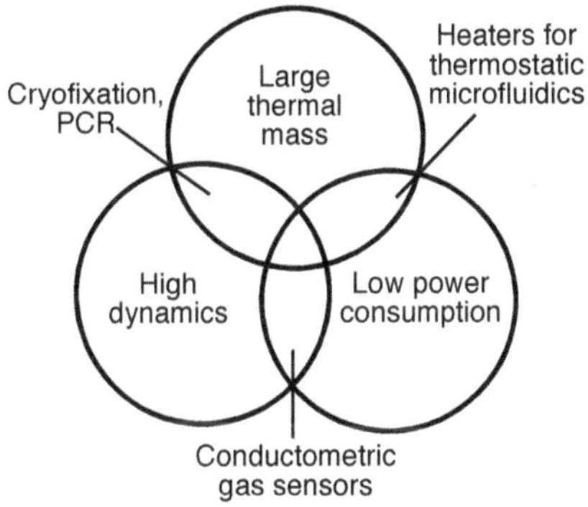

Figure 1. While conductometric gas sensors with micro-hotplates are commonly optimized for low power consumption and high dynamics, their thermal mass can be small. Cryofixation systems, on the other hand, aim to have a high dynamic behavior. Their thermal mass scales with the size of the sample. For cryofixation systems, the power consumption is not a strategic property.

In this work, we focus on high cooling rates of defined thermal masses on micro-hotplates, which are important for preparing biological samples using cryofixation in the field of structural biology. Here, the sample is rapidly cooled down to preserve its ultrastructure in amorphous, or vitreous, ice. This is enabled by the high cooling rates in cryofixation systems, which leave no time for ice crystals to form [29].Cryofixed samples can then be used for investigations using electron microscopy (EM) after freeze substitution, cryogenic electron microscopy (cryo-EM), and X-ray microscopy at cryogenic temperature. It is widely accepted that cooling rates (CRs) over $10^6\,\mathrm{K\,s^{-1}}$ are necessary to enable the full vitrification of pure water. However, for cells and tissues, this requirement is somewhat relaxed depending on the degree of natural cryoprotection imparted by the high concentration of dissolved biomolecules [30,31]. Often, cooling rates of around $10^4\,\mathrm{K\,s^{-1}}$ still allow the vitrification of the sample, sometimes with the formation of locally confined ice crystals of sufficiently small sizes to not perturb the structures of interest [32].

Over the past decades, various methods have been developed to enable the cryofixation of samples of different sizes. Plunge freezing is the most common method for the cryofixation of samples on transmission electron microscopy (TEM) grids [33]. The samples are blotted and plunged into a cryogen with a high cooling efficiency. The thickness of the water layer on top of the grids after blotting is usually less than 1 µm [34]. For samples on the order of 10 µm thick, slam freezing allows vitrification by quickly establishing thermal contact between the sample and a cold block [35]. Samples that are even thicker can only be vitrified using high-pressure freezing at 2000 bar [36].

Although plunge freezing and slam freezing provide reliable and consistent cryofixation, they do not allow optical access to the sample at the precise time of cryofixation due to the limitations of the working principle. Live imaging during cryofixation is often desirable for studying biological processes with sub-second dynamics, such as intraflagellar transport and neuronal firing [37]. In recent years, two design principles for cryofixation systems with optical access have emerged. In the first principle, which we refer to as *equilibrium systems*, the environment is initially warm and the sample is in thermal equilibrium with the environment. The sample is mounted to an optical microscope on a thermally well-conducting substrate in thermal equilibrium with the warm environment (see Figure 2 left). Freezing is accomplished by spraying a cryogenic gas or liquid jet (e.g., liquid nitrogen or cold Helium gas) on the backside of the sample holder and then immersing the sample in liquid nitrogen [38]. In the second principle, which we refer to as *steady-state systems*, the sample is initially maintained far from thermal equilibrium through a micro-hotplate that is counter-cooled with liquid nitrogen (see Figure 2 right) [39]. When the micro-hotplate is turned off, the heat in the sample is quickly absorbed by the cold heat sink below.

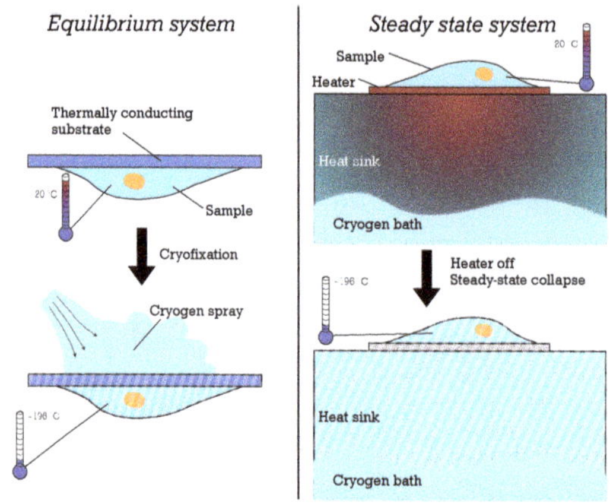

Figure 2. Two different principles for cryofixation with optical access have emerged. In an equilibrium system (**left**), the sample and the system are in a thermal equilibrium at room temperature until the cryofixation is started by spraying a cryogen on the substrate which rapidly cools the sample. A steady-state system (**right**) holds the sample at room temperature while a cryogen cools the backside of the system. This steady state collapses as soon as the heater is turned off and the sample is cryofixed.

This paper proposes a modeling approach for thermal conduction in a steady-state system. We also examine the possibility of increasing the size limits of this system while maintaining cooling rates $>10^6\,\mathrm{K\,s^{-1}}$, as well as the maximum thickness of aqueous samples for cryofixation with cooling rates $>10^4\,\mathrm{K\,s^{-1}}$. For this investigation, the cooling rate is defined as the inverse of the time between the start of the cooling process (20 °C) and the time when the sample reaches a critical temperature of −90 °C.

2. Fundamental Limits for the Cryofixation of Water Layers

If a water layer of any thickness h_{water} is actively cooled from one side, the cooling rate inside the water layer is limited by the thermal diffusivity α_{water}. This imposes a fundamental limit on the cooling rate of a sample with a given size. To calculate this limit, we assume that the water is a layer that is in contact with an ideal heat sink on one side that can instantly change its temperature from room temperature ($T_i = 20\,°\mathrm{C}$) to $T_\infty = -196\,°\mathrm{C}$. The boundary on the opposite side of the layer is assumed to be insulating. The layer is assumed to be unbounded in the other directions so that a one-dimensional heat conduction model can be used [40,41]. Here, using the dimensionless time $\zeta = \alpha_{\text{water}} t / h_{\text{water}}^2$, a series approximation of the temperature T at position y in the channel can be established as follows:

$$T(y,\zeta) = (T_i - T_\infty) \cdot \sum_{n=1}^{\infty} \left[A_n e^{-\lambda_n^2 \zeta} \cos\left(\frac{\lambda_n y}{h_{\text{water}}} \right) \right] + T_\infty \quad (1)$$

where A_n and λ_n are defined by the heat transfer of the interface and the properties of the water layer. Assuming an infinite heat transfer of the interface, this yields the following:

$$A_n = \frac{4 \cdot (-1)^{(n-1)}}{2\pi(n - \frac{1}{2})} \tag{2}$$

$$\lambda_n = \left(n - \frac{1}{2}\right) \cdot \pi \quad n \in \mathbb{N} \tag{3}$$

For $h_{\text{water}} < 50\,\mu\text{m}$, the cooling rate in the water is well above $10^4\,\text{K}\,\text{s}^{-1}$ (see Figure 3). If a cooling rate of at least $10^6\,\text{K}\,\text{s}^{-1}$, as is necessary for the vitrification of pure water, is required, the theoretical maximum thickness is $h_{\text{water}} \leq 7.6\,\mu\text{m}$. Cooling rates higher than $10^6\,\text{K}\,\text{s}^{-1}$ may theoretically be reached in regions up to 3.5 µm inside a thicker water layer. Interestingly, the thickness of the water layer above does not significantly slow down the cooling rate near the boundary.

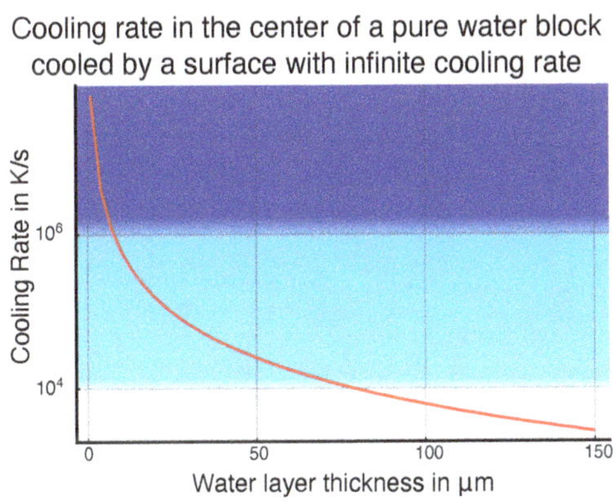

Figure 3. The maximum cooling rate (CR) in a sample at atmospheric pressure is limited by the thermal conduction of the water. $CR > 10^6\,\text{K}\,\text{s}^{-1}$ of water samples cooled from one side are possible for layer heights below 7.6 µm. $CR > 10^4\,\text{K}\,\text{s}^{-1}$ for cryofixation of biological samples is possible for thicknesses up to 50 µm. However, these values assume an infinitely high dissipation rate at the edge of the water layer, which will not be reached by any real system.

3. Domain Model of the Steady-State System

The vitrification of large samples in the steady-state system is only possible by using the correct materials and an optimized geometric design. Therefore, a thermal model of the system must be established. For the systems described in this paper, the dominating effect in determining the cooling rate is heat conduction. Convective heat transfer is neglected due to the small size of the microfludic system, where gravity-driven convection is insignificant ($\text{Ra} \ll 1000$) [42,43]. Also, the rapid nature of cryofixation (rates > $10^3\,\text{K}\,\text{s}^{-1}$) and the large heat transfer rate of conduction ($h_{\text{conduction}} \approx 20\,\text{W} \gg h_{\text{convection,air}} = h * A * \Delta T < 1\,\text{W}$, assuming $h = 30\,\text{W}\,\text{m}^{-2}\,\text{K}^{-1}$ [44], $A = 15\,\text{mm} \times 5\,\text{mm}$ and $\Delta T = 216\,\text{K}$) lead to a low influence of convection in the cryofixation process.

We model the steady-state system as a channel residing on top of a thin heater, an insulation layer, and one or multiple thermally conducting layers below, which act as a heat sink (See Figure 4A—system geometry). We base our model on the two-dimensional heat conduction equation assuming an infinite domain in the z-direction:

$$\nabla(k\nabla \text{T}) = \rho c \frac{\partial T}{\partial t} \tag{4}$$

with initial and boundary conditions:

$$T(y > 0, t = 0) = T_{\mathrm{H}} = 20\,^{\circ}\mathrm{C} \tag{5}$$

$$T(y = -h_{\mathrm{system}}, t) = T_{\mathrm{LN2}} = -196\,^{\circ}\mathrm{C} \tag{6}$$

To model the samples submerged in water, we assume the sample is a microfluidic channel filled with water, with the width b_{water} and height h_{water}. Below the channel there is the heater layer, the insulation layer, and one or multiple heat sink layers. The dimensions of these layers are defined by the width $b_{\mathrm{system}} \gg b_{\mathrm{water}}$ and the height of the bodies below the heater h_{system} (see Figure 4A). As these bodies consist of different materials (insulation layer and heat sink), the thermal conductivity k_{mat}, density ρ_{mat}, and specific heat capacity $c_{p,\mathrm{mat}}$ of the materials are defined accordingly in their geometries. The interface of two layers is modeled as an optimal thermal contact with no contact resistance. Here, both temperature and heat flow are equal at the interface surfaces.

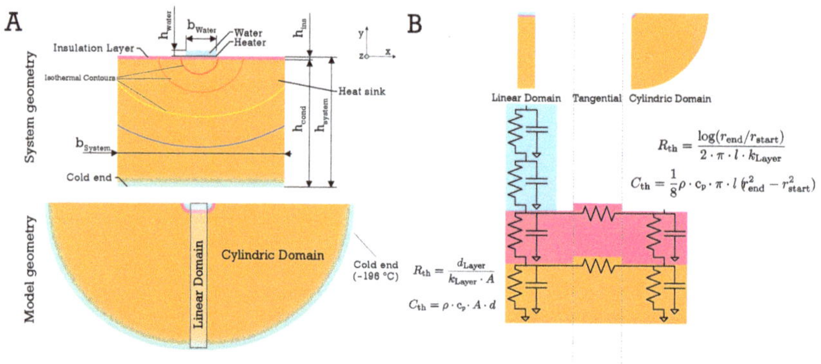

Figure 4. (**A**) The linear temperature distribution below the water channel and the logarithmic temperature distribution around the edge of the water channel can be modeled with a combined model represented by a network of thermal resistances and capacitances. (**B**) Each layer consists of a resistor and a capacitor for the linear domain, a resistor and a capacitor for the semi-cylindrical domain, and a resistor modeling the tangential thermal resistance between the two domains. The resistance and capacitance models for the semi-cylindrical domain are derived from the static solution of the heat conduction equation.

Since the width of the water channel is significantly smaller than that of the underlying layers, we can model the system in a cylindrical representation. Here, the water channel is defined by a cylinder with a radius equal to the real channel height. The water channel is separated from the heat sink by a thin thermal insulation layer, which will be discussed in more detail below. In the model, the insulation layer is represented by a cylindrical shell of thickness h_{ins} (Figure 4A). The outer surface of the heat sink, which is represented by the outermost cylindrical shell, is set constant at $-196\,^{\circ}\mathrm{C}$. Since the system has an insulating boundary on the top and only removes heat downward, the model only incorporates half a cylinder, sectioned from 0 to π. The steady state with the heater heating may be analytically solved (see Appendix A). In this steady state, the heater does not only heat the water channel but also the layers below the heater. Consequently, the insulation layer below the heater decreases the temperature rise in the heat sink below (see Figure 5). This is critical for two reasons: first, the required heater power is reduced and second, the energy stored in the cooled layers below the insulation is reduced, improving their ability to dissipate heat once the heater is turned off.

In addition to the semi-cylindrical domain, a linear domain is introduced to model the system with wider water channels (increase in b_{water}). Here, the water layer is no longer represented by a half-cylinder but through a rectangular domain defined by the width of the water channel b_{water} and its height h_{water}. The layers below are represented by their layer thickness and the width of the water channel. Both geometries are fused to a model using a lumped-element thermal equivalent circuit in the Cauer-type configuration. Each domain of each layer is modeled by a resistor R_{th} and a capacitor C_{th} (see Figure 4B). In each layer, a tangential resistor connects the domains. We implemented this model using ngspice and automated the model preparation, simulation, and data interpretation process in Julia [45]. The model parameters, material, and computational implementation are further described in Appendix C.

A further simplification of the model is presented in Appendix B. Here, the time constants τ_{water} and τ_{ins} of the different layers are used to estimate the expected cooling rate using an analytical equation. This estimator is also used below to explain the influence of the heat sink material and geometry.

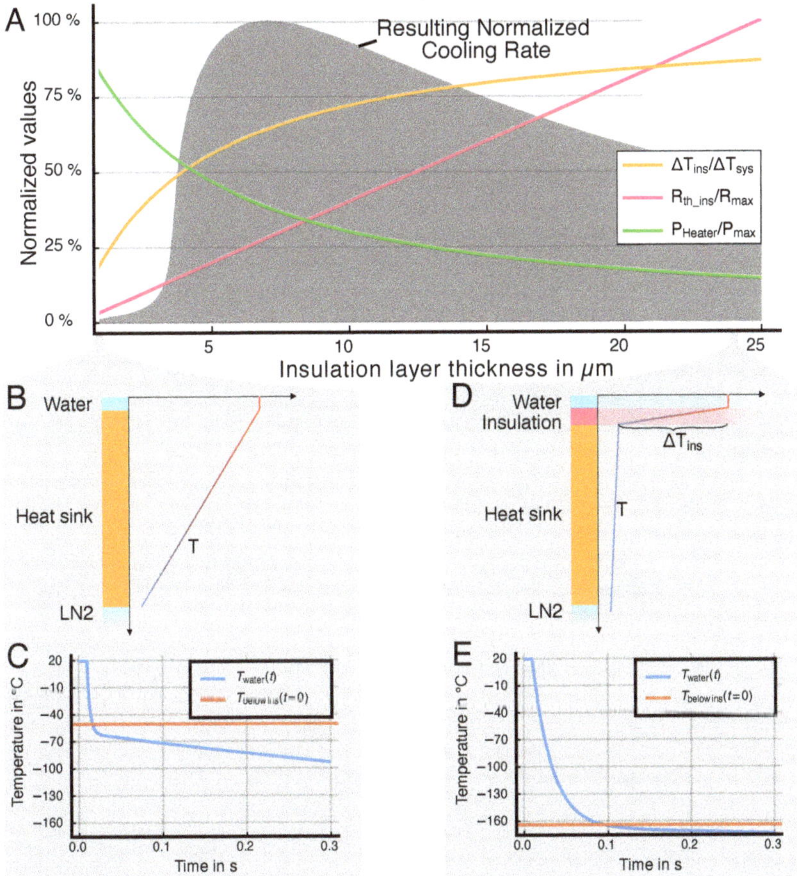

Figure 5. Introducing a thermal insulation layer below the heater reduces the temperature on top of the silicon layer, as the insulation layer now has a significant thermal gradient ΔT (**D**) and therefore reduces the temperature below the insulation layer $T_{below\ ins}$ while the heater is turned on (**A**). Accordingly, the insulation layer increases the cooling rate in the water channel, with the optimal height determined by a trade-off between temperature gradient and time constant. This is because the layers below store less energy before the cryofixation itself starts (and thus are able to conduct more energy away from the channel as soon as the heater is turned off) (**A**,**C**,**E**). The thermal resistance of the insulation layer will increase as the layer height increases. Consequently, the optimal insulation

layer forms a compromise between a high temperature gradient in the insulation layer and the sufficiently short time constant $\tau = R_{ins} * C_{ins}$ of this layer. If there is no insulation layer, the temperature gradient is spread over the whole system, which reduces the cooling rate (**B**,**C**). The plots show the results of simulations using the domain model developed in this paper.

4. Device Fabrication and Assembly of the Steady-State Cryofixation System

The manufacturing process involves the fabrication of the heater on the insulation layer, the fabrication of the microfluidic channel, establishment of micro-to-macro fluidic interfaces, and the assembly of the system to ensure proper thermal contact between the channel and the heater [39,46].

The heater is fabricated on a double-sided polished silicon wafer using standard microfabrication techniques. First, the insulation layer, made of polyimide (PI-2610 from HD Microsystems), is spin-coated on the silicon wafer and cured. Next, the polyimide surface is roughened using a CF_4 plasma to improve the adhesion of the metallization for the heater [47–49]. The Ti heater is then deposited by electron-beam evaporation and patterned using photolithography and wet etching. Electrical contact with the heater is established using an electron-beam-evaporated Au layer on top of the Ti heater. The backside of the heater is metallized with Cr (20 nm), Cr (100 nm), and Cu (300 nm) and soldered to the copper heat sink using indium.

In parallel to the heater fabrication, the microfluidic channel is created using a separate microfabrication process [50]. First, the negative of the microfluidic channel is patterned on a silicon wafer using SU-8 photoresist to create a master mold. A parylene coat on the master mold ensures easy demolding of the PDMS microfluidic channel. Next, the PDMS is spin-coated on the master mold, partially cured on a hotplate, and then cure-bonded to a mechanically strong PDMS structure. This enables an improved peeling process from the wafer. After peeling, the open PDMS channels are closed by plasma bonding a spin-coated 4 µm thick PDMS layer. The low thickness of this layer is critical to ensure a low thermal resistance between water and the heater. The backside of the PDMS channel structure is then plasma-bonded to a silicon chip with contact holes, which allows fluidic connection to the system.

The fabrication method offers flexibility in customizing the water channel dimensions. To modify the channel thickness, a new SU-8 master mold is created with the desired specifications. The thickness of the resulting water channel is controlled by varying the spin speed during the SU-8 deposition process. Additionally, the channel's width and length can be tailored by altering the layout design of the master mold. This versatility allows for precise control over the channel geometry to meet specific experimental requirements.

The thermal connection between heater and channel is enabled by placing the PDMS channel on top of the heater. The channel is connected to a fluidic system that is fixed in position to ensure the stability of the thermal contact during operation. The high thermal conductivity of indium guarantees a low thermal resistance between the heater and the copper heat sink, which is placed in a reservoir that can be filled with liquid nitrogen to establish the steady state. The dimensions of the system are listed in Table 1.

Table 1. Dimensions of the fabricated and assembled steady-state cryofixation system.

Parameter		Value
Water channel thickness	h_{water}	20 µm
Water channel width	b_{water}	100 µm
Water channel length	l_{water}	1.8 mm
PDMS layer between heater and water	h_{pdms}	4 µm
Insulation (PI) layer thickness	h_{ins}	4 µm
Silicon wafer thickness	h_{si}	500 µm
Heater width	b_{heater}	400 µm
Heater length	l_{heater}	3.5 mm

5. Results

5.1. Validating the Domain Model

To validate the domain model, we compare its predictions to a geometrically accurate representation of the system in a finite element modeling simulation. Here, we compare both approaches with a set of 100 randomly generated geometries within the parameter bounds given in Table 2.

Table 2. Parameters and variation range for the validation of the domain model in comparison with the FEM model.

Parameter		Min	Max
Water layer thickness	h_{water}	1 µm	50 µm
Insulation layer thickness	h_{ins}	0.1 µm	15 µm
Silicon layer thickness	h_{si}	1 µm	600 µm
Copper layer thickness	h_{cu}	0.5 mm	15 mm
Heater and water width	b_{water}	0.1 mm	5 mm

The domain model demonstrated strong agreement with the finite element model (FEM) results, with an R-squared value of 0.9279 (see Figure 6). This high correlation extends across the entire range of input parameters tested, suggesting robust performance under varied conditions. The close alignment between the domain model and FEM simulations indicates that the simplified approach effectively captures the key thermal conduction processes in the cryofixation system. Consequently, the domain model could serve as an efficient alternative to full FEM simulations for many applications, offering a balance between computational speed and accuracy. Several outliers can be explained by the geometry parameter combinations in these cases leading to a configuration where the given assumptions of the domain model are less valid (e.g., the assumption of a concentric temperature distribution no longer holds when the thickness of the heat sink is too small).

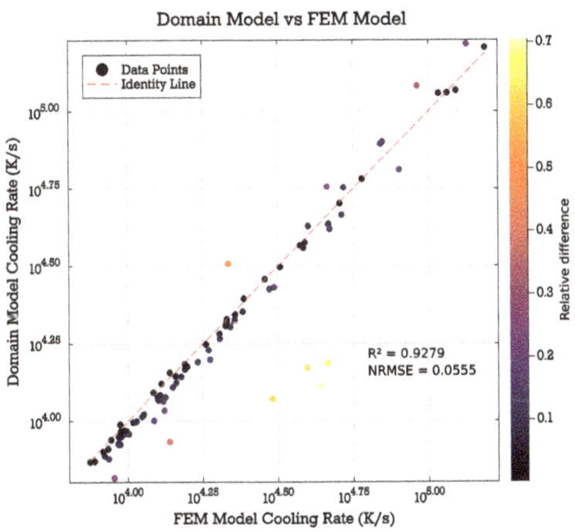

Figure 6. Correlation between calculated cooling rates of the domain model vs. the FEM model. An R-squared value of 0.9279 and a normalized root mean squared error (NRMSE) of 0.0555 indicate a good correlation between both models.

5.2. Experimental Validation of the Domain Model

In [46], Fuest et al. measured the cooling rate in a steady-state cryofixation system. Here, the cooling rate was measured by observing the change in intensity of a temperature-dependent fluorescent dye (Rhodamine B) during cooldown. The cooling rate was measured to be in the range of $2 \times 10^4\,\mathrm{K\,s^{-1}}$. To properly model the thermal conduction in the water channel, the bottom layer of the channel made of PDMS ($h_{pdms} = 4\,\mu m$ as described in [50]) is incorporated into the model with an equivalent electrical network layer similar to the water layer. The domain model estimates a cooling rate of $2.5 \times 10^4\,K\,s^{-1}$ for this system. This is in good agreement with the measured cooling rate. Neglecting the PDMS layer in the model would result in a significantly higher predicted cooling rate, leading to a substantial deviation from the experimental observations. The remaining discrepancies between the model and the measurements can be attributed to non-ideal thermal contacts in the experimental setup. Overall, the presented model can estimate the cooling rate in the system with sufficient accuracy. Therefore, the insights from the domain model can guide the optimization of the system for a high cooling rate with a sufficiently low heater power.

To further qualify the results, we measured the temperature change in the fabricated steady-state cryofixation system (as described in Table 1) using two experimental methods:

- **Fluorescent dye method:** Similar to the method in [46], the fluorescent intensity of the fluorescent dye Rhodamine B is used to indicate the temperature changes in the water channel. This intensity is measured using an sCMOS Camera (Andor Neo) with a frame rate of 413 Hz. A filter cube with a 546/10 nm excitation filter and a 575 nm emission filter reduces the influence of ambient light and improves the signal-to-noise ratio of the fluorescence measurements. As the temperature-dependency of the intensity of the dye is only indicative for temperatures above −21 °C [51], the initial intensity change is used to indicate the cooling rate in the channel.
- **Heater resistance method:** The resistance of the actively powered heater changes with its temperature. This is possible as the resistance of the heater depends on the bulk temperature of the heater with the temperature coefficient of resistance α_R. Using titanium as the heater material results in a robust heater with a significant temperature coefficient $\alpha_R = 0.003\,K^{-1}$ [52]. We measured both the current and voltage with 100 kHz. After an initial calibration run, the resistance is then translated to a temperature at the heater. This temperature is then used to derive the heat flow into the water channel and calculate the temperature in the channel. Here, a model of just the PDMS channel containing the water layer is used (see Figure 7).

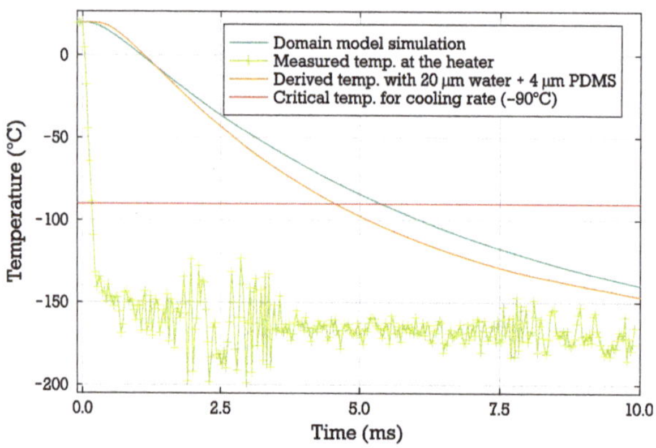

Figure 7. The measured temperature at the heater below the water channel is extracted by measuring the resistance during cryofixation. This temperature measurement is then used to derive the temperature at the top of the water channel by modeling the water channel as a layer structure of a 4 µm PDMS layer and a 20 µm water layer. Compared to the domain model, the cooling transient results in a comparable cooling rate in the water.

We extract a cooling rate of 23,782 K s^{-1} from the heater resistance measurement with a channel height $h_{water} = 20\,\mu m$, while the domain model yields an expected cooling rate of 20,407 K s^{-1} (see Table 3), which shows a good agreement. Optical measurements using the temperature-dependent fluorescent dye suggest a cooling rate in the range of $2 \times 10^4\,\mathrm{K\,s^{-1}}$, which is consistent with the expected cooling rate of the system.

Table 3. Measured heater power and cooling rates of the cryofixation system.

	Domain Model	Measured
Heater power	10.2 W	12.0 W
Cooling rate at heater	867,009 $\mathrm{K\,s^{-1}}$	367,853 $\mathrm{K\,s^{-1}}$
Cooling rate at top of water	20,407 $\mathrm{K\,s^{-1}}$	23,782 $\mathrm{K\,s^{-1}}$

5.3. *Effect of the Heat Sink*

The heat sink does not ideally transport the heat away from the heater. Therefore, we need to consider the impact of the solid support underneath the micro-hotplate on the cooling rate. The solid support is assumed to comprise a silicon substrate and a heat sink made of copper or other materials with high heat conductivity (e.g., diamond). Due to the finite thermal conductivity of these materials, a thermal gradient arises between the insulation layer and the LN2 reservoir.

Both the numerical results of the domain model and the analytical representation of the cooling rate estimator (Appendix B, Equation (A23)) show that the cooling rate cannot exceed a given maximum by changing the insulation layer alone. If the time constant of the insulation layer τ_{ins} is very small compared to the time constant of the water layer τ_{w} and the temperature difference in the insulation layer ΔT_{ins} is sufficiently large, the dominating effect limiting the cooling rate is defined through the material properties of water. Therefore, maximizing ΔT_{ins} is critical for attaining high cooling rates with the steady-state system. This is achieved by optimizing the design and the choice of materials. As the combined thickness of the silicon and the heat sink, h_{heatsink}, only influences ΔT_{ins} through a logarithmic function, its effect is very diminished compared to their effective heat conductivity k_{heatsink}. Here, a 1000-fold value of $k_{\mathrm{heatsink}}/k_{\mathrm{ins}}$ does have a significantly stronger effect than a change in geometry. This effect can also be observed in Figure 8.

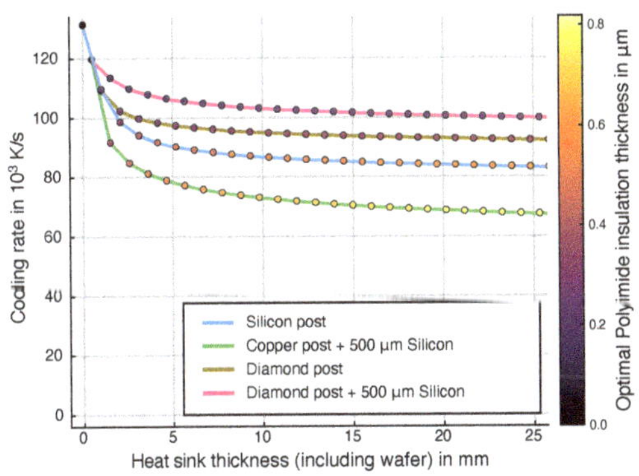

Figure 8. Changing the height of the heat sink below the silicon wafer only influences the achievable cooling rate in a water channel of 20 µm height significantly if the heat sink heights are comparable. A more significant effect may be seen if the material of the heat sink is changed to a material with better conduction and more optimal specific heat capacity (e.g., silicon or diamond). The simulations were performed using the domain model described in Section 3.

The thermal conductivity of silicon increases drastically at low temperatures. This can increase the maximum cooling rate significantly. With an optimized insulation layer, the silicon layer can reach temperatures down to the range of −180 °C to −196 °C while the heater is turned on. According to [53], the thermal conductivity of silicon reaches values of

950 $W\,m^{-1}\,K^{-1}$ to 1670 $W\,m^{-1}\,K^{-1}$ in this temperature range compared to 130 $W\,m^{-1}\,K^{-1}$ at room temperature. As the heat conduction is modeled with resistors, the model incorporates this effect by replacing the resistors for the silicon layer with voltage-dependent resistors. The resulting increase in cooling rate depends on the geometry of the water channel and heater and the heat sink length. The increase ranges from 115 % to 360 % for silicon.

Overall, we see that the heat sink can influence the cooling rate significantly if a high thermal conductivity is provided. Changing the thickness of this layer does have a negligible influence as soon as the thickness is on a larger scale than that of the water layer.

5.4. Size Limitations for Cryofixation in the Steady State System

The cooling rates in the steady-state system reflect the strong influence of the water channel height (Figure 9A). In an optimized system with a diamond heat sink, samples up to 5.4 μm would reach a cooling rate higher than $10^6\,K\,s^{-1}$. If a copper heat sink is used, this limit is reduced to 2.8 μm (Figure 9B). The insulation layer thickness needs to be optimized for each water channel configuration to reach the most optimal results. Other parameters, such as the dimensions of the geometries below the insulation layer, only have a minimal influence on the cooling rate of the system compared to the water channel thickness.

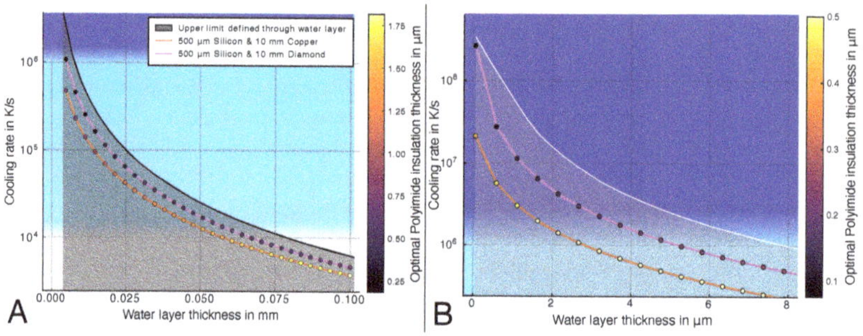

Figure 9. Modeled cooling rates for thick (**A**) and thin (**B**) water layer thicknesses (using the domain model) with varying heat sink materials and thicknesses. Changes in the water and sample layer thickness result in the most significant change in cooling rate in the steady-state system. The resulting curves follow the shape of the natural upper limit imposed by the cooling rate of the water layer itself, which is limited by the thermal conductivity of water.

In contrast, increasing the width of the water channel b_{water} leads to a linear increase in necessary heater power and a decrease in the available cooling rate in the water channel. For heater and channel widths $b_{water} \leq 1\,mm$ and a heat sink length of $h_{heatsink} = 11\,mm$, a cooling rate $> 10^6\,K\,s^{-1}$ can be achieved (e.g., with $h_{water} = 2\,\mu m$). For larger water channel widths up to a few mm, the cooling rate stays above $10^4\,K\,s^{-1}$. The heater power is linearly proportional to the width of the heater if the insulation layer height is fixed. However, if this layer height is optimized for maximum cooling rate, the necessary heater power for wider heaters increases substantially. As all heat heater power from the heater needs to be dissipated by the LN2, the critical heat flux [54] of the liquid nitrogen can impose a power limit to the system. Therefore, it can be necessary to increase the insulation layer thickness further from the cooling rate optimum until the resulting heater power is small enough. The resulting decreased cooling rate is smaller but within the same magnitude as the system with an insulation layer height optimized for maximum cooling rate.

If the cooling rate in a cryofixation system is limited by only the thermal conductivity of the water layer, the system is able to absorb a heat flow exceeding the maximum heat flow through the water layer under the most optimal cryofixation conditions. This limiting heat flow can be estimated with the following equation:

$$\dot{Q} = k_{\text{water}} \cdot b_{\text{water}} \cdot l_{\text{water}} \cdot \frac{T_H - T_{N2}}{h_{\text{water}}} \quad (7)$$

where l_{water} describes the length of the water channel in the z-direction.

In the steady-state system, the thermal resistance of the insulation layer is small enough so that this heat flow can be sustained. At the same time, this same layer guarantees a sufficiently large ΔT_{ins}. On the other hand, equilibrium systems dissipate this heat rapidly through transfer to a cryogenic gas or liquid. This is usually achieved by applying a cryogenic fluid directly to the sample or to the backside of the sample support. Here, the heat conductor needs to have a thermal resistance significantly lower than the thermal resistance of water to avoid impacting the cooling rate. Other established methods like plunge freezing need to not only dissipate heat from the water layer but also from a supporting structure (e.g., EM grids), which has a negative impact on achievable cooling rates.

6. Conclusions and Discussion

Our results demonstrate that the performance of microfluidic cryofixation systems based on solid-supported micro-hotplates can be optimized by tuning three key parameters: the insulation layer thickness (h_{ins}), the thermal conductivity of the solid support and heat sink materials (k_{heatsink}), and the width of the water channel and heater (b_{water}). When these parameters are optimized, the cooling rate becomes limited only by the thermal time constants of the water layer and the insulation layer. Equilibrium systems based on cryo-spraying or plunging are mainly limited by the thickness of the sample carrier and its thermal conductivity. Here, the thermal conductivity, heat capacity, and boiling point of the cryogenic fluid (e.g., pressurized LN2) will have an increased impact compared to in microfluidic steady-state systems.

The most fundamental limit with regard to the size of the sample is dictated by the thermal conductivity of water. Including a steady-state system with limited thermal conduction further decreases the cooling rate in the sample. Using a diamond heat sink, the theoretical maximum cooling rate of an aqueous sample is diminished to 22 % (for $h_{\text{water}} = 1\,\mu\text{m}$) and 66 % (for $h_{\text{water}} = 50\,\mu\text{m}$). The maximum cooling rate is reduced to 13 % (for $h_{\text{water}} = 1\,\mu\text{m}$) and 55 % (for $h_{\text{water}} = 50\,\mu\text{m}$) for copper heat sinks. This attenuation in the cooling rate is at its maximum for low channel heights and decreases for increasing channel heights.

The heat sink can be scaled in size without significantly reducing the available cooling rate in the water layer if the water channel width is significantly smaller than the overall system size. This is possible because the insulation layer below the heater prevents substantial heating of the underlying structures while the system is in its steady state with the sample at room temperature. As the thermal resistance of the heat sink increases, the power in the steady state must be reduced by increasing the thickness of the insulation layer. The best results for large specimen sizes will be reached with an intelligent choice of the materials in this setup. As shown in Figure 9, structures up to 50 µm should be able to reach a sufficient cooling rate using a diamond–silicon–polyimide system.

Using a very wide microfluidic channel ($b_{\text{water}} < 1\,\text{mm}$) or reservoir on a polyimide–silicon–copper system, we expect that full vitrification of pure water samples $< 2\,\mu\text{m}$ is possible according to our models. If larger sample sizes are desired, the copper heat sink needs to be replaced with a material of increased thermal conductivity. Similarly, increasing the width of the reservoir and, consequently, the heater will have an impact on the cooling rate in the water channel, but it will stay above $10^4\,\text{K}\,\text{s}^{-1}$ for heater widths of the same magnitude. The linear increase in heater power consumption with the heater width may be a practical limit in the steady-state system. Therefore, these more extreme configurations need an optimized system with highly conducting materials like diamond or silicon carbide.

The principle described for the steady-state system could also be interesting for other applications requiring a high cooling rate. Some examples are PCR analysis, microactuation, and micromixing as described in the introduction. Additionally, the steady-state system has the advantage of being able to create steep temperature gradients at the edge of the heater. These gradients can be used to improve the control of effects in the microfluidic domain. To actuate pico- and nano-droplets in digital microfludics, these steep gradients could increase thermocapillary effects [24,55–57]. Additionally, the steep temperature gradients achievable with this system could potentially enable temperature gradient focusing (TGF) techniques for separating ionic species in solution [58,59]. Furthermore, an array of micro-heaters could be used to create a pattern of locally controlled temperature zones, potentially allowing for selective deactivation of organisms in specific areas while maintaining optimal conditions for others in the same system.

Author Contributions: Conceptualization, D.B.T. and T.P.B.; methodology, D.B.T., T.P.B. and G.S.; software, D.B.T.; formal analysis, D.B.T.; investigation, D.B.T. and G.S.; funding aquisition, T.P.B.; visualization, D.B.T., writing—original draft, D.B.T.; writing—review and editing, T.P.B. and G.S. All authors have read and agreed to the published version of the manuscript.

Funding: This project was funded by the European Research Council under the European Union's Horizon 2020 research and innovation program (No. 772441).

Data Availability Statement: Data is contained within the article.

Conflicts of Interest: The authors declare no conflicts of interest.

Appendix A. Cylindrical Static Temperature Distribution

If we assume the width of the heater and the water channel to be significantly smaller than the width of the underlying structure, we can simplify the geometry as a semi-cylindrical system. Here, we start with the heat equation in cylindrical coordinates with constant temperature in the z-direction:

$$\frac{1}{r}\frac{\partial}{\partial r}\left(rk\frac{\partial T}{\partial r}\right) = \rho c\frac{\partial T}{\partial t} \tag{A1}$$

The water layer is represented as the innermost cylinder with h_{water}. The subsequent layers are layered around this cylinder with their respective thicknesses.

$$r_{\text{ins}} = h_{\text{water}} \tag{A2}$$

$$r_{\text{si}} = r_{\text{ins}} + h_{\text{ins}} \tag{A3}$$

$$r_{\text{cu}} = r_{\text{si}} + h_{\text{si}} \tag{A4}$$

$$r_{\text{system}} = r_{\text{cu}} + h_{\text{cu}} \tag{A5}$$

Each layer has its own temperature distribution. The boundary conditions are as follows:

$$T_{\text{water}}(r) = T_{\text{Heater}} = 20\,^{\circ}\text{C} \tag{A6}$$

$$T_{\text{cu}}(r) = T_{\text{LN2}} = -196\,^{\circ}\text{C} \tag{A7}$$

Each layer interface between two layers at r_{int} is modeled with a perfect contact:

$$k_1 \cdot \left.\frac{\partial T_1(r)}{\partial r}\right|_{r=r_{\text{int}}} = k_2 \cdot \left.\frac{\partial T_2(r)}{\partial r}\right|_{r=r_{\text{int}}} \tag{A8}$$

The general solution for the given PDE is as follows:

$$T_i(r) = \frac{C_{i_1}}{k_i} \ln(r) + C_{i_2} \tag{A9}$$

Solving with the boundary conditions yields the following:

$$T_{\text{water}}(r) = T_{\text{heater}} \tag{A10}$$

$$T_{\text{ins}}(r) = \frac{T_{c1} - T_{\text{heater}}}{\ln\left(\frac{r_{\text{si}}}{r_{\text{ins}}}\right)} \ln\left(\frac{r}{r_{\text{ins}}}\right) + T_{\text{heater}} \qquad (r_{\text{ins}} \leq r < r_{\text{si}}) \tag{A11}$$

$$T_{\text{si}}(r) = \frac{T_{c2} - T_{c1}}{\ln\left(\frac{r_{\text{cu}}}{r_{\text{si}}}\right)} \ln\left(\frac{r}{r_{\text{si}}}\right) + T_{c1} \qquad (r_{\text{si}} \leq r < r_{\text{cu}}) \tag{A12}$$

$$T_{\text{cu}}(r) = \frac{T_{\text{LN2}} - T_{c2}}{\ln\left(\frac{r_{\text{system}}}{r_{\text{cu}}}\right)} \ln\left(\frac{r}{r_{\text{cu}}}\right) + T_{c2} \qquad (r_{\text{cu}} \leq r < r_{\text{system}}) \tag{A13}$$

with

$$T_{c1} = -\frac{\xi_2 T_{\text{heater}} \ln\left(\frac{r_{\text{cu}}}{r_{\text{si}}}\right) + \xi_1 T_{\text{heater}} \ln\left(\frac{r_{\text{system}}}{r_{\text{cu}}}\right) + \xi_3 T_{\text{LN2}} \ln\left(\frac{r_{\text{si}}}{r_{\text{ins}}}\right)}{\xi_2 \ln\left(\frac{r_{\text{cu}}}{r_{\text{si}}}\right) + \xi_1 \ln\left(\frac{r_{\text{system}}}{r_{\text{cu}}}\right) + \xi_3 \ln\left(\frac{r_{\text{si}}}{r_{\text{ins}}}\right)} \tag{A14}$$

$$T_{c2} = -\frac{\xi_2 T_{\text{LN2}} \ln\left(\frac{r_{\text{cu}}}{r_{\text{si}}}\right) + \xi_1 T_{\text{heater}} \ln\left(\frac{r_{\text{system}}}{r_{\text{cu}}}\right) + \xi_3 T_{\text{LN2}} \ln\left(\frac{r_{\text{si}}}{r_{\text{ins}}}\right)}{\xi_2 \ln\left(\frac{r_{\text{cu}}}{r_{\text{si}}}\right) + \xi_1 \ln\left(\frac{r_{\text{system}}}{r_{\text{cu}}}\right) + \xi_3 \ln\left(\frac{r_{\text{si}}}{r_{\text{ins}}}\right)} \tag{A15}$$

$$\xi_1 = k_{\text{ins}} k_{\text{si}} \quad \xi_2 = k_{\text{cu}} k_{\text{ins}} \quad \xi_1 = k_{\text{cu}} k_{\text{si}} \tag{A16}$$

Using this equation system, the temperature distribution during heater operation can be modeled as cylindrical. To make the system more compliant with the actual setup, we only take the top half section of the cylinder, which does not influence the temperature distribution but the calculated power:

$$P = -k_{\text{si}} \pi L \frac{T_{c2} - T_{c2}}{\ln\left(\frac{r_{\text{cu}}}{r_{\text{si}}}\right)} \tag{A17}$$

This system can also be used to establish the equivalent resistances and capacitances for a Cauer-type network:

$$R_{th,si} = \frac{\Delta T}{P} = \frac{\ln\left(\frac{r_{\text{cu}}}{r_{\text{si}}}\right)}{k_{\text{si}} \pi L} \tag{A18}$$

$$C_{th,si} = \frac{Q_{stored}}{\Delta T} = c_p m = \frac{1}{2} c_p \rho \pi L (r_{cu}^2 - r_{si}^2) \tag{A19}$$

Appendix B. Limiting Case: Estimator for a Narrow Microchannel on a Micro-Hotplate with an Ideal Heat Sink

In the case of a narrow microfluidic channel and an ideal heat sink, a simple circuit model provides some interesting insights. Therefore, in this section we consider a model that includes only the water layer and the insulation layer. From the static cylindrical model (see Appendix A) we know that the temperature difference in the insulation layer can be expressed as

$$\Delta T_{\text{ins}} = \frac{\frac{k_{\text{heatsink}}}{k_{\text{ins}}} \ln\left(\frac{h_{\text{ins}} + h_{\text{w}}}{h_{\text{w}}}\right)}{\ln\left(\frac{h_{\text{heatsink}} + h_{\text{ins}} + h_{\text{w}}}{h_{\text{ins}} + h_{\text{w}}}\right) + \frac{k_{\text{heatsink}}}{k_{\text{ins}}} \ln\left(\frac{h_{\text{ins}} + h_{\text{w}}}{h_{\text{w}}}\right)} \cdot (T_{\text{H}} - T_{\text{LN2}}) \tag{A20}$$

where $T_{\mathrm{H}} = 20\,°\mathrm{C}$ is the temperature of the heater in the steady state and $T_{\mathrm{LN2}} = -196\,°\mathrm{C}$ is the temperature of the LN2. Using this definition, we reduce the network to the following formulation: We model the water layer (R_{w},C_{w}) and the insulation layer (R_{ins},C_{ins}) both as a single RC network using the Cauer representation. The heater between both layers is represented as a power source $P_h(t)$ and the temperature below the insulation layer is set to $T_H - \Delta T_{\mathrm{ins}}$. We can now express the system using the transfer function of the water temperature T_w and the heater power P_h in the complex frequency domain as follows:

$$\frac{T_w}{P_h}(s) = \frac{R_{\mathrm{ins}}}{1 + (R_{\mathrm{ins}}C_{\mathrm{ins}} + R_{\mathrm{w}}C_{\mathrm{w}} + R_{\mathrm{ins}}C_{\mathrm{w}})s + R_{\mathrm{ins}}R_{\mathrm{w}}C_{\mathrm{ins}}C_{\mathrm{w}}s^2} = \frac{R_{\mathrm{ins}}}{1 + \xi_1 s + \xi_2 s^2} \quad \text{(A21)}$$

As the layers described in this system are very thin, ξ_1 is dominant compared to ξ_2. Omitting ξ_2 and renaming $\xi_1 = \tau_{\mathrm{sys}}$ as a time constant yields the transient temperature of the water layer after an inverse transform:

$$T_w(t) = \Delta T_{\mathrm{ins}} * e^{-\frac{t}{\tau_{\mathrm{sys}}}} - \Delta T_{\mathrm{ins}} + T_H \quad \text{(A22)}$$

The used time constant τ_{sys} is the sum of the time constants of both layers $\tau_{\mathrm{ins}} = R_{\mathrm{ins}}C_{\mathrm{ins}}$ and $\tau_{\mathrm{w}} = R_{\mathrm{w}}C_{\mathrm{w}}$ and a combined time constant $\tau_{\mathrm{c}} = R_{\mathrm{ins}}C_{\mathrm{w}}$ that describes the time it takes for the stored heat (as described by C_{w}) to flow through the resistance of the insulation layer R_{ins}. This third time constant becomes relevant if the insulation layer height increases and slows down the heat flow through the insulation layer.

These time constants now form the estimator for the cooling rate. At $t = \tau_1$, 63 % of the temperature difference ΔT_{ins} is surpassed. Therefore, we extract the cooling rate with

$$CR \approx \frac{0.63 \cdot \Delta T_{\mathrm{ins}}}{\tau_{\mathrm{sys}}} = \frac{0.63 \cdot \Delta T_{\mathrm{ins}}}{\tau_{\mathrm{ins}} + \tau_{\mathrm{w}} + \tau_{\mathrm{c}}} \quad \text{(A23)}$$

Appendix C. Model Parameters and Computational Methods

The domain model allows the following parameters to be set before computation:

Table A1. Computational input parameters for the domain model.

Parameter	Symbol	Unit
System width (x-direction)	b_{system}	m
System length (z-direction)	L_{system}	m
For each layer (Water, PDMS, insulation, heat sink):		
Layer thickness	h_{water}	m
Thermal conductivity of water	k_{water}	$\mathrm{W\,m^{-1}\,K^{-1}}$
Specific heat capacity of water	$c_{\mathrm{p,water}}$	$\mathrm{J\,kg^{-1}\,K^{-1}}$
Density of water	ρ_{water}	$\mathrm{kg\,m^{-3}}$

Table A2. Thermal material parameters used in the simulations with the domain model and the reference FEM model [60–64].

	Water	PDMS	Polyimide
Thermal conductivity	$0.560\,\mathrm{W\,m^{-1}\,K^{-1}}$	$0.15\,\mathrm{W\,m^{-1}\,K^{-1}}$	$0.14\,\mathrm{W\,m^{-1}\,K^{-1}}$
Density	$999.9\,\mathrm{kg\,m^{-3}}$	$970\,\mathrm{kg\,m^{-3}}$	$700\,\mathrm{kg\,m^{-3}}$
Specific heat capacity	$4219.9\,\mathrm{J\,kg^{-1}\,K^{-1}}$	$1460\,\mathrm{J\,kg^{-1}\,K^{-1}}$	$2329\,\mathrm{J\,kg^{-1}\,K^{-1}}$
	Silicon	Copper	Diamond
Thermal conductivity	$130\,\mathrm{W\,m^{-1}\,K^{-1}}$	$401\,\mathrm{W\,m^{-1}\,K^{-1}}$	$2900\,\mathrm{W\,m^{-1}\,K^{-1}}$
Density	$2329\,\mathrm{kg\,m^{-3}}$	$8960\,\mathrm{kg\,m^{-3}}$	$3500\,\mathrm{kg\,m^{-3}}$
Specific heat capacity	$700\,\mathrm{J\,kg^{-1}\,K^{-1}}$	$384\,\mathrm{J\,kg^{-1}\,K^{-1}}$	$509\,\mathrm{J\,kg^{-1}\,K^{-1}}$

The domain model was implemented using ngspice (version 34) for circuit simulation. Model preparation, simulation, and data interpretation were automated using Julia (version

1.6.3) [45]. The parameters were converted into resistances and capacitors using the given formulae of the Cauer model and then inserted into the netlist for the ngspice simulation. Temperature-dependent thermal conductivites are represented as voltage-dependent functions that are based on an approximated analytic function of the thermal conductivity dependent on the temperature.

Finite element modeling (FEM) simulations for model validation were performed using COMSOL Multiphysics (version 6.0). The Heat Transfer in Solids interface was used for both a static and a time-dependent study.

References

1. Cavicchi, R.; Suehle, J.; Kreider, K.; Gaitan, M.; Chaparala, P. Fast temperature programmed sensing for micro-hotplate gas sensors. *IEEE Electron. Device Lett.* **1995** , *16*, 286–288. [CrossRef]
2. Simon, I.; Bârsan, N.; Bauer, M.; Weimar, U. Micromachined metal oxide gas sensors: Opportunities to improve sensor performance. *Sens. Actuators B Chem.* **2001**, *73*, 1–26. [CrossRef]
3. Semancik, S.; Cavicchi, R.E.; Kreider, K.G.; Suehle, J.S.; Chaparala, P. Selected-area deposition of multiple active films for conductometric microsensor arrays. *Sens. Actuators B Chem.* **1996**, *34*, 209–212. [CrossRef]
4. Xu, L.; Li, T.; Gao, X.; Wang, Y. Development of a Reliable Micro-Hotplate with Low Power Consumption. *IEEE Sens. J.* **2010**, *11*, 913–919. [CrossRef]
5. Korotcenkov, G.; Cho, B.K. Engineering approaches to improvement of conductometric gas sensor parameters. Part 2: Decrease of dissipated (consumable) power and improvement stability and reliability. *Sens. Actuators B Chem.* **2014**, *198*, 316–341. [CrossRef]
6. Astié, S.; Gué, A.M.; Scheid, E.; Lescouzères, L.; Cassagnes, A. Optimization of an integrated SnO_2 gas sensor using a FEM simulator. *Sens. Actuators A Phys.* **1998**, *69*, 205–211. [CrossRef]
7. Tsamis, C.; Nassiopoulou, A.G.; Tserepi, A. Thermal properties of suspended porous silicon micro-hotplates for sensor applications. *Sens. Actuators B Chem.* **2003**, *95*, 78–82. [CrossRef]
8. Sheng, L.y.; Tang, Z.; Wu, J.; Chan, P.C.H.; Sin, J.K.O. A low-power CMOS compatible integrated gas sensor using maskless tin oxide sputtering. *Sens. Actuators B Chem.* **1998**, *49*, 81–87. [CrossRef]
9. Jung, G.; Hong, Y.; Hong, S.; Jang, D.; Jeong, Y.; Shin, W.; Park, J.; Kim, D.; Jeong, C.B.; Kim, D.U.; et al. A low-power embedded poly-Si micro-heater for gas sensor platform based on a FET transducer and its application for NO_2 sensing. *Sens. Actuators B Chem.* **2021**, *334*, 129642. [CrossRef]
10. Fung, S.K.H.; Tang, Z.; Chan, P.C.H.; Sin, J.K.O.; Cheung, P.W. Thermal analysis and design of a micro-hotplate for integrated gas-sensor applications. *Sens. Actuators A Phys.* **1996**, *54*, 482–487. [CrossRef]
11. Suehle, J.; Cavicchi, R.; Gaitan, M.; Semancik, S. Tin oxide gas sensor fabricated using CMOS micro-hotplates and in-situ processing. *IEEE Electron Device Lett.* **1993**, *14*, 118–120. [CrossRef]
12. Graf, M.; Frey, U.; Taschini, S.; Hierlemann, A. Micro Hot Plate-Based Sensor Array System for the Detection of Environmentally Relevant Gases. *Anal. Chem.* **2006**, *78*, 6801–6808. [CrossRef] [PubMed]
13. Liu, Q.; Ding, G.; Wang, Y.; Yao, J. Thermal Performance of Micro Hotplates with Novel Shapes Based on Single-Layer SiO_2 Suspended Film. *Micromachines* **2018**, *9*, 514. [CrossRef] [PubMed]
14. Majhi, S.M.; Mirzaei, A.; Kim, H.W.; Kim, S.S.; Kim, T.W. Recent advances in energy-saving chemiresistive gas sensors: A review. *Nano Energy* **2021**, *79*, 105369. [CrossRef]
15. Domènech-Gil, G.; Samà, J.; Fàbrega, C.; Gràcia, I.; Cané, C.; Barth, S.; Romano-Rodríguez, A. Highly sensitive SnO_2 nanowire network gas sensors. *Sens. Actuators B Chem.* **2023**, *383*, 133545. [CrossRef]
16. Avraham, M.; Stolyarova, S.; Blank, T.; Bar-Lev, S.; Golan, G.; Nemirovsky, Y. A Novel Miniature and Selective CMOS Gas Sensor for Gas Mixture Analysis—Part 2: Emphasis on Physical Aspects. *Micromachines* **2020**, *11*, 587. [CrossRef]
17. Zhou, Q.; Sussman, A.; Chang, J.; Dong, J.; Zettl, A.; Mickelson, W. Fast response integrated MEMS microheaters for ultra low power gas detection. *Sens. Actuators A Phys.* **2015**, *223*, 67–75. [CrossRef]
18. Lee, S.H.; Park, S.m.; Kim, B.N.; Kwon, O.S.; Rho, W.Y.; Jun, B.H. Emerging ultrafast nucleic acid amplification technologies for next-generation molecular diagnostics. *Biosens. Bioelectron.* **2019**, *141*, 111448. [CrossRef]
19. Shaw, K.J.; Docker, P.T.; Yelland, J.V.; Dyer, C.E.; Greenman, J.; Greenway, G.M.; Haswell, S.J. Rapid PCR amplification using a microfluidic device with integrated microwave heating and air impingement cooling. *Lab Chip* **2010**, *10*, 1725–1728. [CrossRef]
20. Dong, X.; Liu, L.; Tu, Y.; Zhang, J.; Miao, G.; Zhang, L.; Ge, S.; Xia, N.; Yu, D.; Qiu, X. Rapid PCR powered by microfluidics: A quick review under the background of COVID-19 pandemic. *TrAC Trends Anal. Chem.* **2021**, *143*, 116377. [CrossRef]
21. Neuzil, P.; Zhang, C.; Pipper, J.; Oh, S.; Zhuo, L. Ultra fast miniaturized real-time PCR: 40 cycles in less than six minutes. *Nucleic Acids Res.* **2006**, *34*, e77. [CrossRef]
22. Yin, H.; Wu, Z.; Shi, N.; Qi, Y.; Jian, X.; Zhou, L.; Tong, Y.; Cheng, Z.; Zhao, J.; Mao, H. Ultrafast multiplexed detection of SARS-CoV-2 RNA using a rapid droplet digital PCR system. *Biosens. Bioelectron.* **2021**, *188*, 113282. [CrossRef] [PubMed]
23. Chen, X.; Liu, Y.; Zhan, X.; Gao, Y.; Sun, Z.; Wen, W.; Zheng, W. Ultrafast PCR Detection of COVID-19 by Using a Microfluidic Chip-Based System. *Bioengineering* **2022**, *9*, 548. [CrossRef] [PubMed]

24. Nguyen, N.T.; Huang, X. Thermocapillary Effect of a Liquid Plug in Transient Temperature Fields. *Jpn. J. Appl. Phys.* **2005**, *44*, 1139. [CrossRef]
25. Yin, Z.; Prosperetti, A. 'Blinking bubble' micropump with microfabricated heaters. *J. Micromech. Microeng.* **2005**, *15*, 1683. [CrossRef]
26. Guo, G.; Wu, X.; Liu, D.; Liao, L.; Zhang, D.; Zhang, Y.; Mao, T.; He, Y.; Huang, P.; Wang, W.; et al. A Self-Regulated Microfluidic Device with Thermal Bubble Micropumps. *Micromachines* **2022**, *13*, 1620. [CrossRef]
27. Qaiser, N.; Khan, S.M.; Babatain, W.; Nour, M.; Joharji, L.; Shaikh, S.F.; Elatab, N.; Hussain, M.M. A thermal microfluidic actuator based on a novel microheater. *J. Micromech. Microeng.* **2023**, *33*, 035001. [CrossRef]
28. Kim, S.J.; Wang, F.; Burns, M.A.; Kurabayashi, K. Temperature-Programmed Natural Convection for Micromixing and Biochemical Reaction in a Single Microfluidic Chamber. *Anal. Chem.* **2009**, *81*, 4510–4516. [CrossRef]
29. Mazur, P. Stopping Biological Time. *Ann. N. Y. Acad. Sci.* **1988**, *541*, 514–531. [CrossRef]
30. Fahy, G.M.; Wowk, B. Principles of Ice-Free Cryopreservation by Vitrification. In *Cryopreservation and Freeze-Drying Protocols*; Wolkers, W.F., Oldenhof, H., Eds.; Methods in Molecular Biology; Springer: New York, NY, USA, 2021; pp. 27–97. [CrossRef]
31. Yavin, S.; Arav, A. Measurement of essential physical properties of vitrification solutions. *Theriogenology* **2007**, *67*, 81–89. [CrossRef]
32. Sitte, H.; Edelmann, L.; Neumann, K. Cryofixation Without Pretreatment at Ambient Pressure. In *Cryotechniques in Biological Electron Microscopy*; Steinbrecht, R.A., Zierold, K., Eds.; Springer: Berlin/Heidelberg, Germany, 1987; pp. 87–113. [CrossRef]
33. Dobro, M.J.; Melanson, L.A.; Jensen, G.J.; McDowall, A.W. Chapter Three—Plunge Freezing for Electron Cryomicroscopy. In *Methods in Enzymology*; Jensen, G.J., Ed.; Cryo-EM Part A Sample Preparation and Data Collection; Academic Press: Cambridge, MA, USA, 2010; Volume 481, pp. 63–82. [CrossRef]
34. Dubochet, J. The Physics of Rapid Cooling and Its Implications for Cryoimmobilization of Cells. In *Methods in Cell Biology*; Cellular Electron Microscopy; Academic Press: Cambridge, MA, USA, 2007; Volume 79, pp. 7–21. [CrossRef]
35. Escaig, J. New instruments which facilitate rapid freezing at 83 K and 6 K. *J. Microsc.* **1982**, *126*, 221–229. [CrossRef]
36. Dahl, R.; Staehelin, L.A. High-pressure freezing for the preservation of biological structure: Theory and practice. *J. Electron Microsc. Technol.* **1989**, *13*, 165–174. [CrossRef]
37. Szabo, G.V.; Burg, T.P. Time Resolved Cryo-Correlative Light and Electron Microscopy. *Adv. Funct. Mater.* **2024**, *34*, 2313705. [CrossRef]
38. Huebinger, J.; Grecco, H.; Masip, M.E.; Christmann, J.; Fuhr, G.R.; Bastiaens, P.I.H. *Ultrarapid Cryo-Arrest of Living Cells on a Microscope Enables Multiscale Imaging of Out-of-Equilibrium Molecular Patterns*; American Association for the Advancement of Science: Washington, DC, USA, 2021. [CrossRef]
39. Fuest, M.; Nocera, G.M.; Modena, M.M.; Riedel, D.; Mejia, Y.X.; Burg, T.P. Cryofixation during live-imaging enables millisecond time-correlated light and electron microscopy. *J. Microsc.* **2018**, *272*, 87–95. [CrossRef] [PubMed]
40. Çengel, Y.A.; Ghajar, A.J. *Heat and Mass Transfer: Fundamentals & Applications*, 5th ed.; McGraw Hill Education: New York, NY, USA, 2015.
41. Zasadzinski, J.A.N. A new heat transfer model to predict cooling rates for rapid freezing fixation. *J. Microsc.* **1988**, *150*, 137–149. [CrossRef]
42. Ozsun, O.; Alaca, B.E.; Yalcinkaya, A.D.; Yilmaz, M.; Zervas, M.; Leblebici, Y. On heat transfer at microscale with implications for microactuator design. *J. Micromech. Microeng.* **2009**, *19*, 045020. [CrossRef]
43. Guo, Z.Y.; Li, Z.X. Size effect on microscale single-phase flow and heat transfer. *Int. J. Heat Mass Transf.* **2003**, *46*, 149–159. [CrossRef]
44. Hu, X.J.; Jain, A.; Goodson, K.E. Investigation of the natural convection boundary condition in microfabricated structures. *Int. J. Therm. Sci.* **2008**, *47*, 820–824. [CrossRef]
45. Bezanson, J.; Edelman, A.; Karpinski, S.; Shah, V.B. Julia: A Fresh Approach to Numerical Computing. *SIAM Rev.* **2017**, *59*, 65–98. [CrossRef]
46. Fuest, M.; Schaffer, M.; Nocera, G.M.; Galilea-Kleinsteuber, R.I.; Messling, J.E.; Heymann, M.; Plitzko, J.M.; Burg, T.P. In situ Microfluidic Cryofixation for Cryo Focused Ion Beam Milling and Cryo Electron Tomography. *Sci. Rep.* **2019**, *9*, 19133. [CrossRef]
47. Ranucci, E.; Sandgren, Å.; Andronova, N.; Albertsson, A.C. Improved polyimide/metal adhesion by chemical modification approaches. *J. Appl. Polym. Sci.* **2001**, *82*, 1971–1985. [CrossRef]
48. Oh, T.; Kowalczyk, S.; Hunt, D.; Kim, J. Adhesion enhancement of Cr/polyimide interfaces using RF pre-sputtering of polyimide surfaces. *Adhes. Sci. Technol.* **1990**, *4*, 119–129. [CrossRef]
49. Kim, S.H.; Na, S.W.; Lee, N.E.; Nam, Y.W.; Kim, Y.H. Effect of surface roughness on the adhesion properties of Cu/Cr films on polyimide substrate treated by inductively coupled oxygen plasma. *Surf. Coatings Technol.* **2005**, *200*, 2072–2079. [CrossRef]
50. Fuest, M.; Nocera, G.M.; Modena, M.M.; Burg, T.P. Fabrication Advances for Microfluidic Cryofixation. In Proceedings of the MikroSystemTechnik 2017, Congress, Proceedings of VDE, Munich, Germany, 23–25 October 2017; pp. 1–4.
51. Ross, D.; Gaitan, M.; Locascio, L.E. Temperature Measurement in Microfluidic Systems Using a Temperature-Dependent Fluorescent Dye. *Anal. Chem.* **2001**, *73*, 4117–4123. [CrossRef] [PubMed]
52. Singh, B.; Surplice, N.A. The electrical resistivity and resistance-temperature characteristics of thin titanium films. *Thin Solid Films* **1972**, *10*, 243–253. [CrossRef]

53. Silva, E.C.F.d.; Strauch, D.; Rößler, U. (Eds.) *Numerical Data and Functional Relationships in Science and Technology; Subvol. E: Group 3: Condensed Matter; Volume 44. Semiconductors New Data and Updates for Several III–V (Including Mixed Crystals) and II–VI Compounds/ed.: U. Rössler. Authors: E.C.F. da Silva and D. Strauch*; Springer: Berlin/Heidelberg, Germany, 2012; Volume 44.
54. Liang, G.; Mudawar, I. Pool boiling critical heat flux (CHF)—Part 1: Review of mechanisms, models, and correlations. *Int. J. Heat Mass Transf.* **2018**, *117*, 1352–1367. [CrossRef]
55. Choi, K.; Ng, A.H.; Fobel, R.; Wheeler, A.R. Digital Microfluidics. *Annu. Rev. Anal. Chem.***2012**, *5*, 413–440. [CrossRef]
56. Karbalaei, A.; Kumar, R.; Cho, H.J. Thermocapillarity in Microfluidics—A Review. *Micromachines* **2016**, *7*, 13. [CrossRef]
57. Darhuber, A.A.; Valentino, J.P.; Troian, S.M. Planar digital nanoliter dispensing system based on thermocapillary actuation. *Lab Chip* **2010**, *10*, 1061–1071. [CrossRef]
58. Ross, D.; Locascio, L.E. Microfluidic Temperature Gradient Focusing. *Anal. Chem.* **2002**, *74*, 2556–2564. [CrossRef]
59. Matsui, T.; Franzke, J.; Manz, A.; Janasek, D. Temperature gradient focusing in a PDMS/glass hybrid microfluidic chip. *Electrophoresis* **2007**, *28*, 4606–4611. [CrossRef] [PubMed]
60. COMSOL Inc. *COMSOL Multiphysics Material Library (Version 6.0)*; COMSOL Inc.: Burlington, MA, USA, 2021.
61. MatWeb LLC. *Online Materials Information Resource—MatWeb*; MatWeb LLC: Blacksburg, VA, USA, 2011.
62. HD MicroSystems GmbH. *PI-2600 Series—Low Stress Applications*; HD MicroSystems GmbH: San Jose, CA, USA, 2009.
63. Fulkerson, W.; Moore, J.P.; Williams, R.K.; Graves, R.S.; McElroy, D.L. Thermal Conductivity, Electrical Resistivity, and Seebeck Coefficient of Silicon from 100 to 1300°K. *Phys. Rev.* **1968**, *167*, 765–782. [CrossRef]
64. Berman, R.; Simon, F.E.; Ziman, J.M. The Thermal Conductivity of Diamond at Low Temperatures. *Proc. R. Soc. Lond. Ser. A Math. Phys. Sci.* **1953**, *220*, 171–183.

micromachines

Review

Advances in Microfluidic Single-Cell RNA Sequencing and Spatial Transcriptomics

Yueqiu Sun [1,2], Nianzuo Yu [1,2,*], Junhu Zhang [1,2,*] and Bai Yang [1,2]

1 State Key Laboratory of Supramolecular Structure and Materials, College of Chemistry, Jilin University, Changchun 130000, China
2 Joint Laboratory of Opto-Functional Theranostics in Medicine and Chemistry, The First Hospital of Jilin University, Jilin University, Changchun 130000, China
* Correspondence: ynz@jlu.edu.cn (N.Y.); zjh@jlu.edu.cn (J.Z.)

Abstract: The development of micro- and nano-fabrication technologies has greatly advanced single-cell and spatial omics technologies. With the advantages of integration and compartmentalization, microfluidic chips are capable of generating high-throughput parallel reaction systems for single-cell screening and analysis. As omics technologies improve, microfluidic chips can now integrate promising transcriptomics technologies, providing new insights from molecular characterization for tissue gene expression profiles and further revealing the static and even dynamic processes of tissues in homeostasis and disease. Here, we survey the current landscape of microfluidic methods in the field of single-cell and spatial multi-omics, as well as assessing their relative advantages and limitations. We highlight how microfluidics has been adapted and improved to provide new insights into multi-omics over the past decade. Last, we emphasize the contributions of microfluidic-based omics methods in development, neuroscience, and disease mechanisms, as well as further revealing some perspectives for technological advances in translational and clinical medicine.

Keywords: microfluidics; single cell; single-cell RNA sequencing (scRNA-seq); spatial transcriptome

1. Introduction

Tissue is defined as a unit composed of cells and varying amounts of extracellular matrix, which is the structural and functional part of an organ [1]. Therefore, understanding the tissue microenvironment is the key to unraveling the complex flow of genetic information in living systems. In eukaryotes, revealing the high degree of tissue heterogeneity, both cellular and spatial, in physiological and biochemical phenomena is being increasingly regarded as essential for exploring developmental and disease mechanisms. Various single-cell sequencing (SCS) technologies enable the acquisition of gene expression data at the single-cell level, designed to increase the understanding of cellular diversity and interactions in a physiological and specifically immunological context [2–7]. Spatial transcriptome technologies are capable of distinguishing the gene expression of cells in their original locations, thus compensating for the loss of spatial location information during tissue dissociation in SCS technologies [8–10]. Notably, the requirements for resolution, throughput, and multiplexing have strongly motivated the development of new solutions for both single-cell and spatial transcriptome sequencing, aiming to generate more detailed molecular cellular landscapes in tissues to unravel cell diversity and explore tissue architectures [11–13].

 https://doi.org/10.3390/mi16040426

Microfluidics, a tool that integrates microfabrication processes such as microchannels and microvalves, enables the processing of individual cells and their components through the precise control and manipulation of microscale fluids [14]. In the last decade, microfluidics has redefined the experimental platform for SCS and spatial omics sequencing by parallelization, miniaturization, and automatic methodological steps [15,16]. First, assay miniaturization in microfluidics improves enzyme reaction activity and reduces assay costs significantly. While traditional high-throughput single-cell analysis requires a reaction system of at least ~1 μL [17–19], each parallel reaction system in a microfluidic system ranges from a few picoliters to a few nanoliters, thereby improving the concentrations of reagents and reducing reagent consumption [20–22]. The water-in-oil system in Drop-seq is capable of generating approximately 3000 droplets per second with a volume of around 1 nL [23]. DroNc-seq, a modification of the Drop-seq method, can generate droplets with a volume of 300 pL at a higher rate (roughly 4500 droplets per second) [24]. Such improvement allows for scRNA-seq to be performed on cell nuclei. Second, parallel microchannels, microarrays, and microvalves can be used simultaneously for the precise manipulation of fluids to minimize batch effects. When combined with sequencing tools, massive parallel sequencing microdevices are capable of capturing and analyzing at the single-cell scale, which provides a chance for low-cost, high-throughput-based transcriptome analysis and opens up unprecedented opportunities to reveal cellular heterogeneity and its effects on cellular function. For example, Seq-well was able to profile thousands of primary human macrophages in a polydimethylsiloxane (PDMS) chip consisting of ~86,000 subnanoliter wells [21]. Similarly, when used for spatial transcriptome technology, microdevices allow for yielding spatial barcodes unbiasedly through parallel microchannels, resulting in thousands of different spatial spots [25]. Third, a single microfluidic chip enables the integration of multiple operations and compatibility with downstream measuring techniques. With the development of micro/nano manufacturing technology, experimenters can design and fabricate customized microfluidic chips for different clinical research applications. Collectively, as efficient, high-throughput integrated platforms, microfluidics can be applied to SCS and spatial omics sequencing to reveal the diversity of cell types in highly heterogeneous tissues, with further broad applications in developmental, neurological disease research, and more [26,27].

In this review, we survey the current landscape of microfluidic methods in the field of single-cell and spatial multi-omics and their operational principles, as well as assessing their relative advantages and limitations. Additionally, the applications of microfluidic-based methods in a wide range of fields including neuroscience, developmental biology, cancer, and more will be addressed. Last, we discuss future opportunities in the field of translational and clinical medicine.

2. Advances in Single-Cell RNA Sequencing Technologies

Traditional bulk sequencing methods obtain the average data for the gene expression profiles of cell populations [28,29], resulting in the consequent concealment of information at the single-cell level, such as molecular distributions and drug–target interactions. Thus, to understand the single-cell different expression (SCDE), new tools need to be developed to advance single-cell sorting and analysis. In 2009, Tang et al. published the pioneering work of single-cell genome-wide mRNA sequencing (scRNA-seq) [30], thereafter, the field exploded and received great attention (Table 1). Specifically, a significant increase in the number of cell types that can be processed with the throughput has increased from a few to several thousand.

SCS technologies can be distinguished as being either plate-based methods or microfluidic-based methods. Plate-based methods [19], represented by Smart-seq2 [17], are capable of obtaining full-length transcripts. However, the lack of suitable cell segregation strategies results in the quantity of cells available for analysis limiting the throughput of such methods [31]. Compared to conventional plate-based methods, the probability of sample cross-contamination can be reduced with microfluidic-based scRNA-seq methods, based on both microfluidics [20,23] and high-density microwells [21,32], resulting from the cell suspension being partitioned into nanoliter reaction chambers [20–22]. Importantly, this enables microfluidic platforms to be tailored to be compatible with different cell sizes [33,34], providing an expandable framework for a wide range of biological samples [20]. Given the parallelization, massive parallel omics sequencing, adopting microfluidics as a platform, provides a better performance in terms of consistency and throughput [21,35,36].

Table 1. Summary of Mmcrofluidic-based scRNA-seq technologies.

Technology	Year	Method	Sample Type	Fit for Rare Cell (Y/N)	Resolution	Ref.
Droplet-based scRNA-seq	2015	Drop-seq	Cell suspension	N	Single cell	[23]
Droplet-based scRNA-seq	2015	inDrop	Cell suspension	N	Single cell	[20]
Droplet-based scRNA-seq	2017	Chromium 10X	Cell suspension	N	Single cell	[37]
Droplet-based scRNA-seq	2017	DropNc-seq	Cell suspension	N	Single cell	[24]
Droplet-based scRNA-seq	2023	SPEAC-seq	Cell suspension	N	Single cell	[38]
Microwell-based scRNA-seq	2015	Cytoseq	Cell suspension	N	Single cell	[39]
Microwell-based scRNA-seq	2017	BD Rhapsody	Cell suspension	N	Single cell	[40]
Microwell-based scRNA-seq	2015	Solid-phase RNA capture	Cell suspension	N	Single cell	[36]
Microwell-based scRNA-seq	2017	Seq-Well	Cell suspension	N	Single cell	[21]
Microwell-based scRNA-seq	2020	Seq-Well S^3	Cell suspension	N	Single cell	[41]
Microwell-based scRNA-seq	2018	scFTD-seq	Cell suspension	N	Single cell	[42]
Valve-based scRNA-seq	2014	Microfluidic Single-cell whole-transcriptome Sequencing	Cell suspension	N	Single cell	[22]
Valve-based scRNA-seq	2019	Hydro-Seq	blood	Y	Single cell	[43]
Valve-based scRNA-seq	2020	Paired-seq	Cell suspension	N	Single cell	[44]

2.1. *Droplet-Based scRNA-Seq Methods*

In droplet microfluidics systems, the cell suspension is partitioned into surfactant-stabilized droplets, followed by oil–water interactions to individually encapsulate aqueous droplets in an inert carrier oil. So far, all droplet-based scRNA-seq protocols can be categorized into the following five steps: the enrichment and assessment of target cells, encapsulation in droplets, reverse transcription (RT) and molecular amplification, sequencing, and data processing. The droplet microfluidic device greatly facilitates the sequencing throughput by the enrichment and encapsulation of live target cells. Moreover, the ability to naturally adapt to subsequent manipulations can provide a scalable framework for resolving single-cell information.

Drop-seq [23] and inDrop [20], both published in 2015, are two representative works based on 3′ poly-A capture methods (Figure 1a,b). The two protocols use the same principle in terms of barcoding transcripts, but differ in other steps. Specifically, Drop-seq uses a non-deformable rigid resin for encapsulation, which means that both cells and beads obey the Poisson distribution, resulting in a low encapsulation efficiency. After the encapsulation, RT is performed in unison by pooling beads, which synthesizes barcoded primers directly for RNA capture. As opposed to Drop-seq, inDrop uses a droplet microfluidic device to generate barcoded hydrogel microspheres (BHMs). After the encapsulation, RT reactions occur in droplets individually, so the microreactors reduce reaction competition while bulk RT reactions reduce the batch effect, which is particularly important in high-throughput sequencing with a high assay precision. Importantly, Drop-seq is based on Smart-Seq [45] utilizing PCR-based template switching amplification, while inDrop is akin to CEL-Seq [46] following an in vitro transcription (IVT) protocol. Hence, being based on two different protocols means that both methods have their own strengths for amplification and DNA library preparation. Drop-seq has a much higher gene detection ability than inDrop in terms of sensitivity, especially for low-abundance transcripts. However, there is a significant bias in the quantitative analysis of RNA expression in the Drop-seq protocol due to PCR-induced non-linear amplification. On the contrary, CEL-Seq can greatly reduce amplification bias by barcoding and pooling samples before linearly amplifying mRNA with the use of one round of IVT. It is worth noting that the mRNA capture of both Smart-Seq and CEL-Seq requires poly (T) primers, resulting in important transcripts such as circRNAs that are without poly (A) tails being filtered out. Chromium 10X is a commercially available platform for scRNA-seq that delivers standardized capture conditions [37]. With a multichannel microfluidic chip, thousands of cells can be encapsulated into ~100,000 gel beads in emulsions (GEMs) in six minutes, resulting in single-cell capture efficiencies between 50% and 65%, with the probability of a single droplet capturing multiple cells being as low as 0.9%/1000 cells. Notably, such a short treatment time and reproducibility can reduce the probability of cell lysis and RNA degradation [37].

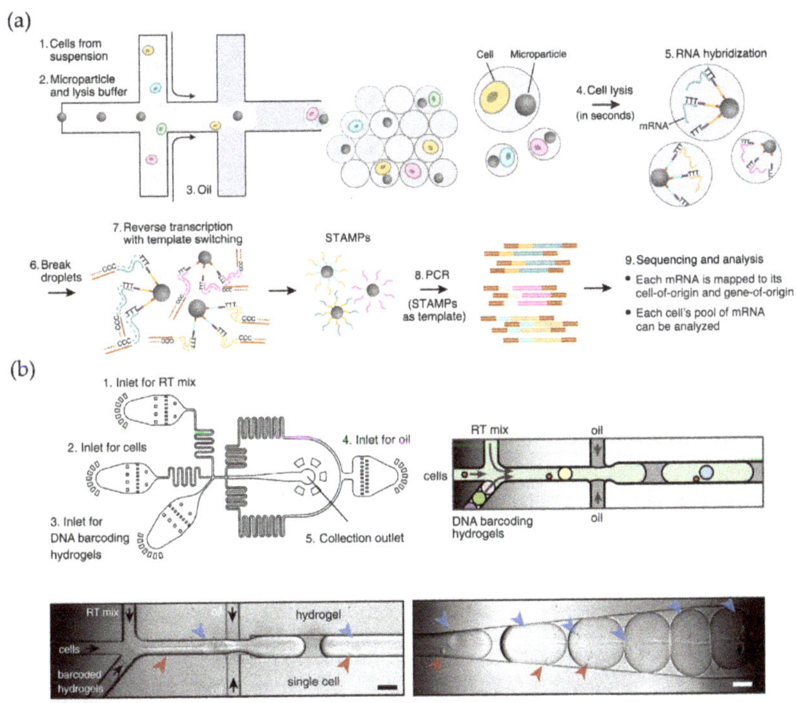

Figure 1. (**a**) Schematic diagram of the workflow of the Drop-seq technique. Reproduced with permission from Ref. [23]. Single cells are encapsulated in tiny droplets along with uniquely barcoded primer beads, enabling large-scale, highly parallel analysis. (**b**) Microfluidic device used in inDrop technology. Reproduced with permission from Ref. [20]. Cells are captured and barcoded in nanoliter droplets with high capture efficiency. Arrows indicate cells (red), hydrogels (blue), and flow direction (black).

Based on the Drop-seq methodology, DropNc-seq achieves a higher capture rate efficiency and further obtains reliable nuclear expression profiles from frozen samples by combining with sNuc-Seq, a low-throughput method utilizing isolated nuclei [24]. In 2023, Wheeler et al. published SPEAC-seq, a forward genetic screening platform [38]. Two-cell encapsulated droplets achieved by microfluidics were co-cultured and then sorted through sequentially gating drops to obtain droplets encapsulating the desired cell–cell pairs, namely activated reporter cells and perturbed cells.

Overall, droplet-based scRNA-seq is severely affected by "doublets", that is, one or more beads and/or cells in a droplet, which occurs for the following two reasons: 1. The presence of two-cell clusters in the suspension resulting from cell–cell interactions. 2. The encapsulation of beads and cells is random, and at least one of them is bound to be governed by the Poisson distribution, even if the variable beads are able to break the Poisson statistics in the case of a tight arrangement [47,48].

2.2. Microwell-Based scRNA-Seq Methods

In comparison to droplet-based scRNA-seq methods, microdevices based on microwell arrays can provide unbiased analysis in SCS at the ensemble level. Specifically, exactly the same nanoliter or picoliter microwells by the tens of thousands are used as stand-alone reactors and can load single cells of comparable sizes by gravity alone. Subsequently, cell lysis, RNA capture, and RT are performed in the original wells [21,39].

In 2015, Fan et al. reported on Cytoseq, initially combining next-generation sequencing (NGS) with the stochastic barcoding of single cells via micropore arrays [39] (Figure 2a). In this strategy, to saturate all wells, the cell suspension concentration and the size of barcoded beads are informed by the geometry and dimensions of the wells. In 2018, Han et al. published a simple method named microwell-seq [32], which can lead to substantial reductions in labor and cost due to the reusability of the PDMS mold, coupled with the low cost of the agarose microwell arrays. Using the microwell arrays, the group characterized cellular heterogeneity and the cross-tissue cellular network with a minimal batch effect, as well as further publishing the first mammalian atlas [32] (Figure 2b). Moreover, given the operations, costs, and turnaround times, researchers may consider BD Rhapsody [40], a proven commercially available platform widely used in immune research and brain sciences [49,50], to observe cell-related information and implement integrative transcriptome and proteome analyses.

To reduce the risk of cross-contamination resulting from the open-top microwell design, the currently preferred method is to seal the microwell array. Bose et al. published a scalable platform for solid-phase RNA capture, consisting of printing RNAs on glass and capturing RNAs with polymer beads [36] (Figure 2c). After loading the cells into the wells and adding lysis buffer, there are different sealing protocols for different modes. The microwell array can be reversibly sealed by a mechanical sealer in the RNA printing mode and by oil injection in the RNA capture mode, respectively. The combination of solid-phase RNA capture with microarrays allows for the exchange of reagents without the need for altering the location of the captured material, which facilitates scalability and miniaturization by eliminating the need for multiple chambers. However, limited by the scale of the bead library and the size of the microwell array, the above platform has a lower throughput compared to Cytoseq. In another report, a portable platform named Seq-Well was developed by Gierahn et al. [21]. By the means of semipermeable membranes and surface-functionalized PDMS arrays, rapid solution exchange and reduced contamination across intercellular mRNAs can be achieved. Also worth mentioning is that Seq-Well uses the same library construction method as Drop-seq, where full-length transcripts are obtained by template switching after RT. However, this method requires a secondary PCR, which may result in the loss of validated products that are not tagged with the handle [51]. Hence, an iterative platform, Seq-Well S^3, was reported in 2020, improving the library construction efficiency and gene detection sensitivity by using a randomly primed second-strand synthesis [41] (Figure 2d). scFTD-seq, published in 2018, proved that freeze–thaw was completely compatible with scRNA-seq [42] (Figure 2e). After the chip was filled with freeze–thaw lysis buffer, the microwell array was sealed using a glass slide, followed by three freeze–thaw cycles and room-temperature incubation. scFTD-seq reduced the cross-contamination, driven, in part, by the seal operation and lack of buffer exchange.

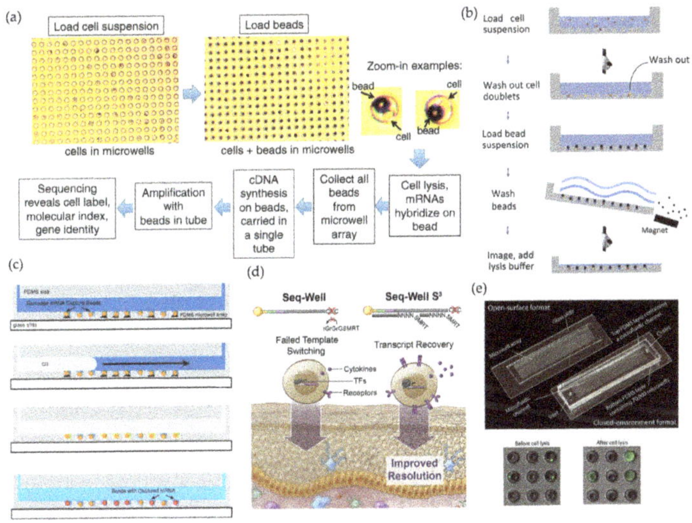

Figure 2. (**a**) Schematic and imaging data for CytoSeq. Reproduced with permission from Ref. [39]. The platform enables routine, digitized gene expression profiling of thousands of single-cell genes. (**b**) Experimental procedure for Microwell-Seq. Reproduced with permission from Ref. [32]. A high-throughput, low-cost platform capable of constructing mouse cellular maps at the single-cell level. (**c**) A schematic of the basic workflow for single-cell RNA printing. Reproduced with permission from Ref. [36]. Microwells in this platform enable low-cost printing of RNA on glass or capturing of RNA on beads. (**d**) Comparison between Seq-Well and Seq-Well S^3. Reproduced with permission from Ref. [41]. Based on Seq-Well, Seq-Well S^3 improves transcript capture and sensitivity. (**e**) scFTD-seq platform and fluorescence imaging data. Reproduced with permission from Ref. [42]. Portable and easy-to-use devices, as well as simplified and modular workflows facilitate clinical use.

In conclusion, microwell-based scRNA-seq methods are theoretically easy to operate and cost-effective due to their fewer complicated peripherals and reagents. However, they are relatively less adaptable to rare cell analysis resulting from the fixed well sizes. Additionally, optimization strategies are still needed to effectively avoid cross-contamination.

2.3. Valve-Based scRNA-Seq Methods

Valve-based microfluidic chips manipulate the fluid flow behavior through pressure-controlled valves [22,43,44]. Specifically, the chip mainly integrates flow channels used for cell suspension flow, valve-controlled channels for changing the pressure, and elastically deformable PDMS films as valves. Changes in pressure control the state of the valve, thus, single-cell isolation can be further realized [52].

In 2014, Streets et al. proposed a semiautomated single-cell high-throughput sequencing program with a high sensitivity and precision [22] (Figure 3a). Individual cells from the cell suspension were sequentially directed by a peristaltic pump into each of the eight trapping chambers and pressurized into the sorting chamber, while multiple cells were discarded through the isolation valve. After sorting, the reagents were filled through the reagent input line and then the reactant was diffused and mixed by controlling the on–off state of the valve above the two chambers. Notably, the integration of valves and pumps had a significant advantage over well-based methods in terms of raw sample contamination removal, which makes it extremely suitable for scarce or easily contaminated samples, such as circulating tumor cells (CTCs). In 2019, Cheng et al. engineered Hydro-Seq chips that could process 10 ml of patient blood containing CTCs [43] (Figure 3b). The customized channel size and valve design captured individual CTCs from the sample into each chamber, while contaminants such as erythrocytes and leukocytes could be transferred. The difference in shape between the bead and capture ports facilitated bead loading. In addition, the valves were closed to the separated chambers, thus allowing for cell lysis and mRNA capture in the separated chambers.

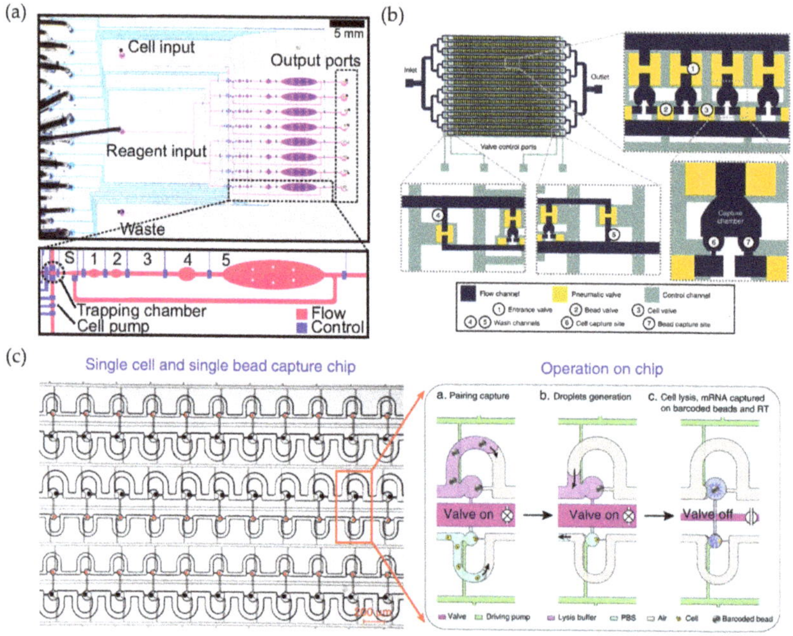

Figure 3. (**a**) Micrograph of the microfluidic device. Reproduced with permission from Ref. [22]. Injection of single-cell suspensions and reagents, and recovery of double-stranded cDNA from the output for single-cell whole transcriptome analysis. Lines indicate the control channels (blue), and the flow channels (purple). (**b**) Schematic diagram of Hydro-Seq technology. Reproduced with permission from Ref. [43]. The platform is capable of cleaning contaminants such as erythrocytes and could be used to isolate rare cells. (**c**) Design of Paired-seq chip. Reproduced with permission from Ref. [44]. The device is capable of efficiently isolating and removing cell-free mRNAs.

In addition, the decontamination capability of the valve-based chip was also demonstrated in the removal of cell-free RNA. Zhang et al. introduced the Paired-seq platform, in which from hundreds to thousands of units were integrated for processing precious samples [44] (Figure 3c). Blocking valves and hydrodynamic differential flow resistance-based units had an immediate effect on the isolation and removal of cell-free mRNAs, which was further validated by subsequent sequencing.

The high-throughput systems above retain fragments of mRNA 3′ end information, while Fluidigm C1 [53], a commercialized platform, employs different protocols, including Smart-Seq [45] and CEL-Seq [46], to generate full-length transcripts. This standardized process reduces the requirements for experimental operations and, therefore, performs robustly and reproducibly in the characterization and data integration of clinical tissues [54] or blood samples [55]. Two prominent examples are the revealing of the intratumoral heterogeneity of the breast cancer transcriptome [54], as well as that of the cellular regulatory pathways in T-cell leukemia [55]. Nevertheless, this calls for caution regarding precious samples and samples that require high-quality gene expression data due to the low cell utilization and the risk of cell damage during the cell capture process [53].

Taking this into account, the valve-based chip structure enables multiple washes of cells and beads, thereby reducing the risk of sample contamination. It is worth noting that the manufacturing process for such SCS platforms, which is highly dependent on the chip structure, is relatively more complex. The integrated distribution of the chip, in particular the shape and design of the valves, greatly affects the efficiency and quality of single-cell capture. For example, the low Reynolds number in channels could be improved by integrating mixing valves, which results in an improved mixing efficiency during DNA amplification [56].

2.4. Single-Cell Multi-Omics Profiling

The central dogma of biology, which tightly links molecular species transcripts to DNA and proteins, implies that single omics information is insufficient to determine the biological function of eukaryotic cells [57,58]. On the one hand, the post-transcriptional status of genes varies from cell to cell within the same organism, or even the tissue, mainly depending on translation and degradation regulatory mechanisms. On the other hand, even in the same cell, the gene expression level fluctuates with time and external stimuli. Therefore, a single-cell multi-omics dataset can capture the static distribution and dynamic information inside and outside a cell to further analyze the mechanisms of gene expression and degradation regulation [59,60]. By mapping cell specificity and tracing cell lineages, the mechanisms of development and disease can be then elucidated [61].

The quantification of proteins as effector molecules of cellular function enables more accurate dynamic modeling to predict gene products and responses to perturbation [62,63]. As a platform capable of precise manipulation in microsystems, the microfluidic device has become a preferred system to integrate multiple quantification technologies at the single-cell level [64–66]. Initially, protein information is read out by the means of fluorescent signals. The relative maturity of imaging technology and analytic software affords immunofluorescence imaging a great potential for directly profiling protein levels [67]. Park et al. used microwell arrays with landmarks to isolate and localize single cells, so that the fluorescence intensity per well was able to quantitatively analyze protein expression dynamics. After cell lysis, the dynamics in transcript levels were obtained by fluorescence intensity via on-chip RT-PCR [68].

However, the throughput and number of fluorescence signal read-outs are limited by spectral overlap and resolution. Therefore, the key to the simultaneous high-throughput detection of the transcriptome and proteome lies in the translation of mRNAs and proteins into readable and unbiased signals. CITE-seq [64] and REAP-seq [69] are both sequencing-based high-throughput methods capable of co-profiling the transcriptome and proteome. Each is represented the following classical barcode-antibody binding mechanisms: streptavidin–biotin interaction and amine chemistry, respectively. Therein, CITE-seq, a widely used tool, is critical for identifying subsets of diseased cells [70], revealing tissue heterogeneity at disease stages [71], and analyzing the mechanisms of immune checkpoint molecular regulations [72]. Such methods are costly due to the use of antibody labeling proteins. In contrast, Delley et al. used oligonucleotide barcode-labeled aptamers, which allowed for the low-cost simultaneous characterization of the transcriptome and intracellular proteins [73].

Another path to responding to genomic expression heterogeneity is to profile the genome and transcriptome in parallel. In 2014, Han et al. used a valve-based platform to isolate the same single-cell mRNA from gDNA. Subsequent RT as well as denaturation and neutralization were performed on the microarray, followed by almost synchronous amplification. Moreover, genes and gene products are not two-by-two linear pathways, and, thus, triple-omics and higher-order single-cell multi-omics are in the process of inferring genomic regulatory models in unprecedented detail [31]. In 2021, Mimitou et al. released ASAP-seq, which is capable of demonstrating chromatin accessibility on a genome-wide scale [66] (Figure 4a). Compatible with multimodal assays such as CITE-seq, ASAP-seq enables a triple- or four-order single-cell multi-omics workflow, revealing stimulus-dependent dynamic changes in T cells. In 2022, Chen et al. profiled CD4 memory T cells with NEAT-seq, a droplet-based microfluidics platform that can simultaneously analyze intracellular proteins, chromatin accessibility, and transcripts through intranuclear staining [74] (Figure 4b).

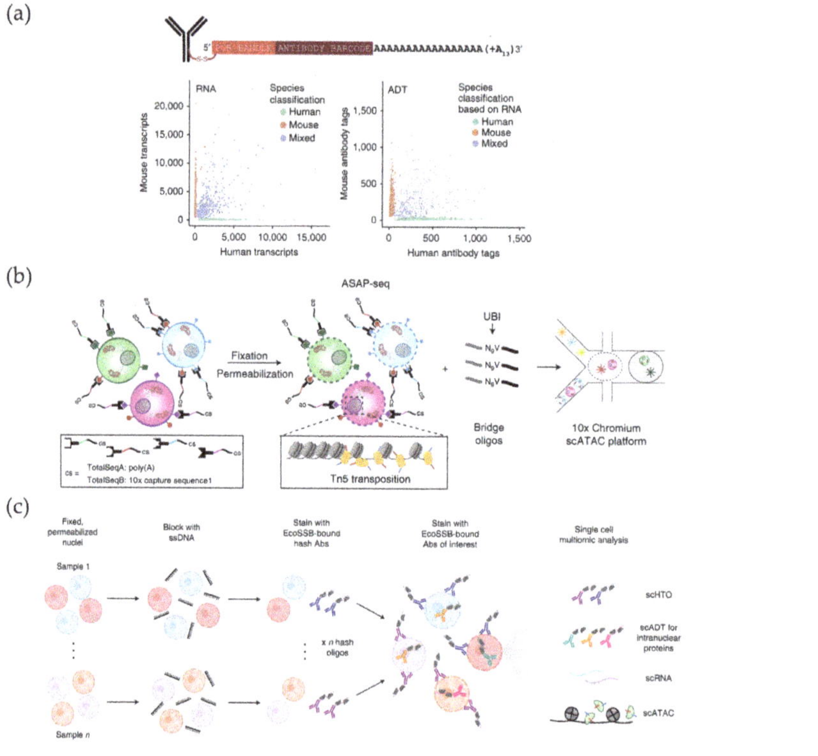

Figure 4. (**a**) Illustration of the DNA-barcoded antibodies used in CITE-seq and analysis of mixtures of mouse and human cells. Reproduced with permission from Ref. [64]. CITE-seq enables simultaneous detection of single cell transcriptomes and protein marks. (**b**) Schematic of ASAP-seq workflow. Reproduced with permission from Ref. [66]. ASAP-seq is capable of demonstrating chromatin accessibility on a genome-wide scale. (**c**) Schematic of NEAT-seq workflow. Reproduced with permission from Ref. [74]. NEAT-seq is a droplet-based microfluidics platform that can simultaneously analyze intracellular proteins, chromatin accessibility, as well as transcripts through intranuclear staining.

In conclusion, due to its precise and unbiased analysis, as well as the comprehensive validation provided on cellular functions and temporal dynamics, microfluidic-platform-based multi-omics sequencing technology is able to reveal the etiology of diseases and highlight disease indicators by integrating multi-omics data. Despite still being in its infancy, the capture of multiple layers provides a time-saving and efficient method for predicting, selecting, and validating regulatory factors of interest. The increased ability to analyze the heterogeneity originating from genetic and environmental stimuli facilitates the further interrogation of molecular regulatory networks spanning the central dogma.

2.5. Technical Challenges and Future Directions

In summary, the challenge in useable yields and data quality in SCS technologies is to reduce cross-contamination so that real single-cell information is obtained [75]. Specifically, each system should ideally (i.e., no cross-contamination) contain both one cell and one bead. However, subject to the Poisson distribution, there may be no or multiple cells and/or beads in the system. In particular, complex components in blood samples such as erythrocytes and cell-free RNA can further introduce additional contamination. Though informatics have been developed to work on this issue from an algorithmic perspective [76,77], microfluidic-based single-cell sequencing methods can minimize cross-contamination in the simplest way possible from a system design. For example, the close-packed arrangement of deformable beads can break the Poisson distribution [47], the sealing of microwell arrays can efficiently avoid inter-system contamination, and the integration of valves and pumps can efficiently scrub contaminants from samples.

3. Advances in Spatial Transcriptomics

While SCS methods provide an exhaustive account of intercellular heterogeneity, they fail to explain the interactions between cells and with the surrounding microenvironment, as well as the spatial distribution [9,78]. Spatially resolved transcriptomics (SRT) technologies have fundamentally reshaped researchers' understanding of tissue homeostasis, tissue development, and even disease mechanisms. SRT technologies can be summarized into the following three categories: imaging-based technologies, laser capture microdissection (LCM)-based technologies, and sequencing-based technologies. Imaging-based technologies are limited by the limited number of fluorescent probes and the complexity of the coding process [79]. In LCM-based technologies, cutting a tissue section with a laser is costly and the analysis area is limited [80]. In contrast, sequencing-based technologies have the advantage of being unbiased, high-throughput, and producing low error [81,82].

Microfluidic systems load cellular information into space through unique array features or microchannels, and further quantify the mRNA expression of a large number of genes in the spatial dimension through high-throughput sequencing (Table 2). The exploration of spatial heterogeneity and the reconstruction of gene expression profiles have deepened the understanding of diseases, neurology, and cancer [83]. Of note, another important implication of SRT technology is to map developmentally relevant spatiotemporal transcriptomics, exploring the developmental trajectory of the embryo, as well as the evolution of life in time and space [84,85].

Table 2. Summary of microfluidic-based spatial transcriptomics technologies.

Technology	Year	Method	Sample Type	Array Substrate	Resolution	Ref.
Array-based SRT	2016	ST	FF	Glass	100 μm	[79]
Array-based SRT	2019	Visium	FF	Glass	55 μm	[86]
Array-based SRT	2019	Slide-Seq	FF	Glass	10 μm	[81]
Array-based SRT	2019	HDST	FF	Silicon	2 μm	[79]
Microchannel-based SRT	2020	DBiT-seq	FF/FFPE	Microfluidic Channel divided tissue	10–50 μm	[82,83]
Microchannel-based SRT	2023	Well-ST-seq	FF	PDMS	10–50 μm	[25]

FF: Fresh-frozen section. FFPE: Formalin-fixed paraffin-embedded section.

3.1. Array-Based SRT Methods

Array-based SRT methods have been widely used to obtain spatially resolved genetic information [9]. Broadly, polyadenylated RNA capture is divided into surface capture and microbead capture [87,88]. The surface capture immobilizes capturing probes by modifying chemical groups, such as epoxy groups, on the surface of glass or silicon wafers [87]. A higher spatial resolution can be achieved by increasing the density of the modification. The microbead capture attaches the capturing probes to microbeads, and the accuracy increases with a decreasing volume of microbeads [89].

Spatial transcriptomics (ST), the first use of surface capture, was proposed by Stamathl et al. [87] and commercialized as the Visium platform (Figure 5a). In total, 1000 + trapping regions 200 μm apart were modified on the slide in ST. After RT of the mRNA from tissue sections captured in situ, cDNA–mRNA complexes with different spatial barcodes were used for library preparation and NGS readout. In the Visium platform, this distance was reduced to 100 μm to achieve a higher spatial resolution. Based on the above platforms, SM-Omics [89] and SPOTS [90] were able to analyze transcriptome information and a panel of protein information from the same tissue section on arrays with a 100 μm and 55 μm resolution, respectively. Slide-Seq technology [88] and HDST technology [89] are representative of microbead capture schemes (Figure 5b,c). Specifically, after modifying magnetic beads with spatially barcoded RT primers, the Slide-Seq technology deposited 10 μm magnetic beads on the plane of the slide, whereas the HDST technology modified 2 μm magnetic beads in the microwells of the slide to capture tissue mRNA. It is worth noting that, although the resolution can be close to that of a single cell, these beads do not capture efficiently enough to support the amount of data required for single-cell analyses. In addition, the spatial barcodes actually map back to adjacent tissue sections, which results in inaccuracy that is quite sensitive to highly heterogeneous samples such as tumors.

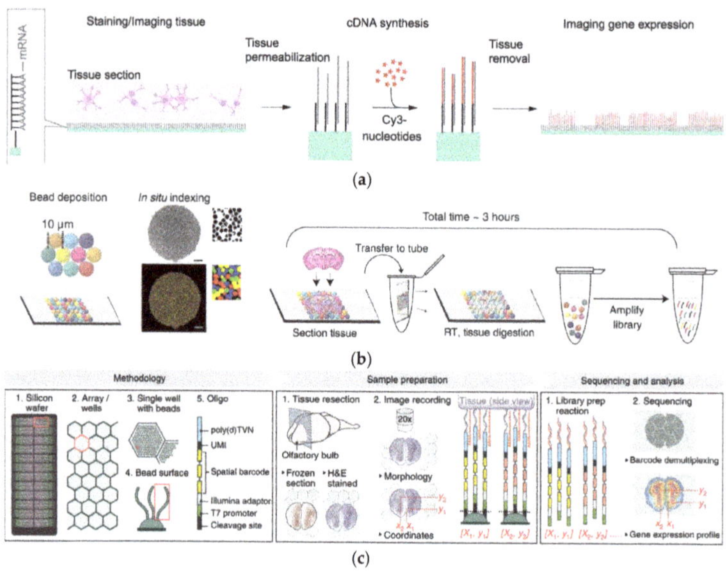

Figure 5. (**a**) Spatially localized cDNA synthesis. Reproduced with permission from Ref. [87]. ST was the first use of surface capture and commercialized as the Visium platform. (**b**) Slide-seq workflow. Reproduced with permission from Ref. [88]. Slide-Seq method deposited 10 μm magnetic beads on the plane of the slide to capture tissue mRNA. (**c**) Schematic diagram of HDST. Reproduced with permission from Ref. [89]. HDST method modified 2 μm magnetic beads in the microwells of the slide to capture tissue mRNA.

3.2. Microchannel-Based SRT Methods

Microchannel-based SRT methods offer an unprecedented perspective on the analysis of biological regulatory networks due to their flexibly adjustable spatially barcoded pixels and their ability to integrate various omics layers. Distinguishing from the solid-phase matrix capture of biomolecules described above, DBiT-seq delivered barcodes for the direct localization of nucleic acids or proteins onto tissue sections [91] (Figure 6a). The key to the method was the sequential use of two PDMS microfluidic chips lined with 50 parallel microchannels. Since the microchannels between the two chips were oriented vertically, the intersecting pixels formed a spatial grid. The resolution of the above pixels could reach up to 10 μm and capture about 2000 genes by adjusting the distance between the microchannels. DBiT-seq is capable of analyzing the proteome, transcriptome, and transcriptome of the same tissue section while maintaining a high resolution [91]. It is also compatible with optical image characterization, which makes the chip capable of an all-round accurate analysis of multi-omics data based on morphological structure. In addition to FF sections, this strategy is also compatible with FFPE tissues with an accuracy of a 50 μm point spacing [92]. Furthermore, microchannel-based PDMS chips can also be used to modify paramagnetic particles, thus avoiding the channel blockage problems associated with tissue deformation while retaining the advantage of flexible resolution adjustment [25] (Figure 6b).

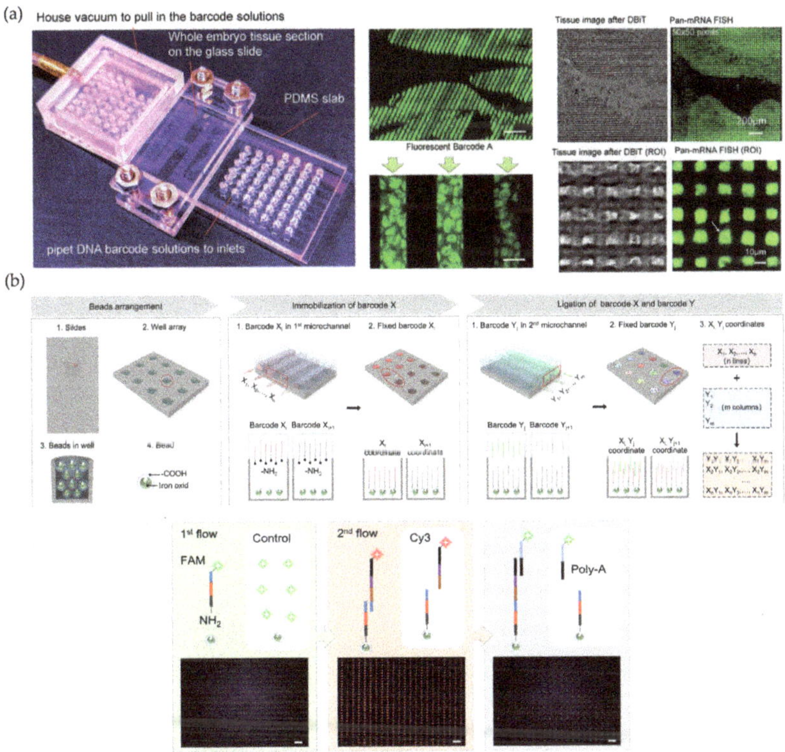

Figure 6. (**a**) Microfluidic device used in DBiT-seq. Reproduced with permission from Ref. [91]. DBiT-seq is capable of analyzing the proteome, transcriptome, and transcriptome of the same tissue section while maintaining a high resolution. (**b**) Schematic diagram and validation of Well-ST-Seq technology. Reproduced with permission from Ref. [25]. Well-ST-Seq avoids channel blockage problems associated with tissue deformation.

As a universal and compatible strategy, microchannel-based SRT methods represented by DBiT-seq extend [93,94] or integrate [95,96] multi-omics data such as proteomes, epigenomes, and so on, thus realizing multiple spatial co-analysis.

3.3. Spatiotemporal Transcriptome

Gene expression and cell-state transitions change continuously in organisms. In order to obtain dynamic multi-omics information, it is necessary to obtain dynamic expression datasets in the temporal and spatial dimensions. The spatiotemporal transcriptome is based on spatial transcriptome technology, which combines spatial information with pseudotime [97,98]. By identifying and tracking key genes during cell migration and differentiation, a deeper understanding of cell fate and processes such as embryonic development and neural differentiation in species including Mus musculus and Homo sapiens can be achieved [84,99]. The size and morphology of fresh-frozen sections under pseudotime are constantly changing with development [100]. Microfluidic devices are capable of adjusting the resolution and size to target regions of interest for comprehensive and detailed analyses [101]. In addition, the low cost and high throughput of microfluidics make it suitable for sequencing large quantities of embryos at different time points [102,103]. Unbiased technical strategies prefer to impute data with spatial and pseudo-temporal contexts as complete cellular differentiation axes. This implies that, when facing problems such as lineage bifurcation decisions, in situ hybridization-based spatiotemporal transcriptomes only target a limited number of target genes, while some capture-based spatiotemporal transcriptomes are unbiased under complex embryonic environments, such as during gastrulation [104].

Higher orders of magnitude of sensitivity and capture efficiency are key to regionalizing spatiotemporal dynamics. To date, researchers have maintained a considerable interest in the development of the mammal post-implantation embryo at all periods of gestation [101,105–107]. Slide-seqV2 identified and recapitulated mouse neocortex developmental trajectories at a near-cellular resolution [80,108]. With nearly an order of magnitude higher sensitivity than the original Slide-seq, more than a thousand key genes during embryonic eye development and genetic drivers of disease could be identified. Subsequently, the spatiotemporal transcriptome has continued to increase in the field of view and resolution. For example, Stereo-seq (Figure 7a), with its cellular resolution and large field of view, was able to track the embryonic organ development trajectory and transcriptional variation in whole mouse embryos spanning E9.5–E16.5 [81], as well as reveal the widest spatiotemporal development of different regions of the human brain across time points to date [100]. sci-Space could elucidate the developmental orientation of nerves in three spatiotemporal dimensions with a single-cell precision by bringing a barcode to each nucleus [102].

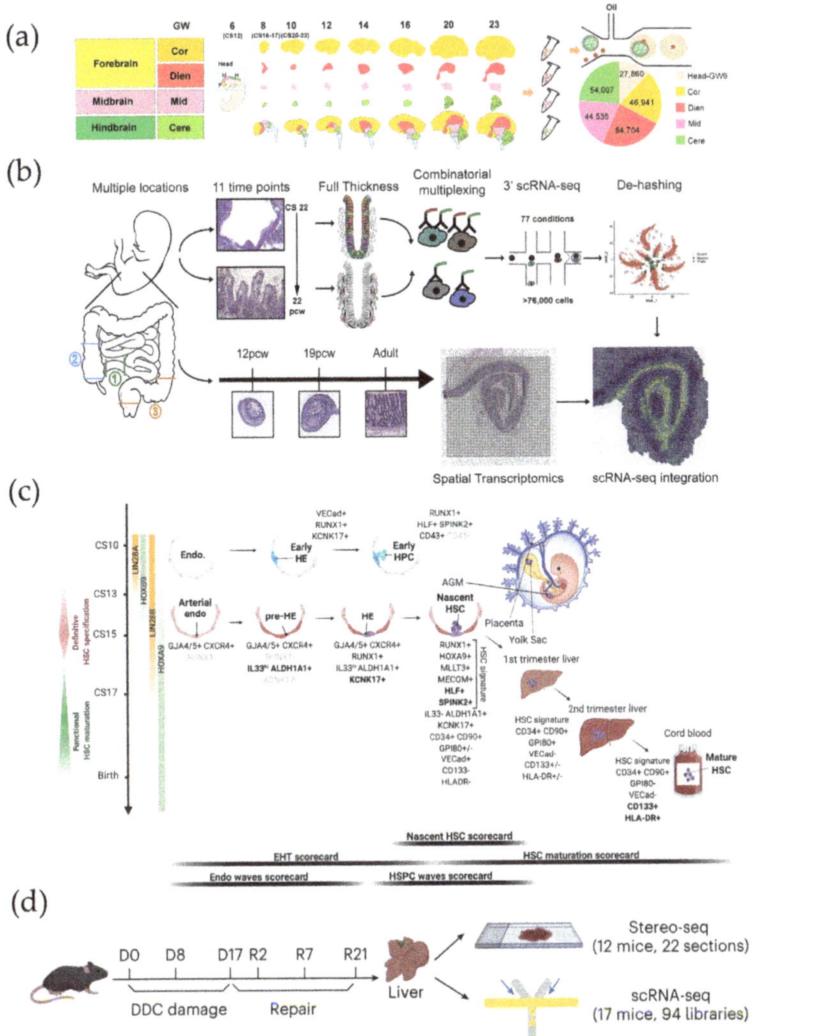

Figure 7. (**a**) Location, period, study design, and number of cells sequenced information. Reproduced with permission from Ref. [100]. (**b**) Overview of study design for intestinal development atlas. Reproduced with permission from Ref. [105]. (**c**) A schematic displaying the key cell types, stages, and markers involved in human hematopoietic stem and progenitor cell (HSPC) specification, emergence, and maturation. Reproduced with permission from Ref. [84]. (**d**) Schematic of DDC-induced injury and repair. Reproduced with permission from Ref. [109]. Spatiotemporal analyses of cholestatic injury and repair processes in mice reveal the potential of the spatiotemporal transcriptome for tissue injury and regeneration studies.

In addition to whole mouse embryos, spatiotemporal transcriptome technology can characterize human organic morphogenesis by linking SCS and SRT. Fawkner-Corbett et al. depicted developmental events in full-thickness intestinal tissue and revealed time-critical transcriptional defects [105] (Figure 7b). Recently, as part of the Human Cell Atlas program, the characterization of the intertemporal development of human embryonic limb organs was made feasible [110]. The understanding of hematopoietic stem cells (HSCs) development has major implications for the diagnosis, prognosis, and treatment of hematologic diseases [111]. The combination of scRNA-seq and Visium commercial platforms elucidated the cellular origins of HSCs and confirmed the sites of HSC emergence and the transcription factor mechanisms. Establishing and visualizing a developmental roadmap for HSCs provides perspectives to guide in vitro transplantable HSC manufacturing efforts [84] (Figure 7c).

The spatiotemporal transcriptome can demonstrate the clinical relevance of developmental cellular mapping [112–115]. Spatiotemporal analyses of cholestatic injury and repair processes in mice reveal the potential of the spatiotemporal transcriptome for tissue injury and regeneration studies [109] (Figure 7d). By characterizing the gene expression for different disease periods, researchers are able to accurately assess the horizontal expression of specific genes during changes in organ morphology versus heterogeneity [116].

3.4. Technical Challenges and Future Directions

Overall, microfluidic-based SRT methods reduce the limitations of cost and laboratory equipment, obtaining thousands of capture spots with different spatial barcodes in an unbiased manner [110]. However, the system performance of such methods is limited by the process of microfluidic devices. First, there is a risk of extremely minute fluid exchange between the adjacent microchannels of the microfluidic device, which leads to unavoidable contamination during spatial barcode probe modification and tissue section encoding. Second, microfluidic-based SRT methods carry the risk of mRNA diffusion during permeabilization compared to SRT methods using hybridization probes. A promising approach is the domain restriction of tissue sections by microwell substrate processing. Finally, while other SRT techniques have achieved a nanometer resolution, the resolution of microfluidic spatial transcriptomic methods is limited to the micrometer level by lithography and nanostructure fabrication. Although efforts have been made to reach up to a 10 μm resolution [25,91], how to obtain spatial information at the subcellular level remains a future challenge [85,117].

4. Clinical Applications and Conclusions

SCS technologies and SRT technologies bridge the gap between clinical histopathology and molecular phenotyping [118], enabling the corroboration of experimentally obtained information on in situ intercellular molecular communication with clinical medical phenomena. When standardized and reproducible sequencing data are combined with clinical precision therapies, SCS technologies and SRT technologies will be able to assist in predicting potential therapeutic targets and even therapeutic effects, thus further advancing researchers' understanding of developmental processes and disease models and reshaping precision therapeutic protocols [59,118–120].

4.1. Early Detection and Disease Classification in Oncology

In translational and clinical medicine, 'one size fits all' therapeutic approaches often result from a lack of the timely subcategorization of diseases and accurate prognoses [121]. In fact, microorganisms may activate a high expression of metabolic pathways [122]. The tumor microenvironment (TME) and highly heterogeneous tumor tissues form a complex immune network [120]. In the above ecosystem, multiple highly variable behaviors, such as immune cell infiltration and immune evasion, further influence the accurate judgment of tumor diagnosis and clinical treatment [123,124].

Spatial transcriptome technology tracks the cancer evolution present in cancerous regions and histologically benign tissues adjacent to cancer in an unsupervised manner, redefining distinct clonal patterns and cancer evolution with an unprecedented spatial resolution [125]. In addition, microfluidics and its expanded platforms have important implications for the deconstruction of hemato-oncological ecosystems in complex microenvironments, such as aberrant regulatory programs and inter-tumoral variability [126]. The results of SCS were able to classify spinal stromal cells and dissect their responses and molecular abnormalities under AML [127]. In a study of mixed-phenotype acute leukemias, single-cell multi-omics analysis inferred a transcription factor, RUNX1, that influences leukemia survival [128] (Figure 8a).

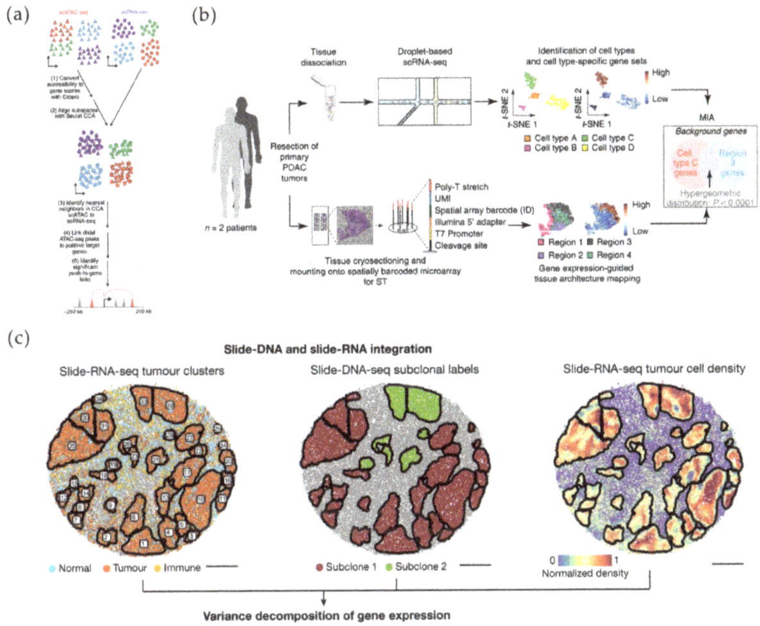

Figure 8. (**a**) Schematic for alignment of scATAC-seq and scRNA-seq data. Reproduced with permission from Ref. [128]. scATAC-seq has simultaneously tagged and fragmented DNA sequences, such as 10x Multiome, in open chromatin regions and further paved the way for other technologies that profile both the epigenome and transcriptome in single nuclei. (**b**) Schematic for alignment of snRNA-seq and Stereo-seq. Reproduced with permission from Ref. [129]. Multimodal intersection analysis (MIA) across the two datasets has revealed the spatial distribution of the cell populations and subpopulations. (**c**) Slide-DNA and slide-RNA integration. Reproduced with permission from Ref. [123]. Slide-DNA-seq has captured spatially resolved DNA sequences from intact tissue sections. The integration of slide-DNA-seq and spatial transcriptomics has uncovered distinct sets of genes that are associated with clone-specific genetics.

Currently, scRNA-seq and ST, either alone or in combination, have been developed as powerful tools for interpreting pathological phenomena and evaluating disease-specific markers in complex or highly heterogeneous tissues. In the process of disease diagnosis and targeted therapies, the information obtained through this emerging technology platform covers the scope and dimensions from cell-to-cell interactions to intracellular transcriptional mechanisms, from holistic networks to dynamic microenvironments, and from tissue heterogeneity to single-cell heterogeneity [16,61,125,130,131]. Multimodal intersection analysis (MIA) enables the integration of scRNA-seq and ST datasets from the same lesion region and the subsequent spatial assessment of the cell types enriched in the region or cluster by the degree of overlap of specifically expressed genes between the two. As an early mode of dataset integration, MIA provides spatial cell specificity, as well as an immune repertoire of lesion regions such as pancreatic ductal adenocarcinoma (PDAC) [129], esophageal squamous cell carcinoma [132], and cervical squamous carcinoma [133] (Figure 8b).

In conclusion, the high degree of tumor heterogeneity poses a great challenge for the prospective validation of precision therapeutic processes. The integration of high-resolution multi-omics analyses will be required to accelerate the analysis of complex biomarkers and therapeutic targets. For example, Slide-DNA-seq enables the spatial analysis of clonal heterogeneity and provides an avenue for analyzing copy number aberrations (CNAs), one of the drivers of cancer progression [123,131] (Figure 8c).

4.2. Cellular Dynamics of the Brain

The central nervous system (CNS) directs the complex activities and interactions of billions of neurons and glial cells [134]. Understanding the nuances between cells through SCS techniques plays a crucial role in the study of neurodevelopment and neurological diseases [135–137].

The highly dynamic brain is the key to the CNS, and analyzing the cerebrospinal fluid (CSF), which protects the CNS, can help in the diagnosis of CNS diseases. Single-cell transcriptomics has been able to determine the cellular profile of cerebrospinal fluid versus blood in an unbiased manner, increasing the diversity of the cellular composition of the CSF and the transcriptional phenotypes of the cells in the bloodstream [138]. In the study of Alzheimer's disease (AD), the integration of snRNA-seq and snATAC-seq datasets is able to identify the transcription factors (TFs) for changes in disease gene expression at a single-nucleus resolution [139]. Furthermore, in addition to exploring disease mechanisms and screening for therapeutic targets, the combination of peptide screening has enabled further analyses of adaptive immune responses in the CSF and blood of patients [99,140].

The connectome, as the key to the functioning of the nervous system, is used to refer to all elements and connections [141]. In animal models, studies of the brain and spinal cord have enriched cell population types and elucidated cell-type-specific responses [103]. Currently, work has been conducted to identify and map cell types with the help of a combination of SCS, SRT, and retrograde labeling. Brain-wide atlases of gene expression constructed by such work reveal the complex topology of the brain connectome and further bridge the gap between the transcriptome and connectome [142]. The Allen Mouse Brain Atlas (AMBA) [143,144] and the Allen Human Brain Atlas (AHBA) [145], the most widely used brain atlases, quantify tens of thousands of genes in the brain. Such comprehensive mapping of the molecular characterization and the spatial patterns of gene expression will help to extend our mechanistic studies of neurodevelopmental and neurodegenerative diseases [142].

4.3. Discussion and Future Directions

Cellular heterogeneity among tissues poses significant challenges for molecular mechanisms and pathology analyses [146,147]. The controllable strategies of microfluidic devices provide new horizons for single-cell, spatial, and spatiotemporal histology. For example, the high-precision manipulation of single cells for SCS [20,21,32] and reliably and reproducibly unbiased capture for SRT [87,88,91] make microfluidics the most promising high-throughput platform. With cost and efficiency in mind, a key question is undoubtedly how researchers can match the most appropriate sequencing method, whether SCS or SRT, for existing biological samples to accurately detect and quantify biomolecules.

First, the sample type plays a decisive role in the selection of sequencing methods. Highly heterogeneous samples such as tumors require deeper and more precise scRNA-seq methods to reveal signaling pathways and stemness, etc. [148]. The introduction of microfluidics has led to a great improvement in the throughput and precision of scRNA-seq compared to traditional tubes and plate-based sequencing. On this basis, more and more characterization modalities are focusing on clinical applications and presenting unprecedented opportunities for precision therapy [149]. A current challenge is that difficult-to-access disease tissues limit sample types and sample sizes. In this context, stem-cell-derived organoids are a highly promising option [150].

Second, a corollary trend is the characterization of comprehensive molecular phenotypes at the single-cell level [151,152]. While single histologic data can provide predictable information with disease progression, for complex diseases such as type 2 diabetes (T2D), osteoarthritis, etc., the combination of transcriptomics, proteomics, metabolomics, etc., can refine the molecular mechanisms of disease etiology, which can then be used for prediction and diagnosis in the clinic [153–157]. Therefore, how to integrate information across these molecular layers is a key issue. Under this premise, a promising direction is the processing of sequencing data [158]. In the process of biomolecular analysis, it is an inevitable trend to go from the whole to the details and from the details back to the whole. To date, the preferred probabilistic approach for revealing complex organizational structures has been the back-convolution of single-cell data against specified spatial transcriptome data. In addition, for example, one of the effective ways to improve spatial annotation is to superimpose expression signatures on single-nucleus RNA-sequencing data [159]. In addition, identifying and removing low-quality sequencing data is the key to obtaining robust multi-omics analysis results [160]. For microfluidic-based SCS methods, the technical noise/dropout events and the methodological limitations of using single-end sequencing largely affect downstream analysis. In this context, imputation is a useful strategy to replace the missing data, thereby restoring the true expression level of the genes [161,162]. For microfluidic-based SRT methods, the tissue morphology, physical distance between data points, and gene expression need to be considered simultaneously. For example, for both healthy and diseased tissues, Pham et al. developed a spatial graph-based method to robustly interrogate biological processes [163].

Finally, in translational research, solid tumors and CTCs have garnered a high interest in the field of single tumor cell research [121,164]. Analyses targeting highly heterogeneous tumor cells have benefited from the application of highly parallel microfluidic devices. The combination of transcriptomics and proteomics has proven to be a highly successful approach to interrogating complex disease mechanisms [165,166]. It is foreseen that the integration of multi-omics information at a single-cell resolution will greatly provide life science and modern medicine, with the capability for early disease prediction and targeted precision therapy.

Author Contributions: Conceptualization, Y.S., N.Y., J.Z. and B.Y.; figure preparation, Y.S., N.Y., J.Z. and B.Y.; Writing—original draft preparation, Y.S., N.Y., J.Z. and B.Y.; writing—review and editing, Y.S., N.Y., J.Z. and B.Y. All authors have read and agreed to the published version of the manuscript.

Funding: This work was supported by the National Natural Science Foundation of China (Grant no. 22275071, 32401229) and Jilin Province Science and Technology Development Program (202401648ZYTS).

Data Availability Statement: Not applicable.

Conflicts of Interest: The authors declare no conflict of interest.

References

1. Neumann, P.E.; Neumann, E.E. General Histological Woes: Definition and Classification of Tissues. *Clin. Anat.* **2021**, *34*, 794–801. [CrossRef] [PubMed]
2. Davis, M.M.; Tato, C.M.; Furman, D. Systems Immunology: Just Getting Started. *Nat. Immunol.* **2017**, *18*, 725–732. [CrossRef]
3. Lee, J.; Hyeon, D.Y.; Hwang, D. Single-Cell Multiomics: Technologies and Data Analysis Methods. *Exp. Mol. Med.* **2020**, *52*, 1428–1442. [CrossRef]

4. Travaglini, K.J.; Nabhan, A.N.; Penland, L.; Sinha, R.; Gillich, A.; Sit, R.V.; Chang, S.; Conley, S.D.; Mori, Y.; Seita, J.; et al. A Molecular Cell Atlas of the Human Lung from Single-Cell RNA Sequencing. *Nature* **2020**, *587*, 619–625. [CrossRef]
5. Wang, Y.; Navin, N.E. Advances and Applications of Single-Cell Sequencing Technologies. *Mol. Cell* **2015**, *58*, 598–609. [CrossRef]
6. Williams, J.W.; Winkels, H.; Durant, C.P.; Zaitsev, K.; Ghosheh, Y.; Ley, K. Single Cell RNA Sequencing in Atherosclerosis Research. *Circ. Res.* **2020**, *126*, 1112–1126. [CrossRef]
7. Nawy, T. Single-Cell Sequencing. *Nat. Methods* **2014**, *11*, 18. [CrossRef]
8. Moses, L.; Pachter, L. Museum of Spatial Transcriptomics. *Nat. Methods* **2022**, *19*, 534–546. [CrossRef]
9. Asp, M.; Bergenstråhle, J.; Lundeberg, J. Spatially Resolved Transcriptomes—Next Generation Tools for Tissue Exploration. *BioEssays* **2020**, *42*, 1900221. [CrossRef]
10. Rao, A.; Barkley, D.; França, G.S.; Yanai, I. Exploring Tissue Architecture Using Spatial Transcriptomics. *Nature* **2021**, *596*, 211–220. [CrossRef]
11. Bach, K.; Pensa, S.; Grzelak, M.; Hadfield, J.; Adams, D.J.; Marioni, J.C.; Khaled, W.T. Differentiation Dynamics of Mammary Epithelial Cells Revealed by Single-Cell RNA Sequencing. *Nat. Commun.* **2017**, *8*, 2128. [CrossRef] [PubMed]
12. Shalek, A.K.; Satija, R.; Shuga, J.; Trombetta, J.J.; Gennert, D.; Lu, D.; Chen, P.; Gertner, R.S.; Gaublomme, J.T.; Yosef, N.; et al. Single-Cell RNA-Seq Reveals Dynamic Paracrine Control of Cellular Variation. *Nature* **2014**, *510*, 363–369. [CrossRef] [PubMed]
13. Pal, B.; Chen, Y.; Vaillant, F.; Jamieson, P.; Gordon, L.; Rios, A.C.; Wilcox, S.; Fu, N.; Liu, K.H.; Jackling, F.C.; et al. Construction of Developmental Lineage Relationships in the Mouse Mammary Gland by Single-Cell RNA Profiling. *Nat. Commun.* **2017**, *8*, 1627. [CrossRef] [PubMed]
14. Whitesides, G.M. The Origins and the Future of Microfluidics. *Nature* **2006**, *442*, 368–373. [CrossRef]
15. Prakadan, S.M.; Shalek, A.K.; Weitz, D.A. Scaling by Shrinking: Empowering Single-Cell "omics" with Microfluidic Devices. *Nat. Rev. Genet.* **2017**, *18*, 345–361. [CrossRef]
16. Tavakoli, H.; Zhou, W.; Ma, L.; Perez, S.; Ibarra, A.; Xu, F.; Zhan, S.; Li, X. Recent Advances in Microfluidic Platforms for Single-Cell Analysis in Cancer Biology, Diagnosis and Therapy. *TrAC Trends Anal. Chem.* **2019**, *117*, 13–26. [CrossRef]
17. Picelli, S.; Faridani, O.R.; Björklund, Å.K.; Winberg, G.; Sagasser, S.; Sandberg, R. Full-Length RNA-Seq from Single Cells Using Smart-Seq2. *Nat. Protoc.* **2014**, *9*, 171–181. [CrossRef]
18. Rodriguez-Meira, A.; Buck, G.; Clark, S.-A.; Povinelli, B.J.; Alcolea, V.; Louka, E.; McGowan, S.; Hamblin, A.; Sousos, N.; Barkas, N.; et al. Unravelling Intratumoral Heterogeneity Through High-Sensitivity Single-Cell Mutational Analysis and Parallel RNA Sequencing. *Mol. Cell* **2019**, *73*, 1292–1305.e8. [CrossRef]
19. Cao, J.; Cusanovich, D.A.; Ramani, V.; Aghamirzaie, D.; Pliner, H.A.; Hill, A.J.; Daza, R.M.; McFaline-Figueroa, J.L.; Packer, J.S.; Christiansen, L.; et al. Joint Profiling of Chromatin Accessibility and Gene Expression in Thousands of Single Cells. *Science* **2018**, *361*, 1380–1385. [CrossRef]
20. Klein, A.M.; Mazutis, L.; Akartuna, I.; Tallapragada, N.; Veres, A.; Li, V.; Peshkin, L.; Weitz, D.A.; Kirschner, M.W. Droplet Barcoding for Single-Cell Transcriptomics Applied to Embryonic Stem Cells. *Cell* **2015**, *161*, 1187–1201. [CrossRef]
21. Gierahn, T.M.; Wadsworth, M.H., II; Hughes, T.K.; Bryson, B.D.; Butler, A.; Satija, R.; Fortune, S.; Love, J.C.; Shalek, A.K. Seq-Well: Portable, Low-Cost RNA Sequencing of Single Cells at High Throughput. *Nat. Methods* **2017**, *14*, 395–398. [CrossRef] [PubMed]
22. Streets, A.M.; Zhang, X.; Cao, C.; Pang, Y.; Wu, X.; Xiong, L.; Yang, L.; Fu, Y.; Zhao, L.; Tang, F.; et al. Microfluidic Single-Cell Whole-Transcriptome Sequencing. *Proc. Natl. Acad. Sci. USA* **2014**, *111*, 7048–7053. [CrossRef] [PubMed]
23. Macosko, E.Z.; Basu, A.; Satija, R.; Nemesh, J.; Shekhar, K.; Goldman, M.; Tirosh, I.; Bialas, A.R.; Kamitaki, N.; Martersteck, E.M.; et al. Highly Parallel Genome-Wide Expression Profiling of Individual Cells Using Nanoliter Droplets. *Cell* **2015**, *161*, 1202–1214. [CrossRef] [PubMed]
24. Habib, N.; Avraham-Davidi, I.; Basu, A.; Burks, T.; Shekhar, K.; Hofree, M.; Choudhury, S.R.; Aguet, F.; Gelfand, E.; Ardlie, K.; et al. Massively Parallel Single-Nucleus RNA-Seq with DroNc-Seq. *Nat. Methods* **2017**, *14*, 955–958. [CrossRef]
25. Yu, N.; Jin, Z.; Liang, C.; Zhang, J.; Yang, B. Well-ST-Seq: Cost-Effective Spatial Transcriptomics at Cellular Level and High RNA Capture Efficiency. *bioRxiv* **2023**, 546974.
26. Ahmed, R.; Zaman, T.; Chowdhury, F.; Mraiche, F.; Tariq, M.; Ahmad, I.S.; Hasan, A. Single-Cell RNA Sequencing with Spatial Transcriptomics of Cancer Tissues. *Int. J. Mol. Sci.* **2022**, *23*, 3042. [CrossRef]
27. Longo, S.K.; Guo, M.G.; Ji, A.L.; Khavari, P.A. Integrating Single-Cell and Spatial Transcriptomics to Elucidate Intercellular Tissue Dynamics. *Nat. Rev. Genet.* **2021**, *22*, 627–644. [CrossRef]
28. Jiao, Y.; Gao, L.; Ji, Y.; Liu, W. Recent Advances in Microfluidic Single-Cell Analysis and Its Applications in Drug Development. *TrAC Trends Anal. Chem.* **2022**, *157*, 116796. [CrossRef]
29. Kalisky, T.; Blainey, P.; Quake, S.R. Genomic Analysis at the Single-Cell Level. *Annu. Rev. Genet.* **2011**, *45*, 431–445. [CrossRef]
30. Tang, F.; Barbacioru, C.; Wang, Y.; Nordman, E.; Lee, C.; Xu, N.; Wang, X.; Bodeau, J.; Tuch, B.B.; Siddiqui, A.; et al. mRNA-Seq Whole-Transcriptome Analysis of a Single Cell. *Nat. Methods* **2009**, *6*, 377–382. [CrossRef]
31. Baysoy, A.; Bai, Z.; Satija, R.; Fan, R. The Technological Landscape and Applications of Single-Cell Multi-Omics. *Nat. Rev. Mol. Cell Biol.* **2023**, *24*, 695–713. [CrossRef] [PubMed]

32. Han, X.; Wang, R.; Zhou, Y.; Fei, L.; Sun, H.; Lai, S.; Saadatpour, A.; Zhou, Z.; Chen, H.; Ye, F.; et al. Mapping the Mouse Cell Atlas by Microwell-Seq. *Cell* **2018**, *172*, 1091–1107.e17. [CrossRef] [PubMed]
33. Schmid, L.; Franke, T. SAW-Controlled Drop Size for Flow Focusing. *Lab A Chip* **2013**, *13*, 1691–1694. [CrossRef] [PubMed]
34. Schmid, L.; Franke, T. Acoustic Modulation of Droplet Size in a T-Junction. *Appl. Phys. Lett.* **2014**, *104*, 133501. [CrossRef]
35. Fan, H.C.; Wang, J.; Potanina, A.; Quake, S.R. Whole-Genome Molecular Haplotyping of Single Cells. *Nat. Biotechnol.* **2011**, *29*, 51–57. [CrossRef]
36. Bose, S.; Wan, Z.; Carr, A.; Rizvi, A.H.; Vieira, G.; Pe'er, D.; Sims, P.A. Scalable Microfluidics for Single-Cell RNA Printing and Sequencing. *Genome Biol.* **2015**, *16*, 120. [CrossRef]
37. Zheng, G.X.Y.; Terry, J.M.; Belgrader, P.; Ryvkin, P.; Bent, Z.W.; Wilson, R.; Ziraldo, S.B.; Wheeler, T.D.; McDermott, G.P.; Zhu, J.; et al. Massively Parallel Digital Transcriptional Profiling of Single Cells. *Nat. Commun.* **2017**, *8*, 14049. [CrossRef]
38. Tang, Y.; Liu, F. SPEAC-Seq: A New Method to Investigate Astrocyte-Microglia Crosstalk. *Brain-X* **2023**, *1*, e22. [CrossRef]
39. Fan, H.C.; Fu, G.K.; Fodor, S.P.A. Combinatorial Labeling of Single Cells for Gene Expression Cytometry. *Science* **2015**, *347*, 1258367. [CrossRef]
40. Birey, F.; Andersen, J.; Makinson, C.D.; Islam, S.; Wei, W.; Huber, N.; Fan, H.C.; Metzler, K.R.C.; Panagiotakos, G.; Thom, N.; et al. Assembly of Functionally Integrated Human Forebrain Spheroids. *Nature* **2017**, *545*, 54–59. [CrossRef]
41. Hughes, T.K.; Wadsworth, M.H.; Gierahn, T.M.; Do, T.; Weiss, D.; Andrade, P.R.; Ma, F.; Silva, B.J.d.A.; Shao, S.; Tsoi, L.C.; et al. Second-Strand Synthesis-Based Massively Parallel scRNA-Seq Reveals Cellular States and Molecular Features of Human Inflammatory Skin Pathologies. *Immunity* **2020**, *53*, 878–894.e7. [CrossRef] [PubMed]
42. Dura, B.; Choi, J.-Y.; Zhang, K.; Damsky, W.; Thakral, D.; Bosenberg, M.; Craft, J.; Fan, R. scFTD-Seq: Freeze-Thaw Lysis Based, Portable Approach Toward Highly Distributed Single-Cell 3′ mRNA Profiling. *Nucleic Acids Res.* **2019**, *47*, e16. [CrossRef] [PubMed]
43. Cheng, Y.-H.; Chen, Y.-C.; Lin, E.; Brien, R.; Jung, S.; Chen, Y.-T.; Lee, W.; Hao, Z.; Sahoo, S.; Min Kang, H.; et al. Hydro-Seq Enables Contamination-Free High-Throughput Single-Cell RNA-Sequencing for Circulating Tumor Cells. *Nat. Commun.* **2019**, *10*, 2163. [CrossRef] [PubMed]
44. Zhang, M.; Zou, Y.; Xu, X.; Zhang, X.; Gao, M.; Song, J.; Huang, P.; Chen, Q.; Zhu, Z.; Lin, W.; et al. Highly Parallel and Efficient Single Cell mRNA Sequencing with Paired Picoliter Chambers. *Nat. Commun.* **2020**, *11*, 2118. [CrossRef]
45. Grün, D.; Kester, L.; van Oudenaarden, A. Validation of Noise Models for Single-Cell Transcriptomics. *Nat. Methods* **2014**, *11*, 637–640. [CrossRef]
46. Hashimshony, T.; Wagner, F.; Sher, N.; Yanai, I. CEL-Seq: Single-Cell RNA-Seq by Multiplexed Linear Amplification. *Cell Rep.* **2012**, *2*, 666–673. [CrossRef]
47. Abate, A.R.; Chen, C.-H.; Agresti, J.J.; Weitz, D.A. Beating Poisson Encapsulation Statistics Using Close-Packed Ordering. *Lab Chip* **2009**, *9*, 2628–2631. [CrossRef]
48. Zhang, X.; Li, T.; Liu, F.; Chen, Y.; Yao, J.; Li, Z.; Huang, Y.; Wang, J. Comparative Analysis of Droplet-Based Ultra-High-Throughput Single-Cell RNA-Seq Systems. *Mol. Cell* **2019**, *73*, 130–142.e5. [CrossRef]
49. Meng, Y.; Ye, F.; Nie, P.; Zhao, Q.; An, L.; Wang, W.; Qu, S.; Shen, Z.; Cao, Z.; Zhang, X.; et al. Immunosuppressive CD10+ALPL+ Neutrophils Promote Resistance to Anti-PD-1 Therapy in HCC by Mediating Irreversible Exhaustion of T Cells. *J. Hepatol.* **2023**, *79*, 1435–1449. [CrossRef]
50. Yu, J.; Xiao, K.; Chen, X.; Deng, L.; Zhang, L.; Li, Y.; Gao, A.; Gao, J.; Wu, C.; Yang, X.; et al. Neuron-Derived Neuropeptide Y Fine-Tunes the Splenic Immune Responses. *Neuron* **2022**, *110*, 1327–1339.e6. [CrossRef]
51. Ramsköld, D.; Luo, S.; Wang, Y.-C.; Li, R.; Deng, Q.; Faridani, O.R.; Daniels, G.A.; Khrebtukova, I.; Loring, J.F.; Laurent, L.C.; et al. Full-Length mRNA-Seq from Single-Cell Levels of RNA and Individual Circulating Tumor Cells. *Nat. Biotechnol.* **2012**, *30*, 777–782. [CrossRef] [PubMed]
52. Unger, M.A.; Chou, H.-P.; Thorsen, T.; Scherer, A.; Quake, S.R. Monolithic Microfabricated Valves and Pumps by Multilayer Soft Lithography. *Science* **2000**, *288*, 113–116. [CrossRef] [PubMed]
53. Xin, Y.; Kim, J.; Ni, M.; Wei, Y.; Okamoto, H.; Lee, J.; Adler, C.; Cavino, K.; Murphy, A.J.; Yancopoulos, G.D.; et al. Use of the Fluidigm C1 Platform for RNA Sequencing of Single Mouse Pancreatic Islet Cells. *Proc. Natl. Acad. Sci. USA* **2016**, *113*, 3293–3298. [CrossRef] [PubMed]
54. Chung, W.; Eum, H.H.; Lee, H.-O.; Lee, K.-M.; Lee, H.-B.; Kim, K.-T.; Ryu, H.S.; Kim, S.; Lee, J.E.; Park, Y.H.; et al. Single-Cell RNA-Seq Enables Comprehensive Tumour and Immune Cell Profiling in Primary Breast Cancer. *Nat. Commun.* **2017**, *8*, 15081. [CrossRef]
55. Satpathy, A.T.; Saligrama, N.; Buenrostro, J.D.; Wei, Y.; Wu, B.; Rubin, A.J.; Granja, J.M.; Lareau, C.A.; Li, R.; Qi, Y.; et al. Transcript-Indexed ATAC-Seq for Precision Immune Profiling. *Nat. Med.* **2018**, *24*, 580–590. [CrossRef]
56. Yang, Y.; Swennenhuis, J.F.; Rho, H.S.; Le Gac, S.; Terstappen, L.W.M.M. Parallel Single Cancer Cell Whole Genome Amplification Using Button-Valve Assisted Mixing in Nanoliter Chambers. *PLoS ONE* **2014**, *9*, e107958. [CrossRef]

57. Guo, Y.; Xiao, P.; Lei, S.; Deng, F.; Xiao, G.G.; Liu, Y.; Chen, X.; Li, L.; Wu, S.; Chen, Y.; et al. How Is mRNA Expression Predictive for Protein Expression? A Correlation Study on Human Circulating Monocytes. *Acta Biochim. Biophys. Sin.* **2008**, *40*, 426–436. [CrossRef]
58. Maier, T.; Güell, M.; Serrano, L. Correlation of mRNA and Protein in Complex Biological Samples. *FEBS Lett.* **2009**, *583*, 3966–3973. [CrossRef]
59. Hasin, Y.; Seldin, M.; Lusis, A. Multi-Omics Approaches to Disease. *Genome Biol.* **2017**, *18*, 83. [CrossRef]
60. He, L.; Wang, W.; Dang, K.; Ge, Q.; Zhao, X. Integration of Single-cell Transcriptome and Proteome Technologies: Toward Spatial Resolution Levels. *View* **2023**, *4*, 20230040. [CrossRef]
61. Kreitmaier, P.; Katsoula, G.; Zeggini, E. Insights from Multi-Omics Integration in Complex Disease Primary Tissues. *Trends Genet.* **2023**, *39*, 46–58. [CrossRef] [PubMed]
62. Vistain, L.F.; Tay, S. Single-Cell Proteomics. *Trends Biochem. Sci.* **2021**, *46*, 661–672. [CrossRef] [PubMed]
63. Sun, W.; Nan, J.; Xu, H.; Wang, L.; Niu, J.; Zhang, J.; Yang, B. Neural Network Enables High Accuracy for Hepatitis B Surface Antigen Detection with a Plasmonic Platform. *Nano Lett.* **2024**, *24*, 8784–8792. [CrossRef] [PubMed]
64. Stoeckius, M.; Hafemeister, C.; Stephenson, W.; Houck-Loomis, B.; Chattopadhyay, P.K.; Swerdlow, H.; Satija, R.; Smibert, P. Simultaneous Epitope and Transcriptome Measurement in Single Cells. *Nat. Methods* **2017**, *14*, 865–868. [CrossRef]
65. Grosselin, K.; Durand, A.; Marsolier, J.; Poitou, A.; Marangoni, E.; Nemati, F.; Dahmani, A.; Lameiras, S.; Reyal, F.; Frenoy, O.; et al. High-Throughput Single-Cell ChIP-Seq Identifies Heterogeneity of Chromatin States in Breast Cancer. *Nat. Genet.* **2019**, *51*, 1060–1066. [CrossRef]
66. Mimitou, E.P.; Lareau, C.A.; Chen, K.Y.; Zorzetto-Fernandes, A.L.; Hao, Y.; Takeshima, Y.; Luo, W.; Huang, T.-S.; Yeung, B.Z.; Papalexi, E.; et al. Scalable, Multimodal Profiling of Chromatin Accessibility, Gene Expression and Protein Levels in Single Cells. *Nat. Biotechnol.* **2021**, *39*, 1246–1258. [CrossRef]
67. Tan, W.C.C.; Nerurkar, S.N.; Cai, H.Y.; Ng, H.H.M.; Wu, D.; Wee, Y.T.F.; Lim, J.C.T.; Yeong, J.; Lim, T.K.H. Overview of Multiplex Immunohistochemistry/Immunofluorescence Techniques in the Era of Cancer Immunotherapy. *Cancer Commun.* **2020**, *40*, 135–153. [CrossRef]
68. Park, S.-M.; Lee, J.Y.; Hong, S.; Lee, S.H.; Dimov, I.K.; Lee, H.; Suh, S.; Pan, Q.; Li, K.; Wu, A.M.; et al. Dual Transcript and Protein Quantification in a Massive Single Cell Array. *Lab Chip* **2016**, *16*, 3682–3688. [CrossRef]
69. Frei, A.P.; Bava, F.-A.; Zunder, E.R.; Hsieh, E.W.Y.; Chen, S.-Y.; Nolan, G.P.; Gherardini, P.F. Highly Multiplexed Simultaneous Detection of RNAs and Proteins in Single Cells. *Nat. Methods* **2016**, *13*, 269–275. [CrossRef]
70. Cadot, S.; Valle, C.; Tosolini, M.; Pont, F.; Largeaud, L.; Laurent, C.; Fournie, J.J.; Ysebaert, L.; Quillet-Mary, A. Longitudinal CITE-Seq Profiling of Chronic Lymphocytic Leukemia during Ibrutinib Treatment: Evolution of Leukemic and Immune Cells at Relapse. *Biomark. Res.* **2020**, *8*, 72. [CrossRef]
71. Antunes, A.R.P.; Scheyltjens, I.; Lodi, F.; Messiaen, J.; Antoranz, A.; Duerinck, J.; Kancheva, D.; Martens, L.; De Vlaminck, K.; Van Hove, H.; et al. Single-Cell Profiling of Myeloid Cells in Glioblastoma across Species and Disease Stage Reveals Macrophage Competition and Specialization. *Nat. Neurosci.* **2021**, *24*, 595–610. [CrossRef] [PubMed]
72. Liu, H.-J.; Du, H.; Khabibullin, D.; Zarei, M.; Wei, K.; Freeman, G.J.; Kwiatkowski, D.J.; Henske, E.P. mTORC1 Upregulates B7-H3/CD276 to Inhibit Antitumor T Cells and Drive Tumor Immune Evasion. *Nat. Commun.* **2023**, *14*, 1214. [CrossRef] [PubMed]
73. Delley, C.L.; Liu, L.; Sarhan, M.F.; Abate, A.R. Combined Aptamer and Transcriptome Sequencing of Single Cells. *Sci. Rep.* **2018**, *8*, 2919. [CrossRef] [PubMed]
74. Chen, A.F.; Parks, B.; Kathiria, A.S.; Ober-Reynolds, B.; Goronzy, J.J.; Greenleaf, W.J. NEAT-Seq: Simultaneous Profiling of Intra-Nuclear Proteins, Chromatin Accessibility and Gene Expression in Single Cells. *Nat. Methods* **2022**, *19*, 547–553. [CrossRef]
75. Armand, E.J.; Li, J.; Xie, F.; Luo, C.; Mukamel, E.A. Single-Cell Sequencing of Brain Cell Transcriptomes and Epigenomes. *Neuron* **2021**, *109*, 11–26. [CrossRef]
76. Participants in the 1st Human Cell Atlas Jamboree; Lun, A.T.L.; Riesenfeld, S.; Andrews, T.; Dao, T.P.; Gomes, T.; Marioni, J.C. EmptyDrops: Distinguishing Cells from Empty Droplets in Droplet-Based Single-Cell RNA Sequencing Data. *Genome Biol.* **2019**, *20*, 63. [CrossRef]
77. Wolock, S.L.; Lopez, R.; Klein, A.M. Scrublet: Computational Identification of Cell Doublets in Single-Cell Transcriptomic Data. *Cell Syst.* **2019**, *8*, 281–291.e9. [CrossRef]
78. Bressan, D.; Battistoni, G.; Hannon, G.J. The Dawn of Spatial Omics. *Science* **2023**, *381*, eabq4964. [CrossRef]
79. Chen, K.H.; Boettiger, A.N.; Moffitt, J.R.; Wang, S.; Zhuang, X. Spatially Resolved, Highly Multiplexed RNA Profiling in Single Cells. *Science* **2015**, *348*, aaa6090. [CrossRef]
80. Ding, J.; Ahangari, F.; Espinoza, C.R.; Chhabra, D.; Nicola, T.; Yan, X.; Lal, C.V.; Hagood, J.S.; Kaminski, N.; Bar-Joseph, Z.; et al. Integrating Multiomics Longitudinal Data to Reconstruct Networks Underlying Lung Development. Available online: https://journals.physiology.org/doi/epdf/10.1152/ajplung.00554.2018 (accessed on 29 October 2024).

81. Du, J.; Yang, Y.-C.; An, Z.-J.; Zhang, M.-H.; Fu, X.-H.; Huang, Z.-F.; Yuan, Y.; Hou, J. Advances in Spatial Transcriptomics and Related Data Analysis Strategies. *J. Transl. Med.* **2023**, *21*, 330. [CrossRef]
82. Williams, C.G.; Lee, H.J.; Asatsuma, T.; Vento-Tormo, R.; Haque, A. An Introduction to Spatial Transcriptomics for Biomedical Research. *Genome Med.* **2022**, *14*, 68. [CrossRef] [PubMed]
83. Canozo, F.J.G.; Zuo, Z.; Martin, J.F.; Samee, M.A.H. Cell-Type Modeling in Spatial Transcriptomics Data Elucidates Spatially Variable Colocalization and Communication Between Cell-Types in Mouse Brain. *Cell Syst.* **2022**, *13*, 58–70.e5. [CrossRef] [PubMed]
84. Calvanese, V.; Capellera-Garcia, S.; Ma, F.; Fares, I.; Liebscher, S.; Ng, E.S.; Ekstrand, S.; Aguadé-Gorgorió, J.; Vavilina, A.; Lefaudeux, D.; et al. Mapping Human Haematopoietic Stem Cells from Haemogenic Endothelium to Birth. *Nature* **2022**, *604*, 534–540. [CrossRef] [PubMed]
85. Chen, A.; Liao, S.; Cheng, M.; Ma, K.; Wu, L.; Lai, Y.; Qiu, X.; Yang, J.; Xu, J.; Hao, S.; et al. Spatiotemporal Transcriptomic Atlas of Mouse Organogenesis Using DNA Nanoball-Patterned Arrays. *Cell* **2022**, *185*, 1777–1792.e21. [CrossRef]
86. 10x Genomics Acquires Spatial Transcriptomics. Available online: https://www.genomeweb.com/sequencing/10x-genomics-acquires-spatial-transcriptomics (accessed on 13 November 2024).
87. Ståhl, P.L.; Salmén, F.; Vickovic, S.; Lundmark, A.; Navarro, J.F.; Magnusson, J.; Giacomello, S.; Asp, M.; Westholm, J.O.; Huss, M.; et al. Visualization and Analysis of Gene Expression in Tissue Sections by Spatial Transcriptomics. *Science* **2016**, *353*, 78–82. [CrossRef]
88. Rodriques, S.G.; Stickels, R.R.; Goeva, A.; Martin, C.A.; Murray, E.; Vanderburg, C.R.; Welch, J.; Chen, L.M.; Chen, F.; Macosko, E.Z. Slide-Seq: A Scalable Technology for Measuring Genome-Wide Expression at High Spatial Resolution. *Science* **2019**, *363*, 1463–1467. [CrossRef]
89. Vickovic, S.; Eraslan, G.; Salmén, F.; Klughammer, J.; Stenbeck, L.; Schapiro, D.; Äijö, T.; Bonneau, R.; Bergenstråhle, L.; Navarro, J.F.; et al. High-Definition Spatial Transcriptomics for in Situ Tissue Profiling. *Nat. Methods* **2019**, *16*, 987–990. [CrossRef]
90. Ben-Chetrit, N.; Niu, X.; Swett, A.D.; Sotelo, J.; Jiao, M.S.; Stewart, C.M.; Potenski, C.; Mielinis, P.; Roelli, P.; Stoeckius, M.; et al. Integration of Whole Transcriptome Spatial Profiling with Protein Markers. *Nat. Biotechnol.* **2023**, *41*, 788–793. [CrossRef]
91. Liu, Y.; Yang, M.; Deng, Y.; Su, G.; Enninful, A.; Guo, C.C.; Tebaldi, T.; Zhang, D.; Kim, D.; Bai, Z.; et al. High-Spatial-Resolution Multi-Omics Sequencing via Deterministic Barcoding in Tissue. *Cell* **2020**, *183*, 1665–1681.e18. [CrossRef]
92. Liu, Y.; Enninful, A.; Deng, Y.; Fan, R. Spatial Transcriptome Sequencing of FFPE Tissues at Cellular Level. *bioRxiv* **2020**, 338475.
93. Deng, Y.; Bartosovic, M.; Ma, S.; Zhang, D.; Kukanja, P.; Xiao, Y.; Su, G.; Liu, Y.; Qin, X.; Rosoklija, G.B.; et al. Spatial Profiling of Chromatin Accessibility in Mouse and Human Tissues. *Nature* **2022**, *609*, 375–383. [CrossRef] [PubMed]
94. Deng, Y.; Bartosovic, M.; Kukanja, P.; Zhang, D.; Liu, Y.; Su, G.; Enninful, A.; Bai, Z.; Castelo-Branco, G.; Fan, R. Spatial-CUT&Tag: Spatially Resolved Chromatin Modification Profiling at the Cellular Level. *Science* **2022**, *375*, 681–686. [CrossRef] [PubMed]
95. Liu, Y.; DiStasio, M.; Su, G.; Asashima, H.; Enninful, A.; Qin, X.; Deng, Y.; Nam, J.; Gao, F.; Bordignon, P.; et al. High-Plex Protein and Whole Transcriptome Co-Mapping at Cellular Resolution with Spatial CITE-Seq. *Nat. Biotechnol.* **2023**, *41*, 1405–1409. [CrossRef] [PubMed]
96. Zhang, D.; Deng, Y.; Kukanja, P.; Agirre, E.; Bartosovic, M.; Dong, M.; Ma, C.; Ma, S.; Su, G.; Bao, S.; et al. Spatial Epigenome-Transcriptome Co-Profiling of Mammalian Tissues. *Nature* **2023**, *616*, 113–122. [CrossRef]
97. Liu, C.; Li, R.; Li, Y.; Lin, X.; Zhao, K.; Liu, Q.; Wang, S.; Yang, X.; Shi, X.; Ma, Y.; et al. Spatiotemporal Mapping of Gene Expression Landscapes and Developmental Trajectories During Zebrafish Embryogenesis. *Dev. Cell* **2022**, *57*, 1284–1298.e5. [CrossRef]
98. Shi, Y.; Huang, L.; Dong, H.; Yang, M.; Ding, W.; Zhou, X.; Lu, T.; Liu, Z.; Zhou, X.; Wang, M.; et al. Decoding the Spatiotemporal Regulation of Transcription Factors During Human Spinal Cord Development. *Cell Res.* **2024**, *34*, 193–213. [CrossRef]
99. Da Mesquita, S.; Papadopoulos, Z.; Dykstra, T.; Brase, L.; Farias, F.G.; Wall, M.; Jiang, H.; Kodira, C.D.; de Lima, K.A.; Herz, J.; et al. Meningeal Lymphatics Affect Microglia Responses and Anti-Aβ Immunotherapy. *Nature* **2021**, *593*, 255–260. [CrossRef]
100. Li, Y.; Li, Z.; Wang, C.; Yang, M.; He, Z.; Wang, F.; Zhang, Y.; Li, R.; Gong, Y.; Wang, B.; et al. Spatiotemporal Transcriptome Atlas Reveals the Regional Specification of the Developing Human Brain. *Cell* **2023**, *186*, 5892–5909.e22. [CrossRef]
101. Asp, M.; Giacomello, S.; Larsson, L.; Wu, C.; Fürth, D.; Qian, X.; Wärdell, E.; Custodio, J.; Reimegård, J.; Salmén, F.; et al. A Spatiotemporal Organ-Wide Gene Expression and Cell Atlas of the Developing Human Heart. *Cell* **2019**, *179*, 1647–1660.e19. [CrossRef]
102. Srivatsan, S.R.; Regier, M.C.; Barkan, E.; Franks, J.M.; Packer, J.S.; Grosjean, P.; Duran, M.; Saxton, S.; Ladd, J.J.; Spielmann, M.; et al. Embryo-Scale, Single-Cell Spatial Transcriptomics. *Science* **2021**, *373*, 111–117. [CrossRef]
103. Shekhar, K.; Lapan, S.W.; Whitney, I.E.; Tran, N.M.; Macosko, E.Z.; Kowalczyk, M.; Adiconis, X.; Levin, J.Z.; Nemesh, J.; Goldman, M.; et al. Comprehensive Classification of Retinal Bipolar Neurons by Single-Cell Transcriptomics. *Cell* **2016**, *166*, 1308–1323.e30. [CrossRef] [PubMed]
104. Lohoff, T.; Ghazanfar, S.; Missarova, A.; Koulena, N.; Pierson, N.; Griffiths, J.A.; Bardot, E.S.; Eng, C.-H.L.; Tyser, R.C.V.; Argelaguet, R.; et al. Integration of Spatial and Single-Cell Transcriptomic Data Elucidates Mouse Organogenesis. *Nat. Biotechnol.* **2022**, *40*, 74–85. [CrossRef] [PubMed]

105. Fawkner-Corbett, D.; Antanaviciute, A.; Parikh, K.; Jagielowicz, M.; Gerós, A.S.; Gupta, T.; Ashley, N.; Khamis, D.; Fowler, D.; Morrissey, E.; et al. Spatiotemporal Analysis of Human Intestinal Development at Single-Cell Resolution. *Cell* **2021**, *184*, 810–826.e23. [CrossRef]
106. Garcia-Alonso, L.; Lorenzi, V.; Mazzeo, C.I.; Alves-Lopes, J.P.; Roberts, K.; Sancho-Serra, C.; Engelbert, J.; Marečková, M.; Gruhn, W.H.; Botting, R.A.; et al. Single-Cell Roadmap of Human Gonadal Development. *Nature* **2022**, *607*, 540–547. [CrossRef]
107. Zhao, L.; Song, W.; Chen, Y.-G. Mesenchymal-Epithelial Interaction Regulates Gastrointestinal Tract Development in Mouse Embryos. *Cell Rep.* **2022**, *40*, 111053. [CrossRef]
108. Stickels, R.R.; Murray, E.; Kumar, P.; Li, J.; Marshall, J.L.; Di Bella, D.J.; Arlotta, P.; Macosko, E.Z.; Chen, F. Highly Sensitive Spatial Transcriptomics at Near-Cellular Resolution with Slide-seqV2. *Nat. Biotechnol.* **2021**, *39*, 313–319. [CrossRef]
109. Wu, B.; Shentu, X.; Nan, H.; Guo, P.; Hao, S.; Xu, J.; Shangguan, S.; Cui, L.; Cen, J.; Deng, Q.; et al. A Spatiotemporal Atlas of Cholestatic Injury and Repair in Mice. *Nat. Genet.* **2024**, *56*, 938–952. [CrossRef]
110. Zhang, B.; He, P.; Lawrence, J.E.G.; Wang, S.; Tuck, E.; Williams, B.A.; Roberts, K.; Kleshchevnikov, V.; Mamanova, L.; Bolt, L.; et al. A Human Embryonic Limb Cell Atlas Resolved in Space and Time. *Nature* **2023**, *635*, 668–678. [CrossRef]
111. Laurenti, E.; Göttgens, B. From Haematopoietic Stem Cells to Complex Differentiation Landscapes. *Nature* **2018**, *553*, 418–426. [CrossRef]
112. Wu, Y.; Yang, S.; Ma, J.; Chen, Z.; Song, G.; Rao, D.; Cheng, Y.; Huang, S.; Liu, Y.; Jiang, S.; et al. Spatiotemporal Immune Landscape of Colorectal Cancer Liver Metastasis at Single-Cell Level. *Cancer Discov.* **2022**, *12*, 134–153. [CrossRef]
113. Qi, J.; Sun, H.; Zhang, Y.; Wang, Z.; Xun, Z.; Li, Z.; Ding, X.; Bao, R.; Hong, L.; Jia, W.; et al. Single-Cell and Spatial Analysis Reveal Interaction of FAP+ Fibroblasts and SPP1+ Macrophages in Colorectal Cancer. *Nat. Commun.* **2022**, *13*, 1742. [CrossRef] [PubMed]
114. Guilliams, M.; Bonnardel, J.; Haest, B.; Vanderborght, B.; Wagner, C.; Remmerie, A.; Bujko, A.; Martens, L.; Thoné, T.; Browaeys, R.; et al. Spatial Proteogenomics Reveals Distinct and Evolutionarily Conserved Hepatic Macrophage Niches. *Cell* **2022**, *185*, 379–396.e38. [CrossRef] [PubMed]
115. Gao, S.; Shi, Q.; Zhang, Y.; Liang, G.; Kang, Z.; Huang, B.; Ma, D.; Wang, L.; Jiao, J.; Fang, X.; et al. Identification of HSC/MPP Expansion Units in Fetal Liver by Single-Cell Spatiotemporal Transcriptomics. *Cell Res.* **2022**, *32*, 38–53. [CrossRef] [PubMed]
116. Egozi, A.; Halpern, K.B.; Farack, L.; Rotem, H.; Itzkovitz, S. Zonation of Pancreatic Acinar Cells in Diabetic Mice. *Cell Rep.* **2020**, *32*, 108043. [CrossRef]
117. Cho, C.-S.; Xi, J.; Si, Y.; Park, S.-R.; Hsu, J.-E.; Kim, M.; Jun, G.; Kang, H.M.; Lee, J.H. Microscopic Examination of Spatial Transcriptome Using Seq-Scope. *Cell* **2021**, *184*, 3559–3572.e22. [CrossRef]
118. Zhang, L. Clinical and Translational Values of Spatial Transcriptomics. *Signal Transduct. Target. Ther.* **2022**, *7*, 111. [CrossRef]
119. Wang, D.C.; Wang, X. Clinical Significance of Spatiotemporal Transcriptional Bursting and Control. *Clin. Transl. Med.* **2021**, *11*, e518. [CrossRef]
120. Pickard, K.; Stephenson, E.; Mitchell, A.; Jardine, L.; Bacon, C.M. Location, Location, Location: Mapping the Lymphoma Tumor Microenvironment Using Spatial Transcriptomics. *Front. Oncol.* **2023**, *13*, 1258245. [CrossRef]
121. Wu, Y.; Cheng, Y.; Wang, X.; Fan, J.; Gao, Q. Spatial Omics: Navigating to the Golden Era of Cancer Research. *Clin. Transl. Med.* **2022**, *12*, e696. [CrossRef]
122. Sun, Y.; Liu, M.; Sun, W.; Tang, X.; Zhou, Y.; Zhang, J.; Yang, B. A Hemoglobin Bionics-Based System for Combating Antibiotic Resistance in Chronic Diabetic Wounds via Iron Homeostasis Regulation. *Adv. Mater.* **2024**, *36*, 2405002. [CrossRef]
123. Zhao, T.; Chiang, Z.D.; Morriss, J.W.; LaFave, L.M.; Murray, E.M.; Del Priore, I.; Meli, K.; Lareau, C.A.; Nadaf, N.M.; Li, J.; et al. Spatial Genomics Enables Multi-Modal Study of Clonal Heterogeneity in Tissues. *Nature* **2022**, *601*, 85–91. [CrossRef] [PubMed]
124. Donoghue, M.T.A.; Schram, A.M.; Hyman, D.M.; Taylor, B.S. Discovery Through Clinical Sequencing in Oncology. *Nat. Cancer* **2020**, *1*, 774–783. [CrossRef] [PubMed]
125. Erickson, A.; He, M.; Berglund, E.; Marklund, M.; Mirzazadeh, R.; Schultz, N.; Kvastad, L.; Andersson, A.; Bergenstråhle, L.; Bergenstråhle, J.; et al. Spatially Resolved Clonal Copy Number Alterations in Benign and Malignant Tissue. *Nature* **2022**, *608*, 360–367. [CrossRef] [PubMed]
126. van Galen, P.; Hovestadt, V.; Wadsworth, M.H., II; Hughes, T.K.; Griffin, G.K.; Battaglia, S.; Verga, J.A.; Stephansky, J.; Pastika, T.J.; Lombardi Story, J.; et al. Single-Cell RNA-Seq Reveals AML Hierarchies Relevant to Disease Progression and Immunity. *Cell* **2019**, *176*, 1265–1281.e24. [CrossRef]
127. Baryawno, N.; Przybylski, D.; Kowalczyk, M.S.; Kfoury, Y.; Severe, N.; Gustafsson, K.; Kokkaliaris, K.D.; Mercier, F.; Tabaka, M.; Hofree, M.; et al. A Cellular Taxonomy of the Bone Marrow Stroma in Homeostasis and Leukemia. *Cell* **2019**, *177*, 1915–1932.e16. [CrossRef]
128. Granja, J.M.; Klemm, S.; McGinnis, L.M.; Kathiria, A.S.; Mezger, A.; Corces, M.R.; Parks, B.; Gars, E.; Liedtke, M.; Zheng, G.X.Y.; et al. Single-Cell Multiomic Analysis Identifies Regulatory Programs in Mixed-Phenotype Acute Leukemia. *Nat. Biotechnol.* **2019**, *37*, 1458–1465. [CrossRef]

129. Moncada, R.; Barkley, D.; Wagner, F.; Chiodin, M.; Devlin, J.C.; Baron, M.; Hajdu, C.H.; Simeone, D.M.; Yanai, I. Integrating Microarray-Based Spatial Transcriptomics and Single-Cell RNA-Seq Reveals Tissue Architecture in Pancreatic Ductal Adenocarcinomas. *Nat. Biotechnol.* **2020**, *38*, 333–342. [CrossRef]
130. Cheng, M.; Jiang, Y.; Xu, J.; Mentis, A.-F.A.; Wang, S.; Zheng, H.; Sahu, S.K.; Liu, L.; Xu, X. Spatially Resolved Transcriptomics: A Comprehensive Review of Their Technological Advances, Applications, and Challenges. *J. Genet. Genom.* **2023**, *50*, 625–640. [CrossRef]
131. Wu, C.-Y.; Lau, B.T.; Kim, H.S.; Sathe, A.; Grimes, S.M.; Ji, H.P.; Zhang, N.R. Integrative Single-Cell Analysis of Allele-Specific Copy Number Alterations and Chromatin Accessibility in Cancer. *Nat. Biotechnol.* **2021**, *39*, 1259–1269. [CrossRef]
132. Guo, W.; Zhou, B.; Yang, Z.; Liu, X.; Huai, Q.; Guo, L.; Xue, X.; Tan, F.; Li, Y.; Xue, Q.; et al. Integrating Microarray-Based Spatial Transcriptomics and Single-Cell RNA-Sequencing Reveals Tissue Architecture in Esophageal Squamous Cell Carcinoma. *EBioMedicine* **2022**, *84*, 104281. [CrossRef]
133. Ou, Z.; Lin, S.; Qiu, J.; Ding, W.; Ren, P.; Chen, D.; Wang, J.; Tong, Y.; Wu, D.; Chen, A.; et al. Single-Nucleus RNA Sequencing and Spatial Transcriptomics Reveal the Immunological Microenvironment of Cervical Squamous Cell Carcinoma. *Adv. Sci.* **2022**, *9*, e2203040. [CrossRef] [PubMed]
134. Habibey, R.; Arias, J.E.R.; Striebel, J.; Busskamp, V. Microfluidics for Neuronal Cell and Circuit Engineering. *Chem. Rev.* **2022**, *122*, 14842–14880. [CrossRef] [PubMed]
135. Rostom, R.; Svensson, V.; Teichmann, S.A.; Kar, G. Computational Approaches for Interpreting scRNA-Seq Data. *FEBS Lett.* **2017**, *591*, 2213–2225. [CrossRef] [PubMed]
136. Stuart, T.; Satija, R. Integrative Single-Cell Analysis. *Nat. Rev. Genet.* **2019**, *20*, 257–272. [CrossRef]
137. Gal, E.; London, M.; Globerson, A.; Ramaswamy, S.; Reimann, M.W.; Muller, E.; Markram, H.; Segev, I. Rich Cell-Type-Specific Network Topology in Neocortical Microcircuitry. *Nat. Neurosci.* **2017**, *20*, 1004–1013. [CrossRef]
138. Schafflick, D.; Xu, C.A.; Hartlehnert, M.; Cole, M.; Schulte-Mecklenbeck, A.; Lautwein, T.; Wolbert, J.; Heming, M.; Meuth, S.G.; Kuhlmann, T.; et al. Integrated Single Cell Analysis of Blood and Cerebrospinal Fluid Leukocytes in Multiple Sclerosis. *Nat. Commun.* **2020**, *11*, 247. [CrossRef]
139. Morabito, S.; Miyoshi, E.; Michael, N.; Shahin, S.; Martini, A.C.; Head, E.; Silva, J.; Leavy, K.; Perez-Rosendahl, M.; Swarup, V. Single-Nucleus Chromatin Accessibility and Transcriptomic Characterization of Alzheimer's Disease. *Nat. Genet.* **2021**, *53*, 1143–1155. [CrossRef]
140. Gate, D.; Saligrama, N.; Leventhal, O.; Yang, A.C.; Unger, M.S.; Middeldorp, J.; Chen, K.; Lehallier, B.; Channappa, D.; De Los Santos, M.B.; et al. Clonally Expanded CD8 T Cells Patrol the Cerebrospinal Fluid in Alzheimer's Disease. *Nature* **2020**, *577*, 399–404. [CrossRef]
141. Sporns, O.; Tononi, G.; Kötter, R. The Human Connectome: A Structural Description of the Human Brain. *PLoS Comput. Biol.* **2005**, *1*, e42. [CrossRef]
142. Fornito, A.; Arnatkevičiūtė, A.; Fulcher, B.D. Bridging the Gap between Connectome and Transcriptome. *Trends Cogn. Sci.* **2019**, *23*, 34–50. [CrossRef]
143. Lein, E.S.; Hawrylycz, M.J.; Ao, N.; Ayres, M.; Bensinger, A.; Bernard, A.; Boe, A.F.; Boguski, M.S.; Brockway, K.S.; Byrnes, E.J.; et al. Genome-Wide Atlas of Gene Expression in the Adult Mouse Brain. *Nature* **2007**, *445*, 168–176. [CrossRef] [PubMed]
144. Ng, L.; Bernard, A.; Lau, C.; Overly, C.C.; Dong, H.-W.; Kuan, C.; Pathak, S.; Sunkin, S.M.; Dang, C.; Bohland, J.W.; et al. An Anatomic Gene Expression Atlas of the Adult Mouse Brain. *Nat. Neurosci.* **2009**, *12*, 356–362. [CrossRef] [PubMed]
145. Hawrylycz, M.J.; Lein, E.S.; Guillozet-Bongaarts, A.L.; Shen, E.H.; Ng, L.; Miller, J.A.; van de Lagemaat, L.N.; Smith, K.A.; Ebbert, A.; Riley, Z.L.; et al. An Anatomically Comprehensive Atlas of the Adult Human Brain Transcriptome. *Nature* **2012**, *489*, 391–399. [CrossRef]
146. Gravina, S.; Dong, X.; Yu, B.; Vijg, J. Single-Cell Genome-Wide Bisulfite Sequencing Uncovers Extensive Heterogeneity in the Mouse Liver Methylome. *Genome Biol.* **2016**, *17*, 150. [CrossRef]
147. Papalexi, E.; Satija, R. Single-Cell RNA Sequencing to Explore Immune Cell Heterogeneity. *Nat. Rev. Immunol.* **2018**, *18*, 35–45. [CrossRef]
148. Scott, D.W.; Gascoyne, R.D. The Tumour Microenvironment in B Cell Lymphomas. *Nat. Rev. Cancer* **2014**, *14*, 517–534. [CrossRef]
149. Stewart, B.J.; Fergie, M.; Young, M.D.; Jones, C.; Sachdeva, A.; Blain, A.; Bacon, C.M.; Rand, V.; Ferdinand, J.R.; James, K.R.; et al. Spatial and Molecular Profiling of the Mononuclear Phagocyte Network in Classic Hodgkin Lymphoma. *Blood* **2023**, *141*, 2343–2358. [CrossRef]
150. He, Z.; Maynard, A.; Jain, A.; Gerber, T.; Petri, R.; Lin, H.-C.; Santel, M.; Ly, K.; Dupré, J.-S.; Sidow, L.; et al. Lineage Recording in Human Cerebral Organoids. *Nat. Methods* **2022**, *19*, 90–99. [CrossRef]
151. Peixoto, L.; Risso, D.; Poplawski, S.G.; Wimmer, M.E.; Speed, T.P.; Wood, M.A.; Abel, T. How Data Analysis Affects Power, Reproducibility and Biological Insight of RNA-Seq Studies in Complex Datasets. *Nucleic Acids Res.* **2015**, *43*, 7664–7674. [CrossRef]

152. Halpern, K.B.; Shenhav, R.; Matcovitch-Natan, O.; Toth, B.; Lemze, D.; Golan, M.; Massasa, E.E.; Baydatch, S.; Landen, S.; Moor, A.E.; et al. Single-Cell Spatial Reconstruction Reveals Global Division of Labour in the Mammalian Liver. *Nature* **2017**, *542*, 352–356. [CrossRef]
153. Madsen, R.; Lundstedt, T.; Trygg, J. Chemometrics in Metabolomics—A Review in Human Disease Diagnosis. *Anal. Chim. Acta* **2010**, *659*, 23–33. [CrossRef] [PubMed]
154. Steuer, R. Review: On the Analysis and Interpretation of Correlations in Metabolomic Data. *Brief. Bioinform.* **2006**, *7*, 151–158. [CrossRef] [PubMed]
155. Joyce, A.R.; Palsson, B.Ø. The Model Organism as a System: Integrating "omics" Data Sets. *Nat. Rev. Mol. Cell Biol.* **2006**, *7*, 198–210. [CrossRef]
156. Uhlén, M.; Fagerberg, L.; Hallström, B.M.; Lindskog, C.; Oksvold, P.; Mardinoglu, A.; Sivertsson, Å.; Kampf, C.; Sjöstedt, E.; Asplund, A.; et al. Proteomics. Tissue-Based Map of the Human Proteome. *Science* **2015**, *347*, 1260419. [CrossRef]
157. Khan, M.A.B.; Hashim, M.J.; King, J.K.; Govender, R.D.; Mustafa, H.; Al Kaabi, J. Epidemiology of Type 2 Diabetes—Global Burden of Disease and Forecasted Trends. *J. Epidemiol. Glob. Health* **2020**, *10*, 107–111. [CrossRef]
158. Ren, X.; Zhong, G.; Zhang, Q.; Zhang, L.; Sun, Y.; Zhang, Z. Reconstruction of Cell Spatial Organization from Single-Cell RNA Sequencing Data Based on Ligand-Receptor Mediated Self-Assembly. *Cell Res.* **2020**, *30*, 763–778. [CrossRef]
159. Maynard, K.R.; Collado-Torres, L.; Weber, L.M.; Uytingco, C.; Barry, B.K.; Williams, S.R.; Catallini, J.L.; Tran, M.N.; Besich, Z.; Tippani, M.; et al. Transcriptome-Scale Spatial Gene Expression in the Human Dorsolateral Prefrontal Cortex. *Nat. Neurosci.* **2021**, *24*, 425–436. [CrossRef]
160. Chen, G.; Ning, B.; Shi, T. Single-Cell RNA-Seq Technologies and Related Computational Data Analysis. *Front. Genet.* **2019**, *10*, 317. [CrossRef]
161. Huang, M.; Wang, J.; Torre, E.; Dueck, H.; Shaffer, S.; Bonasio, R.; Murray, J.I.; Raj, A.; Li, M.; Zhang, N.R. SAVER: Gene Expression Recovery for Single-Cell RNA Sequencing. *Nat. Methods* **2018**, *15*, 539–542. [CrossRef]
162. Li, W.V.; Li, J.J. An Accurate and Robust Imputation Method scImpute for Single-Cell RNA-Seq Data. *Nat. Commun.* **2018**, *9*, 997. [CrossRef]
163. Pham, D.; Tan, X.; Balderson, B.; Xu, J.; Grice, L.F.; Yoon, S.; Willis, E.F.; Tran, M.; Lam, P.Y.; Raghubar, A.; et al. Robust Mapping of Spatiotemporal Trajectories and Cell–Cell Interactions in Healthy and Diseased Tissues. *Nat. Commun.* **2023**, *14*, 7739. [CrossRef] [PubMed]
164. Zhu, S.; Qing, T.; Zheng, Y.; Jin, L.; Shi, L. Advances in Single-Cell RNA Sequencing and Its Applications in Cancer Research. *Oncotarget* **2017**, *8*, 53763–53779. [CrossRef] [PubMed]
165. Steinberg, J.; Ritchie, G.R.S.; Roumeliotis, T.I.; Jayasuriya, R.L.; Clark, M.J.; Brooks, R.A.; Binch, A.L.A.; Shah, K.M.; Coyle, R.; Pardo, M.; et al. Integrative Epigenomics, Transcriptomics and Proteomics of Patient Chondrocytes Reveal Genes and Pathways Involved in Osteoarthritis. *Sci. Rep.* **2017**, *7*, 8935. [CrossRef] [PubMed]
166. Steinberg, J.; Southam, L.; Roumeliotis, T.I.; Clark, M.J.; Jayasuriya, R.L.; Swift, D.; Shah, K.M.; Butterfield, N.C.; Brooks, R.A.; McCaskie, A.W.; et al. A Molecular Quantitative Trait Locus Map for Osteoarthritis. *Nat. Commun.* **2021**, *12*, 1309. [CrossRef]

micromachines

Review

Recent Advances in Microfluidic Impedance Detection: Principle, Design and Applications

Yigang Shen [1,†], Zhenxiao Wang [1,†], Tingyu Ren [1], Jianming Wen [1,2], Jianping Li [1,*] and Tao Tang [3,*]

1 The Institute of Precision Machinery and Smart Structure, College of Engineering, Zhejiang Normal University, Jinhua 321004, China; yigangshen@zjnu.edu.cn (Y.S.); 1029502@zjnu.edu.cn (Z.W.); rentingyu001205@163.com (T.R.); wjming@zjnu.cn (J.W.)
2 College of Mathematical Medicine, Zhejiang Normal University, Jinhua 321004, China
3 Department of Neurosurgery, Chongqing General Hospital, Chongqing University, Chongqing 401147, China
* Correspondence: lijp@zjnu.cn (J.L.); tang.tao@cqu.edu.cn (T.T.)
† These authors contributed equally to this work.

Abstract: Under the dual drivers of precision medicine development and health monitoring demands, the development of real-time biosensing technologies has emerged as a key breakthrough in the field of life science analytics. Microfluidic impedance detection technology, achieved through the integration of microscale fluid manipulation and bioimpedance spectrum analysis, has enabled the real-time monitoring of biological samples ranging from single cells to organ-level systems, now standing at the forefront of biological real-time detection research. This review systematically summarizes the core principles of microfluidic impedance detection technology, modeling methods for cell equivalent circuits, system optimization strategies, and recent research advancements in biological detection applications. We first elucidate the fundamental principles of microfluidic impedance detection technologies, followed by a comprehensive analysis of cellular equivalent circuit model construction and microfluidic system design optimization strategies. Subsequently, we categorize applications based on biological sample types, elaborating on respective research progress and existing challenges. This review concludes with prospective insights into future developmental trajectories. We hope this work will provide novel research perspectives for advancing microfluidic impedance detection technology while stimulating interdisciplinary collaboration among researchers in biology, medicine, chemistry, and physics to propel technological innovation collectively.

Keywords: microfluidics; impedance; cell equivalent circuits; single cell

1. Introduction

Amid frequent global health crises (e.g., COVID-19 pandemic, dengue fever outbreaks), surging demands for precision medicine, and escalating environmental and food safety challenges, the development of rapid, sensitive, and portable detection technologies has become a shared pursuit across scientific and industrial communities. Traditional detection methods (such as polymerase chain reaction (PCR) [1] and enzyme-linked immunosorbent assay (ELISA) [2]), while characterized by high specificity, are constrained by operational complexity, expensive equipment, and time-consuming procedures, making them inadequate for urgent on-site analysis requirements. As a groundbreaking solution, point-of-care testing (POCT) technology enables direct detection at sampling sites through portable devices and integrated reagents, eliminating the need for complex laboratory processing and facilitating the immediate acquisition of test results. This technology aims

 https://doi.org/10.3390/mi16060683

to enhance medical decision-making efficiency through device portability and streamlined workflows that rapidly deliver analytical outcomes.

Microfluidic technology is an advanced methodology that enables the precise manipulation of nano- to pico-liter scale fluids through engineered microscale channels and structures (typically tens to hundreds of micrometers). Its defining characteristic lies in integrating conventional laboratory functions into miniaturized chips, earning the designation "lab on a chip". The concept of microfluidics was first proposed by Eringen et al. in 1964 [3]. In 1979, Terry et al. [4] developed a silicon-based gas chromatographic device that served as the prototype for modern microfluidic systems. Manz et al. introduced the conceptual framework for microfluidic design in 1990 [5]. A milestone advancement occurred in 1998 when the Whitesides group at Harvard University, through iterative material testing and based on microfabrication techniques, innovatively proposed using polydimethylsiloxane (PDMS) and simplified manufacturing methods to fabricate microfluidic chips [6], thereby establishing a comprehensive technical system. Through synergistic innovations in micro/nanofabrication processes (e.g., photolithography, soft lithography) and novel materials (e.g., PDMS, glass), microfluidic technology has evolved into an interdisciplinary field combining fluid dynamics, biomedicine, and sensing technologies. Its core advantages manifest in four dimensions: miniaturization (reducing sample/reagent consumption to 1/1000 of conventional methods), high-throughput capacity (enabling the parallel processing of thousands of reaction units), precise control (regulating particle motion via multi-physical fields including electric and pressure fields), and system integration (achieving unified sample pretreatment, reaction, and detection modules) [7–10].

Impedance detection technology is a universal analytical methodology based on measurements of impedance characteristics in circuits or materials under alternating current excitation. By analyzing the variations in impedance magnitude and phase with frequency, this technique reveals the electrical, physical, and chemical properties of analyzed subjects. In 1975, Kivelson et al. [11] advanced a theoretical model elucidating the relationship between polarization effects and frequency through investigations of dielectric relaxation behaviors in complex systems, providing fundamental theoretical support for the application of alternating current (AC) impedance spectroscopy in heterogeneous material analysis. A pivotal milestone emerged in 1999 when Ayliffe et al. [12] pioneered single-cell impedance measurements within microchannels, utilizing paired microelectrodes with a few-micrometer gap to achieve the impedance detection of human polymorphonuclear leukocytes and fish erythrocytes in 10 μm wide microchannels, thereby demonstrating the first evidence of single-cell electrical property discrimination capabilities. This technology exhibits three core advantages: non-invasiveness (sample preservation), multidimensional information (a comprehensive analysis of parameters including the impedance modulus, phase angle, and spectral curves), and high sensitivity (detectable to minute variations) [13,14].

The deep integration of microfluidics with impedance detection technologies has established revolutionary pathways for point-of-care testing and on-site analysis. This synergistic convergence not only enhances analytical dimensionality and precision but also drives a paradigm shift from laboratory-based to field-deployable platforms, achieving breakthroughs from macroscopic solutions to the microscale detection of single cells and bacteria. The defining features of microfluidic impedance detection technology encompass label-free operation (eliminating fluorescent/radioactive labeling), real-time dynamic monitoring (millisecond-level responsiveness), multiparameter resolution (membrane capacitance, cytoplasmic conductivity, and dielectric constant), and ultrahigh sensitivity (detection capability down to single microorganisms or microscale particles). Currently, three primary application matrices are established: fundamental research (single-cell heterogeneity analysis and organ-on-a-chip development), clinical diagnostics (rapid pathogen screening and bedside testing), and industrial applications (high-throughput drug screening and real-time environmental pollutant monitoring).

This review systematically examines the technical principles, system design methodologies, and recent advancements in microfluidic impedance sensing technology, with particular emphasis on its critical research value in biological detection, spanning from single-cell to organ-level analyses. A rigorous analytical framework is established to dissect the modeling approaches for cellular equivalent circuit models in microfluidic impedance systems, followed by comprehensive discussions on design principles categorized under static and dynamic operational paradigms. From an application perspective, the current implementation domains of microfluidic impedance analysis technology are classified according to biological sample types, delineated into four major categories, tumor detection, blood detection, organ on a chip, and microbial detection, as shown in Figure 1. Finally, this review culminates in a critical examination of prevailing technical bottlenecks (e.g., signal noise suppression, multimodal data fusion) and prospective breakthrough directions (e.g., AI-driven intelligent algorithms, flexible electronics integration), aiming to provide theoretical foundations and technical roadmaps for interdisciplinary innovation in precision medicine, green chemistry, and POCT fields.

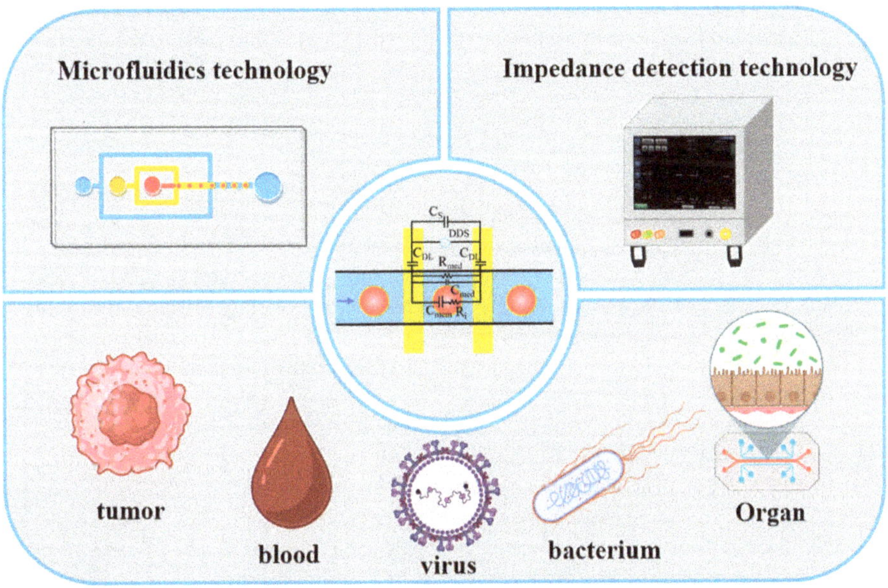

Figure 1. Innovative applications of microfluidic impedance detection technology in various fields.

2. Theory of Microfluidic Impedance Analysis

This chapter establishes a theoretical framework for microfluidic impedance analysis, spanning three domains: hydrodynamics, cellular equivalent circuit modeling, and system design. Hydrodynamic characteristics and microchannel flow behaviors are first elucidated. Equivalent circuit modeling methods for detecting cellular electro-mechanical properties via impedance measurements are subsequently introduced. Design principles for static and dynamic microfluidic impedance detection systems are comprehensively detailed, encompassing electrode configurations and accuracy enhancement strategies through electric field optimization, hydrodynamic focusing, and signal compensation. Progressing systematically from microscopic mechanisms to macroscopic implementations, this hierarchical framework provides the technical foundation for multidimensional single-cell characterization and precision sensing.

2.1. Microchannel Flow Characterization

The essence of microfluidics lies in constructing miniaturized fluidic networks, where microscale channels (specifically width, height, and depth dimensions within the micrometer range) are precisely fabricated via microfabrication techniques (e.g., soft lithography, laser ablation, or injection molding). Multi-physical field coupling effects (pressure, electric fields, and capillary forces) are harnessed to achieve controlled fluid delivery, mixing, separation, and reaction control. Microfluidic chips are typically fabricated using materials such as PDMS, glass, or thermoplastic polymers. Channel wettability is adjusted by surface chemical modifications (e.g., silanization or plasma treatment) to suit different application requirements. Since the dimensions of microfluidic channels are sub-micron to millimeter, the hydrodynamic behavior within these confined spaces is mainly governed by microscale effects, exhibiting four characteristic phenomena: under low Reynolds number conditions, laminar flow dominates, and interlayer material exchange is achieved only by molecular diffusion; due to the high surface area-to-volume ratio, the effects of surface tension, wettability, and wall adsorption are significantly enhanced; at the microscale, synergistic interactions between electric, thermal, and hydrodynamic fields are dramatically enhanced; and the shortened diffusion path and improved mass/heat transfer efficiency in microliter-scale volumes reduce reagent consumption to 1/100–1/1000 of conventional methods.

2.2. Cell Model

The impedance testing technique is a general analytical method that quantifies the internal electrical characteristics and structural features of a test object by measuring the impedance response under alternating electric field excitation in combination with spectral analysis and equivalent circuit modeling. The principle of operation involves applying a frequency-controlled AC excitation signal to the test object, collecting voltage and current waveform data synchronously, and calculating the amplitude ratio and phase difference to construct a frequency-dependent complex impedance spectrum. Subsequently, RLC equivalent circuit modeling or dielectric relaxation theory can establish quantitative correlations between characteristic frequency peaks, capacitive arc radii in the impedance spectrum, and physical/chemical properties (e.g., dielectric constant, conductivity). Impedance detection, characterized by non-invasiveness, high sensitivity, and cost-effectiveness, has become an important tool for linking microscopic biophysical and electrical properties to macroscopic applications in medicine, environmental monitoring, and food safety.

Impedance detection techniques are used in the electrical characterization of cells that can be represented as a uniformly conducting cytoplasm (cytoplasmic conductivity σ_i) wrapped by a dielectric membrane (specific membrane capacitance C_{sm}) according to the widely used spherical monoshell model [15]. When suspended cells are exposed to an AC electric field, the electrical properties of the cell–medium mixture can be represented by Maxwell's mixture theory (MMT) and the corresponding equivalent circuit model (ECM), as shown in Figure 2a. In the simplest analytical model, the impedance consists of solution resistance (R_{med}), solution capacitance (C_{med}), cytoplasmic resistance (R_i), membrane capacitance(C_{mem}), double-layer capacitance (C_{DL}), and stray capacitance (C_s).

$$Z^*_{mix} = \frac{R_{med}\,(1 + j\omega R_i C_{mem})}{j\omega R_{med}\,C_{mem} + (1 + j\omega R_i C_{mem})(1 + j\omega R_{med}\,C_{med})} \tag{1}$$

Considering C_s and C_{DL}, the total impedance is given by

$$Z^*_{total} = \frac{2Z^*_{mix}}{2j\omega C_s Z^*_{mix} + j\omega C_{DL} Z^*_{mix} + 2} \quad (2)$$

Cellular impedance measurements leverage frequency-dependent characteristics: lower frequencies (<0.1 MHz) primarily analyze cell size, membrane properties, and concentration quantification, while higher frequencies (>0.1 MHz) probe intracellular features including organelle status, cytoplasmic composition, and pathophysiological conditions. At different frequencies, the main factors affecting impedance during cell impedance measurements vary, as shown in Figure 2c. At the first frequency dispersion (<0.1 MHz), the presence of the dielectric membrane causes the impedance to be primarily determined by the electric double-layer capacitance and cell size; at the second frequency dispersion (0.1–0.5 MHz), the electric field penetrates the membrane, and the impedance is predominantly determined by its dielectric constant; at the third dispersion (0.5–10 MHz), the dielectric membrane becomes permeable to the electric field, and the impedance is strongly influenced by intracellular conductivity (i.e., the ability of ions or electrons to migrate within the cytoplasm); and at the fourth dispersion (>10 MHz), the impedance is affected by stray capacitance [16]. Intrinsic cellular electrical properties are typically resolved by minimizing the squared discrepancy between measured impedance Z^*_{exp} and model-estimated Z^*_{total}:

$$\min \sum_n \left[Z^*_{exp}(\omega_n) - Z^*_{total}(\omega_n)\right]^2, n = 1, 2, \ldots, N \quad (3)$$

2.3. Cell Deformation Model

In the mechanical characterization of cells using impedance detection technology, resolving intrinsic mechanical parameters necessitates mapping the power-law rheological deformation process of single-cell viscoelasticity to corresponding impedance signals. Within the mechanical sensing zone, where single cells displace equivalent-volume media, an electrical equivalent model can be constructed based on sensing geometry, as shown in Figure 2b. Excitation frequencies are selected within the intermediate range (0.5–10 MHz) to eliminate C_s and C_{DL} influences. Consequently, the total resistance in the mechanical sensing zone is expressed as

$$R_{overall} = R_{cell} + R_{med} = \frac{\rho_{med} L_{channel}}{S} + \frac{(\rho_{cell} - \rho_{med})}{S} \cdot L_p \quad (4)$$

where ρ denotes resistivity (cell or medium), L represents protruding cell length or medium path in constriction channels ($L_{channel} = L_p + L_{med}$), and S indicates the channel cross-sectional area. The measured resistance $R_{overall}$ thus proportionally correlates with cellular protrusion length L_p under constant $L_{channel}$, enabling the real-time tracking of dynamic deformation processes.

Cellular deformation derived from impedance measurements is subsequently utilized to resolve intrinsic mechanical parameters. According to power-law rheology [17], the cell undergoes three distinct phases during its passage through the constriction channel, as shown in Figure 2d. Phase I begins as the cell rapidly enters the constriction channel. Phase II involves the elastic deformation of the cell within the channel, exhibiting power-law deformation behavior consistent with the rheological model; this phase allows the determination of intrinsic mechanical parameters. Phase III occurs when the cell is fully deformed within the constriction channel and accelerates through it.

During Phase II, cellular viscoelastic deformation occurs within the microchannel, where the measured deformation length $L_P(t)$ adheres to a power-law rheological model. This dynamic behavior is governed by intrinsic cellular viscoelasticity, with core parameters including Young's modulus and the power-law exponent. Subsequently, deformation data $L_P(t)$ from Phase II is utilized to generate theoretical deformation curves $L_{cell}(t)$ via the power-law rheological model. Finally, intrinsic mechanical parameters, Young's modulus and the power-law exponent, are determined through least-squares minimization.

$$\min \sum_i \left[L_P(t_i) - L_{cell}(t_i)\right]^2 \quad (5)$$

To characterize background fluid impedance for modeling prerequisites, an equivalent circuit model [18] is established, as shown in Figure 2e:

$$Z_1 = Z_{1r} + jZ_{1i} = R_e + \left[\frac{1}{\frac{1}{R_s} + j\omega C_s}\right] \quad (6)$$

where j denotes the imaginary unit, ω is the angular frequency, and Z_1 represents complex background impedance (real: Z_{1r}, imaginary: Z_{1i}). R_e characterizes electrode interfacial reactance, with the zero-phase element (ZPE) substituting conventional constant phase element (CPE) to model frequency-independent components. R_s and C_s correspond to fluid resistance and capacitance, respectively. This model accurately describes phosphate-buffered solution (PBS) electrochemical properties, providing a baseline for subsequent impedance variation analysis with biochemical additives.

For the precise simulation of experimental electrochemical impedance spectroscopy, Cao et al. [19] developed an equivalent circuit model systematically characterizing oocyte measurements at a microfluidic orifice, as shown in Figure 2f. The 12-parameter model comprises medium resistance R_s, plasma membrane resistance R_M, cytoplasmic capacitance C_P, cytoplasmic resistance R_P, perivitelline space resistance R_{PVS}, zona pellucida resistances (R_1: extra-orifice, R_2: orifice periphery, R_3: intra-orifice), medium resistance R_A in the orifice region, zona pellucida capacitance C, electrode crosstalk capacitance C_E, and the CPE. The CPE accounts for non-ideal capacitive behavior at electrode–medium interfaces due to surface roughness.

2.4. *Microfluidic Impedance System Design*

Microfluidic impedance sensing technology achieves the high-sensitivity real-time dynamic monitoring of cells or biomolecules by integrating the precise fluid manipulation capability of microfluidic chips with the bioelectrical characteristic analysis capacity of impedance detection. The design methodology of microfluidic impedance systems is systematically introduced below through both static and dynamic configurations.

2.4.1. Static System Design—Electrical Impedance Spectroscopy (EIS)

In cell culture or open microfluidic systems, where cells are maintained in static states or prolonged growth phases for extended durations, this operational paradigm is defined as static system design. Under such conditions, the detection sample is fully immersed between the detection electrodes and subjected to impedance analysis to capture temporal impedance variations over defined intervals. For static configurations, critical parameters including the volumetric dimensions of the microfluidic chip and the spatial arrangement of detection electrodes must be optimized. For example, in impedance monitoring throughout cell culture cycles, the detection electrodes are conventionally positioned on opposing sides of the cultivation chamber. During static measurements, intrinsic dielectric properties of cells such as cytoplasmic conductivity, specific membrane capacitance, Young's modulus, cell membrane capacitance, cytoplasmic resistivity, and phase angle are quantitatively resolved.

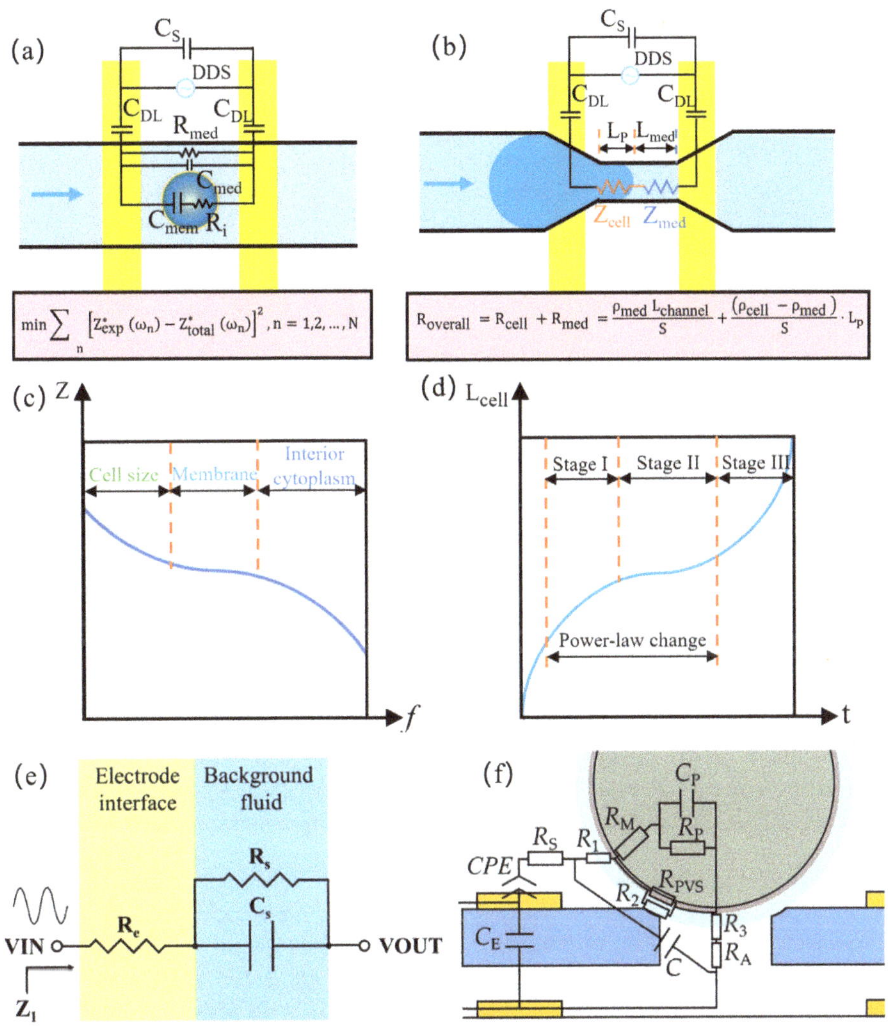

Figure 2. Equivalent circuit models: (**a**) an equivalent circuit diagram when measuring the electrical properties of the cell; (**b**) an equivalent circuit diagram when measuring the mechanical properties of the cell in the constricted channel; (**c**) corresponding impedance spectra showing the different dispersion dominated by R, C_{sm}, and σ_i; (**d**) an equivalent impedance–deformability mapping model of a single cell extruded through the constricted channel; (**e**) an equivalent circuit model of the background fluid (reprinted with permission from [18], published by AIP Publishing, 2025); (**f**) an equivalent circuit model at the orifice (reprinted with permission from [19], published by Elsevier, 2023).

For single cells in a flowing state, developing an effective method to achieve precise capture and in-depth analysis remains a critical challenge to be addressed. A fully electric single-cell manipulation platform was developed by Eeckhoudt et al. [20] using dual-voltage dielectrophoresis technology, which enabled the integrated control of the entire workflow including capture, retention, impedance analysis, and the release of Saccharomyces cerevisiae.

2.4.2. Dynamic System Design—Impedance Flow Cytometry (IFC)

In dynamic system design, this situation refers to the high-speed movement of a biological sample within a microfluidic channel. The core principle of microfluidic impedance sensing technology lies in the fact that as targets (e.g., cells, particles) flow through the microchannel, their dielectric properties (e.g., membrane insulation, intracellular conductivity) perturb the preset AC electric field within the channel. By measuring the impedance changes (the combined effects of resistive and capacitive responses), information on the size, morphology, mechanical properties, and biochemical status of the target can be obtained. The combined advantages of this technology are threefold: real-time capability with dynamic tracking, high throughput with automation, and enhanced sensitivity and stability. This technology paradigm overcomes the static limitations of traditional impedance detection and provides a miniaturized, high-precision solution for applications such as single-cell analysis, the rapid screening of pathogens, and environmental and food monitoring. For single-cell analysis, the size of the microchannel sensing zone should be close to the size of the cell or pathogen under test for optimal accuracy; however, to avoid channel blockage, the channel size is usually slightly larger than the target size. Constricted channels are commonly used when measuring cell deformability, but such channels tend to clog. Adding xanthan gum to mixed samples can effectively reduce clogging [21]. Regarding microchannel size design, when the sensing area size matches the measured particles, the relative error of the electrical characteristic parameters measured by IFC is less than 10%. The microchannel height has the greatest impact on measurement accuracy, with channel heights exceeding the particle size increasing the relative error to 30%, while the maximum relative error caused by the electrode gap is 21.4%, slightly higher than the error caused by the channel width (20.9%) [16]. Therefore, the systematic design optimization of these parameters can further improve the measurement accuracy.

To overcome the limitations of traditional solid microchannels—such as susceptibility to microchannel occlusion and non-adjustable dimensions—Wang et al. [22] recently developed a microfluidic impedance-based flow cytometry technique based on virtual constriction microchannels. This technology constructs virtual constriction zones through the co-flow of conductive samples and insulating sheath fluids, with embedded microelectrodes positioned beneath the microchannel for impedance measurements. Compared to conventional mechanically constricted microchannels, the virtual constriction approach offers three key advantages: the effective avoidance of direct physical contact between cells and channel walls; the maintenance of high-throughput detection; and a significant reduction in the impedance sensing volume, thereby enhancing detection sensitivity. In the design of virtual constriction microchannels, the flow rate ratio of sample to sheath flow is the critical parameter. Adjusting this ratio controls the effective width of the virtual constriction zone, directly influencing both impedance detection sensitivity and throughput. The experimental results indicate that a sample-to-sheath flow ratio of 1:0.5 achieves stable fluidic focusing, with a focused stream width of 24.2 ± 0.2 μm and a theoretical detection throughput of ~1000 cells per second.

2.4.3. Electrode Design

In the design of electrode configurations, two main types are used: coplanar electrodes and opposing electrodes. Coplanar electrodes are located at the bottom of the microchannel and have two modes of operation: absolute measurement (using two electrodes) and differential measurement (using three electrodes). In the differential mode, a voltage is applied through the central electrode, and the differential currents from the two side electrodes are measured to achieve a higher signal-to-noise ratio (SNR) and estimate particle velocity from the peak signal spacing. Opposing electrodes distributed at the top and bottom of the channel produce a more uniform electric field distribution, resulting in higher sensitivity than coplanar designs at the cost of increased manufacturing complexity. Additionally, liquid electrodes positioned in a transverse chamber can generate a vertical equipotential surface distribution, simplifying electric field uniformity design. The increased detection volume of liquid electrodes, however, reduces sensitivity. Each configuration presents a tradeoff between manufacturing simplicity, sensitivity enhancement, and application adaptability.

To improve the accuracy of data measurement in microfluidic impedance sensing systems, researchers have approached the problem from three aspects: electric field optimization, particle focusing, and signal compensation. Kim et al. [23] created a uniform electric field within the sensing channel by making the length of the sensing channel sufficiently long relative to the distance between the electrodes to which the potential was applied, thereby minimizing the edge effect. When focusing particles, researchers can constrain the particle trajectories through active (acoustic or dielectrophoresis) or passive (sheath flow, inertial or viscoelastic focusing) methods to ensure consistent paths through the detection area. However, this may increase the complexity of the system or dilute the samples. For this purpose, Fang et al. [24] designed a unique coplanar differential electrode device. By measuring the changes in the induced current, they successfully resolved the three-dimensional position of a single cell (with a lateral resolution of 2.1 μm and a vertical resolution of 1.2 μm), thereby avoiding the limitations of the detection flux imposed by traditional sheath flow or narrow channels. For signal compensation, calibration curves established by the correlation between electrical parameters (such as the peak width and peak height ratio) and particle position, and multi-electrode designs (such as five pairs of electrodes) or a cross wiring, are used to compensate for the signal deviation caused by position ambiguity. Zhong et al. [25] eliminated the dependence of signal sensitivity on particle height through a unique coplanar electrode configuration. In addition, customized algorithms can also be developed for special signal waveforms (such as multi-peak signals or pulse width variations) to enhance the robustness of single-cell feature extraction. These methods significantly improve the repeatability and accuracy of impedance data by reducing the errors caused by position dependence.

Additionally, static and dynamic systems can be integrated through microchannel design and the application of hydrodynamic principles. A microfluidic device integrating IFC and EIS was developed by Feng et al. [26]. Passive hydrodynamic trapping units (based on the principle of minimum flow resistance path) enabled efficient single-cell capture, while coplanar electrodes achieved the synchronous analysis of cellular electrical properties. The IFC module rapidly detected discrete impedance variations among three cancer cell lines (HeLa, HepG2, and A549), whereas the EIS module quantitatively resolved membrane capacitance and cytoplasmic conductivity. The experimental results demonstrated less than 5% deviation between IFC and EIS measurements for identical cells under flow rates of 10 nL/min to 1000 nL/min (cells were displaced from trapping sites at flow rates exceeding 10 μL/min). The EIS data maintained stability across this broad flow range, verifying IFC's effectiveness as a high-efficiency supplement to EIS. This dual-mode strategy significantly enhanced single-cell electrophysiological characterization efficiency and reliability by synergizing the high-throughput capability of IFC with the high-resolution analytical power of EIS.

3. Application of Microfluidic Impedance Analysis

This chapter focuses on the diversified applications of microfluidic impedance detection technology in fields including tumor diagnostics, hematological analysis, organ-on-a-chip systems, and microbial detection. The broad application prospects spanning molecular to organ levels are comprehensively demonstrated, highlighting its significant translational value and potential from fundamental research to clinical practice. Current technical limitations are critically discussed. First, we delineate the differential characteristics between microfluidic impedance detection and three conventional detection methodologies (Table 1).

Table 1. Comparison of microfluidic impedance detection technology with three other detection methods.

Category	Microfluidic Impedance Detection	Fluorescence Detection [27,28]	PCR [29,30]	ELISA [31,32]
Principle	Label-free electrical detection based on dielectric property differences in cells/particles	Fluorescent dye labeling with optical signal acquisition	Nucleic acid amplification and fluorescence quantification	Antigen–antibody binding with enzymatic chromogenic reaction
Advantages	Label-free operation, real-time dynamic monitoring, miniaturized equipment, low sample consumption, simplified workflow	High sensitivity (single-molecule level), multi-channel parallel detection, mature technology	Ultra-high sensitivity (aM–fM), exceptional specificity, quantitative capability	High specificity, high throughput, standardized protocols, commercial maturity
Limitations	Susceptible to environmental interference, limited multi-target detection capability, complex chip design requirements	Label-dependent (increased cost/time), photobleaching issues, expensive instrumentation	Thermal cycler dependency, complex instrumentation, contamination risks, non-nucleic acid targets undetectable	Antibody quality dependency (potential cross-reactivity), lengthy procedures, limited small-molecule detection
Detection time	Minutes (real time)	Minutes/hours	1–3 h	3–24 h
Sensitivity	pg/mL level	fg/mL level	aM–fM level	pg/mL level
Throughput	Medium–high (chip-dependent)	Medium	Medium	High
Cost	Low equipment cost, high chip development cost	High reagent cost (labels), high equipment cost	High equipment/reagent costs	Moderate equipment/reagent costs
Applications	Point-of-care testing, cellular analysis	Single-molecule detection, imaging analysis	Pathogen detection, gene expression	Protein biomarkers, clinical diagnostics

3.1. Tumor Detection

Tumor detection leverages microfluidic impedance sensing technology to identify, classify, and analyze tumor biomarkers, tumor cells (e.g., HeLa, HL-60, MCF-7, HepG2), or exosomes. This process focuses on elucidating tumor biology and molecular mechanisms to enable early cancer screening and precision medicine. Cancer detection leverages tumor biomarker or circulating tumor cell analysis to enable early cancer discovery and high-risk population screening; the characterization of tumor-specific cellular features clarifies molecular profiles, guiding personalized therapeutic strategies (e.g., targeted therapy, immunotherapy) to optimize efficacy while reducing adverse effects; and liquid biopsy technologies like exosome detection achieve non-invasive dynamic monitoring for evaluating treatment response and recurrence risks. Additionally, these technologies provide critical experimental platforms for drug development and mechanistic studies. Thus, cancer detection is not only pivotal for clinical diagnosis but also drives advancements in personalized oncology and cancer prevention.

In the static design of microfluidic impedance detection systems, addressing the challenge of detecting tumors at very early or asymptomatic stages, Hu et al. [33] developed a porous 3D graphene aerogel biochip and combined it with EIS technology to achieve the ultra-sensitive detection of alpha-fetoprotein (LOD = 7.9 pg/mL) and carcinoembryonic antigen (LOD = 6.2 pg/mL), and the detection limit of tumor exosomes was as low as 10 particles/mL, surpassing the performance of commercial immunoassay kits. When performing dynamic detection on cells, to improve the accuracy of measurement data, it is generally necessary to focus the cells. Zhu et al. [34] used a virtual sensing zone using polyethylene glycol (PEG) as a sheath flow, which increased the detection sensitivity and signal-to-noise ratio by 7.92 times and 1.42 times, respectively, significantly better than the traditional contraction channels. Traditional microfluidic channels are prone to clogging in the contracted part because their size cannot be changed. To address channel clogging, Li et al. [21] developed an anti-clogging microfluidic impedance cytometer with xanthan gum-based cell pretreatment, achieving 240 cells/s throughput and leveraging the triphasic mechanical dynamics (creep, friction, and relaxation) of long constriction channels to analyze the complex mechanical properties of single cells, thereby enhancing the measurement accuracy of mechanical characteristic parameters, as shown in Figure 3a. Additionally, Luan et al. [35] introduced a parallel physical fitting solver to analyze specific membrane capacitance and cytoplasmic conductivity at a high speed of 1613 cells/s, which is 27,000 times faster than traditional methods, as shown in Figure 3b.

However, when analyzing cell parameters and then classifying cells, a single mechanical or electrical characteristic parameter is not sufficient to classify cells more accurately. To solve this problem, by simultaneously analyzing the five-dimensional intrinsic parameters of single cells, including Young's modulus, fluidity, cell radius, cytoplasmic conductivity, and specific membrane capacitance, Feng et al. [36] achieved a high-precision classification rate of 93.4% for three cancer cell subtypes (HepG2/MCF-7/MDA8) and 95.1% for the cytoskeleton perturbation model (fixed/treated with cytochalasin B), as shown in Figure 3c. In pursuit of higher classification performance, Tang et al. [37] developed an asymmetric serpentine microchannel chip and combined various dielectric parameters (cell diameter, impedance amplitude, impedance phase shift and electrical opacity) with machine learning classification models to identify cell types, achieving 96.2–99.6% accuracy in discriminating human leukocytes from four tumor cell lines (A549, MCF7, H226, and H460). This work demonstrated the data-mining potential of machine learning for uncovering subtle cellular signatures hidden in multi-parameter dielectric datasets. To enable the rapid detection of tumor cells in human blood for improving early cancer diagnosis rates or treatment efficacy assessment, Zhu et al. [38] developed an automated liquid biopsy device integrating inertial microfluidics and impedance sensing. Combined with a deep learning algorithm, this system achieved the label-free detection of circulating tumor cells in peripheral blood within 15 min. When analyzing a cell population, measuring the overall viability of the cells is a difficult problem, Xie et al. [39] designed a microfluidic impedance system to quantify the toxicity of pollutants to three-dimensional human liver cancer cell clusters

(HepG2) through the cell clustering index, and proved that it can be used as a measurement indicator of overall cell viability. In addition, mass spectrometry analysis technology based on microfluidic chips has further improved the precision analysis of single cell internal substances. For mass spectrometry analysis technology, it is usually necessary to purify and desalt the samples. Zhu et al. [40] developed a one-step impedance flow cytometry protocol integrating cell sorting and desalting (>99% efficiency), resolving pre-MS purification and salt interference simultaneously, as shown in Figure 3d.

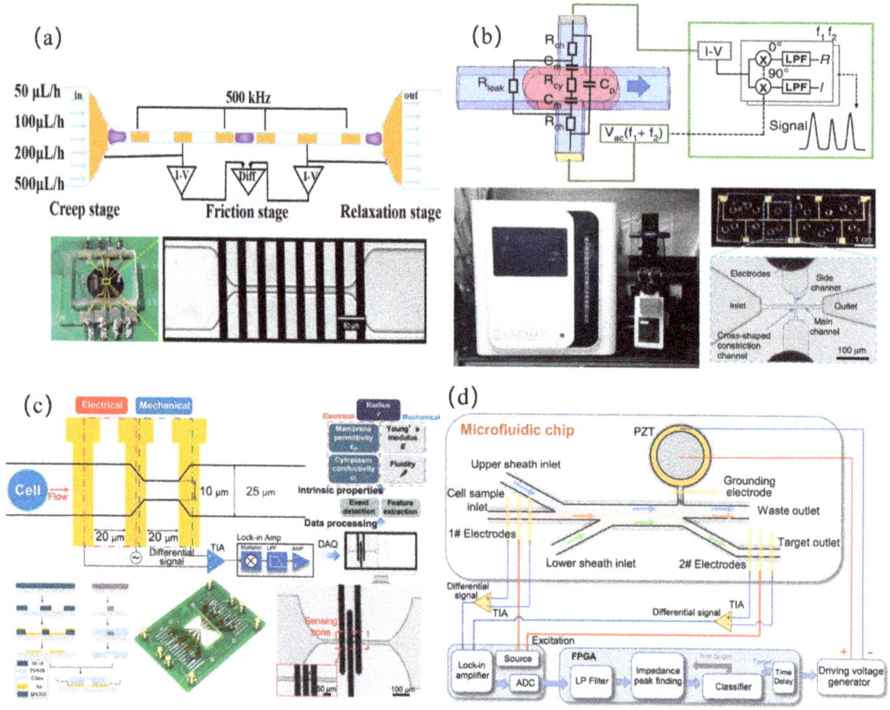

Figure 3. Tumor detection: (**a**) microfluidic impedance flow cytometer featuring xanthan gum pretreatment protocol to mitigate microchannel occlusion in constriction-based cellular interrogation systems (reprinted with permission from [21], published by American Chemical Society, 2024); (**b**) ultrahigh-throughput single-cell biophysical analyzer employing parallel physics-informed solver architecture for concurrent quantification of specific membrane capacitance and cytoplasmic conductivity (reprinted with permission from [35], published by Springer Nature, 2023); (**c**) multimodal microfluidic system with integrated electromechanical sensing modules for concurrent mechanoelectrical phenotyping of cellular intrinsic properties (reprinted with permission from [36], published by Wiley-VCH Verlag, 2023); (**d**) integrated microfluidic impedance cytometer implementing label-free operation mode with single-step analytical process combining cellular characterization and on-chip desalination (reprinted with permission from [40], published by Wiley-VCH Verlag, 2024).

3.2. Blood Detection

Blood detection employs microfluidic impedance sensing technology to identify, classify, and analyze blood components (e.g., red/white blood cells, proteins, biomarkers), enabling the assessment of individual health status, disease diagnosis, or therapeutic efficacy monitoring. Its clinical significance lies in the early detection of conditions such as anemia, infection, diabetes, and hepatic/renal dysfunction while tracking chronic disease progression or treatment outcomes to guide therapeutic adjustments; the dynamic monitoring of blood parameters also identifies latent health risks, supporting disease prevention and health management. As an indispensable core tool in clinical diagnostics and population screening, it integrates precision, non-invasiveness, and real-time capabilities to advance personalized healthcare.

Innovations in single-cell and blood parameter detection have driven progressive breakthroughs across research teams. Caselli et al. [41] pioneered neural network-based raw impedance data parsing to extract intrinsic dielectric properties (cell radius, membrane capacitance, and cytoplasmic permittivity/conductivity) at millisecond resolution, resolving signal overlap distortions in high-throughput single-cell analysis while capturing hidden single-cell signals, as shown in Figure 4a. Advancing multi-parameter blood analysis, Zhbanov et al. [42] developed a flow-mode microfluidic dielectric spectroscopy sensor, combining effective medium theory to simultaneously quantify six parameters (RBC count, hemoglobin, etc.) with <3.5% clinical error, suppressing RBC aggregation via flow impedance and introducing an orientation factor to quantify RBC alignment.

In the pursuit of optimizing blood cell detection performance, research teams have propelled progress by breaking through the limitations of traditional methods through multidimensional innovations. Huang et al. [43], through dielectrophoretic (DEP) force manipulation and a uniquely engineered structural design, achieved high mixing efficiency, enabling the real-time measurement of dynamic permeability changes in single red blood cells within 0.19 s. Zhong et al. [25] eliminated the dependence of sensitivity on cell height via a novel coplanar electrode configuration, achieving high-throughput detection at 1000 cells/s. The position-insensitive characteristics of this design not only enabled the precise classification of three leukocyte subtypes, but also simultaneously acquired red blood cell indices while avoiding traditional hemolysis steps, as shown in Figure 4b. To further enhance classification accuracy, Wang et al. [22] constructed a dynamic detection zone via sheath flow crossover and achieved 99.8% classification accuracy for three leukemia cell types (K562, Jurkat, and HL-60) and $\geq$99.2% recognition precision for four leukocyte subtypes (NEU/EOS/MON/LYM) using a recurrent neural network.

In the development of integrated microfluidic devices, research teams are progressively advancing the practicality and scenario adaptability of detection technologies through modular design and functional integration. Oshabaheebwa et al. [44] developed a fully miniaturized, portable, and wash-free microfluidic impedance detection platform capable of generating an erythrocyte occlusion index within 15 min, enabling the rapid assessment of red blood cell health, functional status, and therapeutic efficacy, as shown in Figure 4c. Fu et al. [45] developed an origami-inspired electrochemical microfluidic paper-based device that integrates plasma separation, buffer absorption, and three-channel EIS detection This system achieved the simultaneous detection of three cardiac protein markers (LODs: 4.6/1.2/146 pg/mL) in fingerprick whole blood, delivering rapid analysis within 46 min—a performance comparable to commercial ELISA kits, as shown in Figure 4d. To address the challenge of stroke subtyping, Sayad et al. [46] designed a magneto-impedance biosensor combined with a microfluidic chip for glial fibrillary acidic protein (GFAP) extraction from plasma. By leveraging Dynabeads magnetic labels, the system achieved a detection sensitivity of 1.0 ng/mL for GFAP in acute stroke patient blood samples, fulfilling the clinical need for subtype differentiation. To address the issues of time-consuming diagnostics and frequent false positives in urinary tract infection (UTI) detection, Petchakup et al. [47] developed a microfluidic inertial impedance sensing platform. This system enables the label-free sorting of neutrophils directly from urine samples in 5 min, followed by impedance spectroscopy analysis, thus facilitating rapid, culture-free urine screening for infection biomarkers.

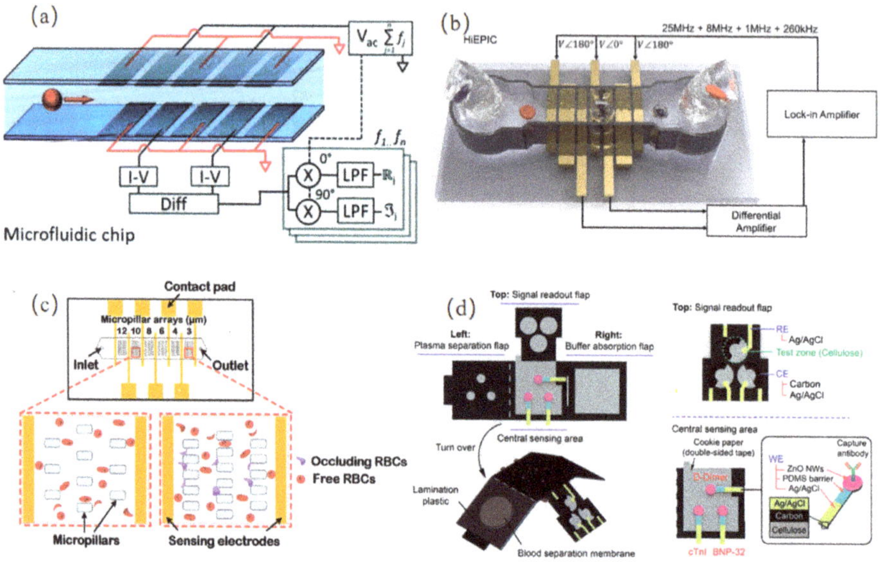

Figure 4. Blood detection: (**a**) neural network-based decoding of raw impedance data streams enabling extraction of single-cell signatures hidden within overlapping cellular measurements (reprinted with permission from [41], published by Royal Society of Chemistry, 2022); (**b**) microfluidic chip with unique coplanar electrode configuration achieving concentration-independent sensitivity through elimination of high-concentration dependency (reprinted with permission from [25], published by Elsevier, 2022); (**c**) portable, wash-free microfluidic platform for multiplex assessment of erythrocyte health, functional characterization and therapeutic efficacy monitoring within 15 min (reprinted with permission from [44], published by Elsevier, 2024); (**d**) origami-inspired electrochemical microfluidic paper-based device for simultaneous quantitation of three cardiac protein biomarkers in fingerprick whole blood without pretreatment (reprinted with permission from [45], published by American Chemical Society, 2023).

3.3. *Organ on a Chip*

Organ on a chip (OOC) is a biomimetic microfluidic platform that replicates the physiological functions and microenvironment of human organs, providing an efficient and cost-effective experimental tool for drug screening, toxicology studies, and disease modeling. Traditional in vitro models fail to simulate the complex structures and dynamic interactions of organs, while animal studies face ethical constraints and interspecies differences. To bridge these gaps, OOC technology integrates microfluidics to precisely regulate fluid flow, mechanical forces, and biochemical signals and incorporates impedance sensing to monitor cellular dynamics and functional responses in real time, enabling a paradigm-changing approach to organ-level physiological and pathological studies. In recent years, OOC systems that integrate microfluidic and impedance analysis technologies have been widely used in vascular, brain, liver, kidney, and intestinal models, establishing high-fidelity in vitro platforms for precision medicine and personalized drug development. The convergence of microfluidic and impedance technologies highlights their transformative potential to advance biomedical research and therapeutic innovation.

The core design of OOC systems lies in the deep integration of impedance sensing and microenvironmental regulation. Such architecture provides a high-fidelity in vitro platform for advancing precision medicine and personalized drug development by recapitulating tissue-specific physiological and pathological dynamics. As the body's central transport network, vascularization on a chip further enables the interconnection of multi-organ chips, thereby reconstructing a more physiologically complete in vitro system. Na et al. [48] proposed a hemodynamic similarity principle that enables the rapid derivation of the input impedance for in vitro microgravitational systems derived from in vivo arterial systems, thereby accurately replicating the desired endothelial hemodynamic microenvironment. To address the challenge of the dynamic monitoring of tumor angiogenesis, Huang et al. [49] developed a 3D biomimetic model based on matrix gel microchannels. By integrating microelectrodes to quantify impedance changes during vascular extension, they demonstrated a linear relationship between the vascular growth rate and vascular endothelial growth factor concentration and revealed the regulatory role of cancer cell count in angiogenesis, as shown in Figure 5a.

The brain, serving as the central control hub for the entire body's functions, plays a critically important role in regulating human physiological activities. Liang et al. [50] developed a bioinspired brain microenvironment chip by integrating interstitial flow regulation and real-time neural circuit monitoring. This system deciphered the dynamic characteristics of synaptic connection/disconnection processes and established a highly biomimetic model for studying the mechanisms of neurodegenerative diseases, as shown in Figure 5b. In addition, Ceccarelli et al. [51] designed a blood–brain barrier (BBB) chip integrated with thin-film electrodes, utilizing EIS to track the integrity and maturation of the BBB in real time. This innovation provides a highly efficient in vitro platform for investigating drug permeation mechanisms in neurological disorders. As the central metabolic hub for both catabolism and anabolism, the liver plays a pivotal role in detoxification and biosynthesis, and studying its function is essential for elucidating disease mechanisms such as metabolic disorders, fibrosis, and hepatocellular carcinoma. Dhwaj et al. [52] developed a liver-on-a-chip system powered by a low-power impedance micropump to enable continuous nutrient perfusion, mimicking cardiac pulsation-driven medium circulation for the real-time monitoring of healthy cell growth over 14 days. The compact design (compatible with standard Petri dishes) eliminates contamination risks associated with conventional perfusion systems.

As the central filtration hub of the blood, the kidneys are pivotal for maintaining systemic homeostasis. The development of in vitro renal models is critical for advancing therapies for uremia and other renal pathologies. Liang et al. [53] developed a microfluidic impedance sensing system to dynamically track the formation and drug response of renal tubular epithelial barriers. They demonstrated that basement membrane matrix (ABM)-supported barriers exhibited significantly superior monolayer stability and growth rates compared to Transwell membrane scaffolds. The platform validated its high sensitivity to biochemical stimuli through calcium switch assays and employed equivalent circuit modeling to correlate barrier impedance with cellular density and junctional tightness, as shown in Figure 5c. As the terminal site for nutrient absorption, intestinal dysfunction has garnered significant attention due to the growing prominence of modern food safety issues (e.g., chemical additives, microbial contamination). Fernandes et al. [54] developed a multi-channel microfluidic system that monitored the growth and polarization processes of human epithelial cells by conducting continuous impedance measurements over five consecutive days, enabling the real-time tracking of epithelial barrier integrity, as shown in Figure 5d. To address the limitations of traditional models in simulating microenvironments with varying oxygen concentrations, Khalid et al. [55] proposed a method using inkjet-printed embedded dissolved oxygen (DO) and reactive oxygen species (ROS) electrochemical sensors. This approach was implemented on a custom-developed bilayer gut-on-a-chip platform to monitor both developmental phases (initially normoxic) and induced hypoxia over 6 days.

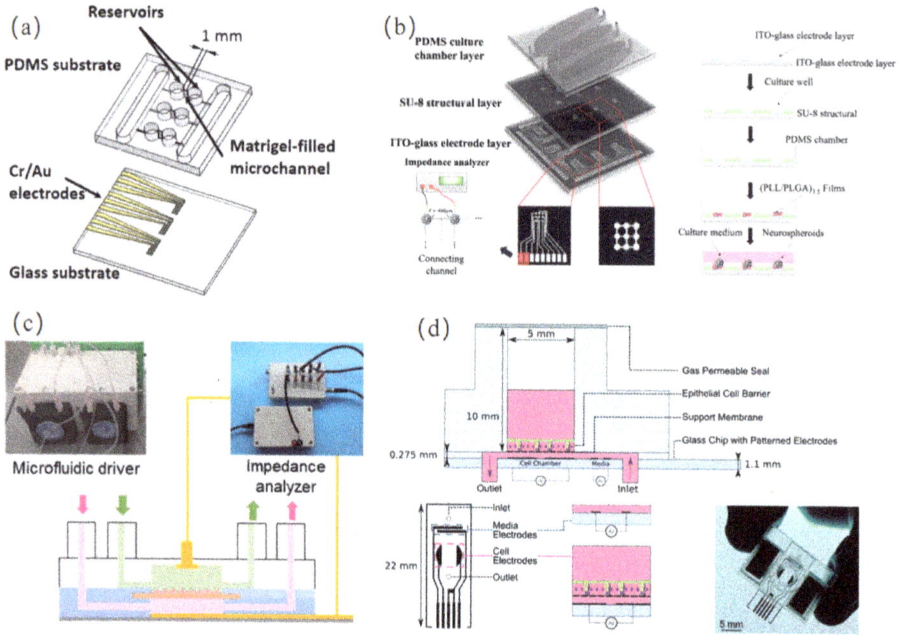

Figure 5. Organ on a chip: (**a**) a 3D biomimetic model based on matrix gel microchannels for the dynamic monitoring of tumor angiogenesis (reprinted with permission from [49], published by Elsevier, 2022); (**b**) a brain-on-chip biomimetic microenvironment with interstitial flow regulation and neural network analysis for the dynamic characterization of synaptic connection/disconnection kinetics (reprinted with permission from [50], published by the Royal Society of Chemistry, 2024); (**c**) a microfluidic platform for the spatiotemporal tracking of renal tubular epithelial barrier formation and pharmacodynamic evaluation (reprinted with permission from [53], published by Elsevier, 2023); (**d**) a multi-channel microfluidic array for the real-time assessment of epithelial barrier integrity (reprinted with permission from [54], published by the Royal Society of Chemistry, 2022).

3.4. Microbial Detection

Microbial detection is defined as the process of qualitative and quantitative analysis for microorganisms (viruses, bacteria, fungi, and algae) using microfluidic impedance sensing technology. Its principal value lies in providing critical data support for disease control, food safety, and environmental protection through the precise identification of microbial characteristics. In healthcare applications, pathogens can be rapidly localized using this technique, enabling the rational guidance of antibiotic usage and the containment of infectious disease transmission. Within the food industry, microbial contamination-induced foodborne illnesses are effectively prevented, thereby ensuring food safety. For environmental monitoring, specific marine algae species are monitored to provide early warnings of ecological imbalance and environmental pollution risks. Furthermore, in biopharmaceutical and agricultural sectors, the dynamic monitoring of microbial variations is implemented through this technology, facilitating process optimization and risk management. These multidimensional capabilities establish it as a pivotal tool for safeguarding human health and achieving sustainable development.

In microbial detection, the detectability of microbial throughput or concentration is recognized as a pivotal performance metric. Regarding throughput metrics, advancements have been systematically pursued. Chen et al. [56] constructed a 3D detection zone with cylindrical electrodes and microchannels, achieving a throughput of 1800 Haematococcus pluvialis cells per second, as shown in Figure 6a. For concentration range optimization, expanding the detectable dynamic range becomes imperative to enhance analytical versatility. Chen et al. [57] developed a microfluidic system with an ultra-wide detection range (615–2.7 $\times$ 10^8 cells/mL), simultaneously differentiating bacterial viability and Gram types (LOD = 10^2 cells/mL). Microfluidic impedance sensing technology enables not only single-bacterium detection but also concurrent multi-pathogen discrimination. Muhsin et al. [58] optimized microfluidic layouts to concurrently detect Salmonella, Legionella, and E.coliO157:H7 (LOD = 3 cells/mL, 30–40 min detection, no pre-enrichment). For rapid bacterial antibiotic resistance testing, Chen et al. [59] created a single-bacterium impedance detection platform. Antibiotic susceptibility tests were completed within 32 min (30 min antibiotic exposure added to 2 min impedance analysis), significantly accelerating detection cycles, as shown in Figure 6b. However, the discrimination between common bacteria and drug-resistant bacteria is recognized as a significant technical challenge. Tang et al. [60] developed parallel impedance cytometers with dual microchannels, where opposing impedance signals in synchronized time sequences enabled high-precision identification. Bacterial classification is recognized as a technically demanding process due to submicron dimensions, presenting greater technical challenges compared to conventional eukaryotic cell analysis. Convolutional neural networks were employed by Zhang et al. [61] to analyze impedance flow cytometry data, achieving >96% accuracy in bacillus/coccus/vibrio classification and >95% precision in Escherichia coli vs. Salmonella enteritidis discrimination, as shown in Figure 6c. However, current methodologies that constrain idealized sample validation, without addressing the complexity of real-sample matrices, may pose critical technical barriers for clinical translation. Piedimonte et al. [62] created a differential impedance system with gold microelectrodes and nanoadsorbed antibodies, achieving 88 pg/mL anti-dengue antibody sensitivity (superior to commercial ELISA). The reconfigurable probe design further enabled the rapid screening of other pathogens and diseases, as shown in Figure 6d. Almalaysha et al. [63] demonstrated 1–2 cells/mL Salmonella typhimurium detection in raw chicken, validating real-sample reliability. An impedance spectroscopy-based electronic tongue was engineered by Coatrini-Soares et al. [64], where optimized sensing units fabricated through layer-by-layer thin films were combined with machine learning algorithms employing decision tree models to process capacitive data, achieving the sensitive detection of Staphylococcus aureus in milk (LOD = 2.01 CFU/mL). Dowdell et al. [65] integrated online flow cytometry into water ozonation systems, successfully predicting microbial anomalies through the 40-day continuous monitoring of total/intact cell counts.

Beyond the detection targets, microfluidic impedance sensing technology enables the high-sensitivity and non-invasive analysis of other microscopic biological samples (e.g., stem cells, oocytes, sweat, and microplastics). Stem cell and oocyte detection is utilized to assess cellular viability and quality in regenerative medicine or assisted reproduction, thereby optimizing therapeutic outcomes. Cam et al. [66] quantified the therapeutic window of bone marrow mesenchymal stem cell exosomes through real-time impedance analysis, with their repair efficacy and functional recovery capabilities for ischemic acute kidney injury structures being precisely evaluated. Due to long-standing limitations in oocyte quality assessment caused by the subjectivity of zona pellucida hardening detection, Cao et al. [19] developed a microfluidic cell capture technique where impedance variations were correlated with zona pellucida aspiration lengths. This approach allowed the calculation of the zona pellucida's Young's modulus and the quantitative characterization of its aspiration process at the opening.

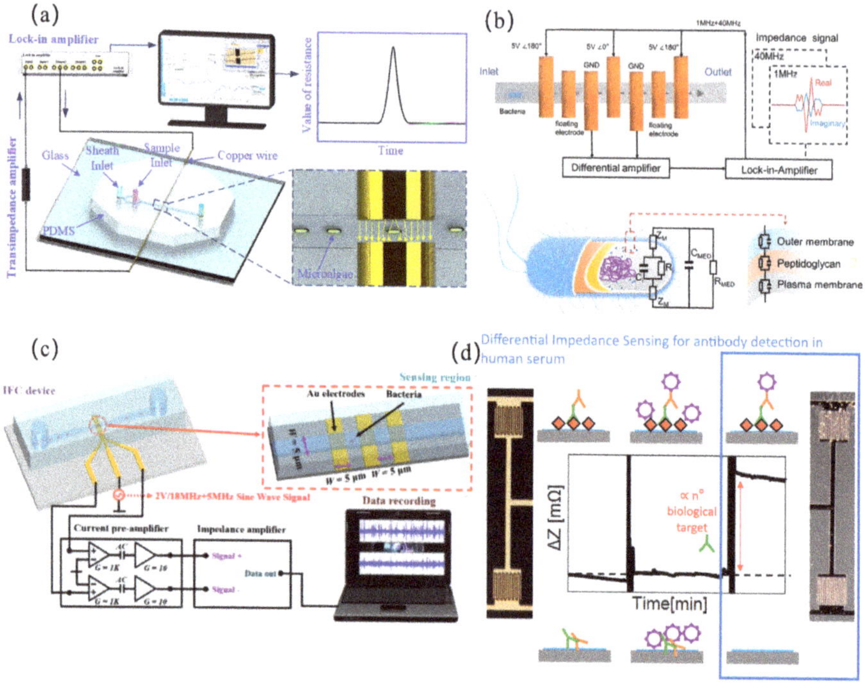

Figure 6. Microbial detection: (**a**) 3D-architected microfluidic impedance cytometer with dual cylindrical electrodes and embedded microchannels for enhanced cellular interrogation (reprinted with permission from [56], published by Royal Society of Chemistry, 2024); (**b**) single-bacterium-resolution impedance cytometry platform enabling rapid antimicrobial susceptibility testing via electrophysiological profiling (reprinted with permission from [59], published by Wiley-VCH Verlag, 2024); (**c**) microfluidic device integrated with convolutional neural network-based deep learning framework for automated interpretation of impedance flow cytometry data to achieve precision bacterial classification (reprinted with permission from [61], published by American Chemical Society, 2024); (**d**) differential impedance biosensing system implementing phase-sensitive signal demodulation for label-free detection of anti-dengue viral antibodies in low-concentration sera (reprinted with permission from [62], published by Elsevier, 2022).

Sweat analysis not only achieves the dynamic monitoring of electrolytes, metabolites, and inflammatory factors but also provides real-time data for sports medicine, dermatological disorders, and chronic disease management. In wearable device research, human sweat analysis is predominantly conducted. Wang et al. [67] engineered a self-powered multifunctional microfluidic sweat analysis system where spontaneous sweat transport was driven by wettability gradient design. By integrating impedance–colorimetric dual-mode sensing, the simultaneous detection of sweat flow rate, electrolyte concentration, glucose, and pH was accomplished with the response time reduced to 100 ms. To enhance wearable system practicality, Zahed et al. [68] constructed a wireless monitoring patch system using flexible materials, integrating microfluidic channels with multimodal sensing. The reduced graphene oxide glucose sensor exhibited an enhanced sensitivity of 19.97 $\mu A/mM/cm^2$ (2.3-fold improvement versus non-microfluidic designs) under microfluidic assistance, achieving the synchronous dynamic monitoring of sweat glucose and ECG signals during exercise while validating stability and bend resistance in complex scenarios. With increasing demands for marine pollution monitoring, microplastic identification in seawater has emerged as a critical challenge. Butement et al. [69] implemented an impedance-based cytometer combined with machine learning algorithms, enabling the direct recognition and counting of microplastic particles within phytoplankton mixtures (1.5–10 μm) in seawater-mimicking media. For practical sample detection, Silva et al. [70] developed an impedance-based microfluidic electronic tongue. Multi-sensing units coated with nanostructured thin films were integrated, and chemometric methods including principal component analysis and partial least squares-discriminant analysis were employed. The accurate discrimination of authentic coconut water samples (classification accuracy >90%) was achieved, along with the high-precision prediction of physicochemical parameters such as soluble solid content and total titratable acidity. Xue et al. [71] developed a copper nanoparticle-enhanced microfluidic flow injection analysis system. Electrochemical impedance spectroscopy was employed to reveal a 20-fold lower charge transfer resistance at copper nanoparticle-modified multi-walled carbon nanotube screen-printed carbon electrodes versus standard counterparts. Microfluidic parameters were optimized and this modified electrode was utilized to achieve highly sensitive dopamine (DA) detection (LOD = 0.33 nM). Across three authentic sample matrices—artificial cerebrospinal fluid, fetal bovine serum, and human urine—recovery rates of 96.5% to 103.8% were demonstrated, proving its practical utility. These technological advancements not only expand conventional detection dimensions but also demonstrate precision and operational efficiency, significantly accelerating the translation of fundamental research into clinical applications.

Notwithstanding the multi-domain application potential demonstrated by microfluidic impedance sensing technology, the following critical bottlenecks persist: Practical application validations remain insufficient for most current technologies. High-sensitivity and high-classification-accuracy performance metrics are predominantly validated under laboratory-standardized conditions using simplified models (e.g., single-cell lines or purified samples). However, in complex biological fluids (e.g., whole blood, interstitial fluid, or clinical samples with high background interference), performance robustness and reliability are compromised by non-target component interference, matrix effects, and dynamic physiological environmental factors. Although machine learning significantly enhances classification efficiency, models exhibit pronounced dataset-specific dependency and limited generalization capability for unknown sample types. The opacity of black-box models hinders biologically interpretable explanations, undermining clinical decision-making credibility. Practical implementation is hindered by complex structural dependencies (e.g., virtual contraction channels, or sheath flow control), resulting in difficulties in device miniaturization, elevated manufacturing costs, and operational complexity. These constraints conflict with the portability and user-friendliness requirements of POCT; Despite advancements in simultaneous multidimensional physical/electrical characterization, accuracy is affected by parameter coupling effects. The absence of unified data fusion standards and cross-platform comparability further impedes large-scale technological deployment. These challenges collectively emphasize the necessity for future research to prioritize clinical translation, technical simplification, multimodal data standardization, and system stability optimization.

4. Conclusions and Outlook

In summary, this review systematically examines recent advancements in the integration of microfluidic and impedance sensing technologies, along with their technological evolution in biosensing applications. Starting from the fundamental principles of microfluidic technology and impedance spectrum analysis, the detection mechanisms based on fluidic characteristics and impedance analysis were introduced, with particular emphasis on the construction logic of cellular impedance equivalent circuit models. Detection systems were classified into static and dynamic categories according to cellular motion states, with their design principles and critical factors for enhancing detection accuracy being elucidated, respectively. Finally, the innovative applications were systematically categorized into four domains: tumor detection, blood detection, OOC, and microbial detection based on biological target dimensions.

Despite critical progress achieved through microstructural and electrode design optimizations, several challenges remain (Figure 7).

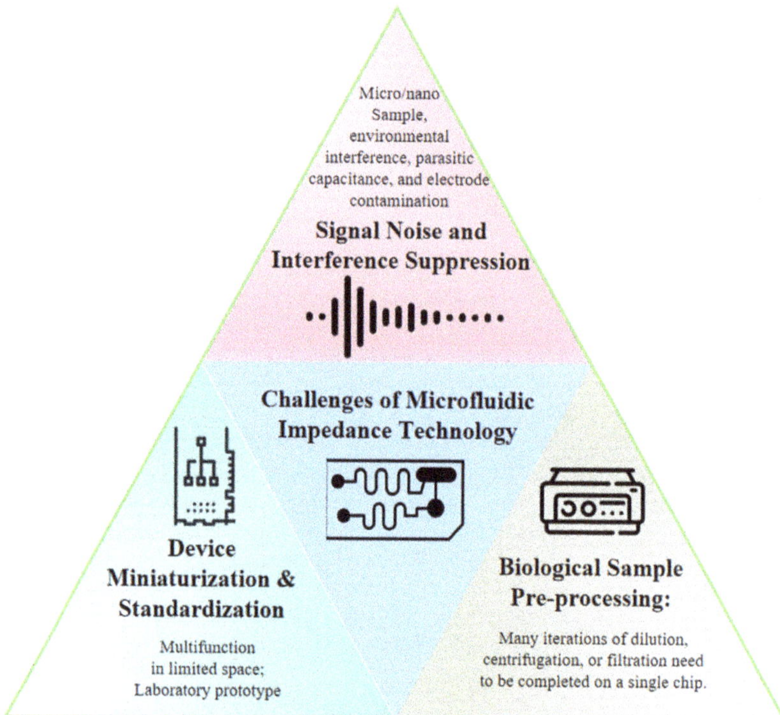

Figure 7. Conceptual map of prevailing challenges in microfluidic impedance detection technology.

(1) Signal Noise and Interference Suppression: Signal noise represents a critical challenge affecting data accuracy and system reliability during microfluidic impedance measurements. Due to the compact size and high sensitivity of microfluidic devices, susceptibility to external electromagnetic interference exists, particularly in open measurement environments. Furthermore, device-intrinsic parasitic capacitance, poor contacts, and electrode contamination introduce additional noise, degrading signal quality. Differential measurement techniques are typically employed to effectively suppress common-mode noise through differential signal extraction for automatic noise balancing. Shielding materials and electromagnetic isolation methods, such as copper foil tapes, are utilized to minimize external interference. Signal processing approaches including filtering algorithms and adaptive noise cancelation algorithms have been implemented for post-processing raw signals to enhance signal quality and stability. Although numerous studies attempt to model and analyze impedance data using statistical methods and machine learning techniques, limitations persist in generalization capabilities and practical applicability due to insufficient data volume and sample imbalance. Particularly, machine learning algorithms require extensive annotated datasets for training, which are often unavailable in real-world scenarios.

(2) Device Miniaturization and Standardization: While microfluidic technology inherently offers miniaturization advantages, achieving further size reduction becomes complex when integrating multi-channel, multi-parametric detection systems. Conventional microfluidic systems typically rely on single electrode arrays for impedance measurements, whereas practical applications necessitate integrated multiple electrodes and intricate microchannel designs to accommodate diverse detection needs. The miniaturization of impedance analyzers remains challenging, impeding portable system development. Additionally, standardization deficiencies exist since most microfluidic chips are laboratory-customized via photolithography without uniform specifications for electrode dimensions/materials. This variability hinders standardized performance evaluation across methodologies.

(3) Biological Sample Pre-processing: Microfluidic impedance technology has significant applications in biomedical detection, particularly for the rapid analysis of biological samples like blood and urine. However, the inherent heterogeneity of biological samples, including cell density, protein concentration, and ionic strength, directly affects impedance measurements. Sample pre-processing is crucial in practice. For instance, whole blood contains abundant cellular components and solutes; direct measurement may cause signal confusion due to conductivity differences between cell membranes and solutions. Accuracy enhancement often requires pre-processing steps such as dilution, centrifugation, or filtration. Consequently, designing integrated pre-processing zones (e.g., filtration/separation units) within microfluidic chips is essential.

Based on our discussion, the future development of microfluidic impedance sensing technology will focus on two core objectives: "practicality optimization" and "clinical adaptability enhancement".

(1) Standardization and manufacturing simplification must be prioritized: Unified chip design specifications (the electrode layout and channel dimensions) and detection protocols should be established to facilitate laboratory-to-industry translation. Cost-effective microfabrication techniques (e.g., lithography injection molding scale production) and novel electrode materials (flexible conductive polymers, carbon-based composites) require development to reduce precious metal dependency while improving batch consistency and mechanical stability.

(2) Anti-interference capability and portability require targeted improvements: Integrated pretreatment modules (e.g., embedded filtration membranes) should be optimized to minimize matrix interference in complex samples (whole blood and saliva). Miniaturized signal processing units combined with low-power circuits must evolve toward handheld/wearable formats to meet POCT demands. Bioresorbable material innovations enabling implantable monitoring devices demand breakthroughs in long-term signal drift correction and biosafety evaluation through fundamental studies on material-biological interface dynamics.

(3) Artificial intelligence (AI) systems and data processing strategies: AI and machine learning (ML) are increasingly being used in the data analysis and processing of microfluidic impedance chip systems [72–74]: Convolutional Neural Networks (CNNs) automatically extract spatial features (e.g., cellular morphology, impedance distribution), where local patterns are captured through convolutional layers while dimensionality reduction and noise resistance are achieved via pooling layers. Raw images/signals are directly processed to identify cell-type differences, and multi-source parameters (temperature/pH) are fused to enable real-time sensor drift correction. Recurrent neural network variants (GRU/LSTM) model temporal dynamics (e.g., impedance changes during cell division, biomarker fluctuations in sweat) through

gating mechanisms, with long-term dependencies captured to predict physiological trends. These are combined with CNNs to form hybrid architectures (CNN-GRU) for spatiotemporal feature co-analysis. These methods further enhance the ability of impedance signals to predict cell behavior and response, accurately predict future dynamic changes in cells through real-time and historical data and greatly enhance the diagnostic performance of microfluidic impedance analysis technology. In addition, AI algorithms can also be used to control microfluidic impedance systems, form intelligent feedback mechanisms, adjust experimental conditions (such as the flow rate, measurement frequency, etc.) in real time, perform high-precision measurements of biological samples of different sizes and types, and achieve closed-loop control and adaptive adjustment. Furthermore, the AI method of integrating impedance analysis with multimodal data fusion such as optical imaging, biosensors, and Raman spectroscopy analysis is expected to further improve the sensitivity and specificity of microfluidic chips in diagnosis and detection.

(4) AI-driven impedance microfluidic chip design: With the development of AI intelligent design microfluidic chips, as well as artificial intelligence expert models based on large language models such as ChatGPT [73,75], the development of impedance microfluidic technology chip design AI and expert models will become the key to the next generation of impedance microfluidic technology. This will greatly lower the threshold for the application of impedance microfluidic chips and make the technology accessible to a wider range of biological research fields.

(5) Multimodal integration: Synergistic combinations with acoustic, magnetoelastic, and Raman spectroscopic techniques should create complementary multidimensional sensing. Self-powered systems, surface functionalization, self-cleaning microstructures, and modular replaceable electrodes require integration to extend chip lifespan and reduce maintenance costs.

These pragmatic improvements will propel microfluidic impedance sensing technology from laboratories into clinical, household, and industrial settings, ultimately establishing it as a practical tool for precision medicine and health monitoring with global health impacts. This review adheres to the PRISMA (Preferred Reporting Items for Systematic Reviews and Meta-Analyses) standards (Supplementary Materials).

Supplementary Materials: The following supporting information can be downloaded at: https://www.mdpi.com/article/10.3390/mi16060683/s1, PRISMA 2020 checklist. Reference [76] are cited in the supplementary materials.

Author Contributions: Conceptualization, Y.S., T.T. and J.W.; methodology, J.L. and Z.W.; validation, J.L. and Z.W.; investigation, J.L. and Z.W.; resources, J.L. and Z.W.; data curation, J.L. and Z.W.; writing—original draft preparation, J.L., Z.W., T.R. and Y.S.; writing—review and editing, Y.S.; visualization, Z.W. and T.R.; supervision, Y.S., T.T. and J.W.; project administration, Y.S.; funding acquisition, Y.S., T.T. and J.L. All authors have read and agreed to the published version of the manuscript.

Funding: This research was funded by the National Natural Science Foundation of China (22304143, 52205075), the Zhejiang Provincial Natural Science Foundation of China (LQN25E050020, LZ24E050008), the Open Fund Project of Key Laboratory of CNC Equipment reliability, the Ministry of Education (JLU-cncr-202407), the Key Science and Technology Plan Project of Jinhua City, China (2023-3-084), the New Chongqing Project of Chongqing (CSTB2024YCJH-KYXM0115), and the Research Funding of Chongqing General Hospital (2025022102).

Acknowledgments: During the preparation of this manuscript, the authors used DeepSeek-R1 to edit the English and correct the grammar. The authors have reviewed and edited the output and take full responsibility for the content of this publication.

Conflicts of Interest: The authors declare no conflicts of interest.

References

1. Du, J.; Wu, S.; Niu, L.; Li, J.; Zhao, D.; Bai, Y. A Gold Nanoparticles-Assisted Multiplex PCR Assay for Simultaneous Detection of *Salmonella typhimurium, Listeria monocytogenes* and *Escherichia coli* O157:H7. *Anal. Methods* **2020**, *12*, 212–217. [CrossRef]
2. Kumar, B.K.; Raghunath, P.; Devegowda, D.; Deekshit, V.K.; Venugopal, M.N.; Karunasagar, I.; Karunasagar, I. Development of Monoclonal Antibody Based Sandwich ELISA for the Rapid Detection of Pathogenic Vibrio Parahaemolyticus in Seafood. *Int. J. Food Microbiol.* **2011**, *145*, 244–249. [CrossRef] [PubMed]
3. Eringen, A.C. Simple Microfluids. *Int. J. Eng. Sci.* **1964**, *2*, 205–217. [CrossRef]
4. Terry, S.C.; Jerman, J.H.; Angell, J.B. A Gas Chromatographic Air Analyzer Fabricated on a Silicon Wafer. *IEEE Trans. Electron Devices* **1979**, *26*, 1880–1886. [CrossRef]
5. Manz, A.; Graber, N.; Widmer, H.M. Miniaturized Total Chemical Analysis Systems: A Novel Concept for Chemical Sensing. *Sens. Actuators B Chem.* **1990**, *1*, 244–248. [CrossRef]
6. Duffy, D.C.; McDonald, J.C.; Schueller, O.J.A.; Whitesides, G.M. Rapid Prototyping of Microfluidic Systems in Poly(Dimethylsiloxane). *Anal. Chem.* **1998**, *70*, 4974–4984. [CrossRef]
7. Shen, Y.; Yalikun, Y.; Aishan, Y.; Tanaka, N.; Sato, A.; Tanaka, Y. Area Cooling Enables Thermal Positioning and Manipulation of Single Cells. *Lab Chip* **2020**, *20*, 3733–3743. [CrossRef]
8. Shen, Y.; Yuan, Y.; Tang, T.; Ota, N.; Tanaka, N.; Hosokawa, Y.; Yalikun, Y.; Tanaka, Y. Continuous 3D Particles Manipulation Based on Cooling Thermal Convection. *Sens. Actuators B Chem.* **2022**, *358*, 131511. [CrossRef]
9. Shen, Y.; Yalikun, Y.; Tanaka, Y. Recent Advances in Microfluidic Cell Sorting Systems. *Sens. Actuators B Chem.* **2019**, *282*, 268–281. [CrossRef]
10. Shen, Y.; Chen, Z.; Zhang, Y.; Hou, X. Liquid Interfacial Tension Design of Building New Concept Materials. *Matter* **2023**, *6*, 2506–2508. [CrossRef]
11. Kivelson, D.; Madden, P. Theory of Dielectric Relaxation. *Mol. Phys.* **1975**, *30*, 1749–1780. [CrossRef]
12. Ayliffe, H.E.; Frazier, A.B.; Rabbitt, R.D. Electric Impedance Spectroscopy Using Microchannels with Integrated Metal Electrodes. *J. Microelectromechanical Syst.* **1999**, *8*, 50–57. [CrossRef]
13. Tang, T.; Liu, X.; Kiya, R.; Shen, Y.; Yuan, Y.; Zhang, T.; Suzuki, K.; Tanaka, Y.; Li, M.; Hosokawa, Y.; et al. Microscopic Impedance Cytometry for Quantifying Single Cell Shape. *Biosens. Bioelectron.* **2021**, *193*, 113521. [CrossRef]
14. Tang, T.; Liu, X.; Yuan, Y.; Kiya, R.; Shen, Y.; Zhang, T.; Suzuki, K.; Tanaka, Y.; Li, M.; Hosokawa, Y.; et al. Dual-Frequency Impedance Assays for Intracellular Components in Microalgal Cells. *Lab Chip* **2022**, *22*, 550–559. [CrossRef]
15. Sun, T.; Green, N.G.; Gawad, S.; Morgan, H. Analytical Electric Field and Sensitivity Analysis for Two Microfluidic Impedance Cytometer Designs. *IET Nanobiotechnol.* **2007**, *1*, 69–79. [CrossRef]
16. Chai, H.; Feng, Y.; Zhu, J.; Meng, X.; Liang, F.; Bai, J.; Wang, W. Evaluating the Accuracy of Impedance Flow Cytometry with Cell-Sized Liposomes. *ACS Sens.* **2023**, *8*, 2681–2690. [CrossRef] [PubMed]
17. Chen, Z.; Zhu, Y.; Xu, D.; Alam, M.; Shui, L.; Chen, H. Cell Elasticity Measurement Using a Microfluidic Device with Real-Time Pressure Feedback. *Lab Chip* **2020**, *20*, 2343–2353. [CrossRef]
18. Tsai, S.-L.; Wey, J.-J.; Lai, S.-C.; Huang, Y.-Y. A Novel Microfluidic Chip Based on an Impedance-Enhanced Electrochemical Immunoassay. *Phys. Fluids* **2025**, *37*, 011920. [CrossRef]
19. Cao, Y.; Floehr, J.; Azarkh, D.; Schnakenberg, U. Microfluidic Aspiration-Assisted Electrical Impedance Spectroscopy System Is a Reliable Tool for the Characterization of Oocyte Hardening. *Sens. Actuators B Chem.* **2023**, *380*, 133316. [CrossRef]
20. Eeckhoudt, R.V.D.; Christiaens, A.-S.; Ceyssens, F.; Vangalis, V.; Verstrepen, K.J.; Boon, N.; Tavernier, F.; Kraft, M.; Taurino, I. Full-Electric Microfluidic Platform to Capture, Analyze and Selectively Release Single Cells. *Lab Chip* **2023**, *23*, 4276–4286. [CrossRef] [PubMed]
21. Li, S.-S.; Xue, C.-D.; Hu, S.-Y.; Li, Y.-J.; Chen, X.-M.; Zhao, Y.; Qin, K.-R. Long-Term Stable and Multifeature Microfluidic Impedance Flow Cytometry Based on a Constricted Channel for Single-Cell Mechanical Phenotyping. *Anal. Chem.* **2024**, *96*, 17754–17764. [CrossRef] [PubMed]
22. Wang, M.; Zhang, J.; Chen, X.; Li, Y.; Huang, X.; Wang, J.; Li, Y.; Huo, X.; Chen, J. Microfluidic Impedance Flow Cytometer Leveraging Virtual Constriction Microchannel and Its Application in Leukocyte Differential. *Microsyst. Nanoeng.* **2024**, *10*, 192. [CrossRef] [PubMed]
23. Kim, B.; Yao, W.; Rhie, J.W.; Chun, H. Microfluidic Potentiometric Cytometry for Size-Selective Micro Dispersion Analysis. *BioChip J.* **2022**, *16*, 471–479. [CrossRef]
24. Fang, Q.; Feng, Y.; Zhu, J.; Huang, L.; Wang, W. Floating-Electrode-Enabled Impedance Cytometry for Single-Cell 3D Localization. *Anal. Chem.* **2023**, *95*, 6374–6382. [CrossRef]
25. Zhong, J.; Tang, Q.; Liang, M.; Ai, Y. Accurate Profiling of Blood Components in Microliter with Position-Insensitive Coplanar Electrodes-Based Cytometry. *Sens. Actuators B Chem.* **2022**, *367*, 132068. [CrossRef]

26. Feng, Y.; Huang, L.; Zhao, P.; Liang, F.; Wang, W. A Microfluidic Device Integrating Impedance Flow Cytometry and Electric Impedance Spectroscopy for High-Efficiency Single-Cell Electrical Property Measurement. *Anal. Chem.* **2019**, *91*, 15204–15212. [CrossRef] [PubMed]
27. Wu, G.; Qiu, H.; Liu, X.; Luo, P.; Wu, Y.; Shen, Y. Nanomaterials-Based Fluorescent Assays for Pathogenic Bacteria in Food-Related Matrices. *Trends Food Sci. Technol.* **2023**, *142*, 104214. [CrossRef]
28. Guo, Y.; Wang, R.; Wei, C.; Li, Y.; Fang, T.; Tao, T. Carbon Quantum Dots for Fluorescent Detection of Nitrite: A Review. *Food Chem.* **2023**, *415*, 135749. [CrossRef]
29. Wang, Y.; Zhu, L.; Guo, P.; Zhang, Y.; Lan, X.; Xu, W. Research Progress of All-in-One PCR Tube Biosensors Based on Functional Modification and Intelligent Fabrication. *Biosens. Bioelectron.* **2024**, *246*, 115824. [CrossRef]
30. Zhao, X.; Peng, H.; Hu, J.; Wang, L.; Zhang, F. Nanotechnology-Enabled PCR with Tunable Energy Dynamics. *Jacs Au* **2024**, *4*, 3370–3382. [CrossRef]
31. Aly, N.S.; Kim, H.-S.; Marei, Y.M.; Elhamshary, A.S.; Bayoumi, I.R.; Omar, R.E.; Mohammed, D.A.; Miyoshi, S.-I.; Rashed, G.A. Diagnosis of Toxoplasmosis Using Surface Antigen Grade 1 Detection by ELISA, Nano-Gold ELISA, and PCR in Pregnant Women. *Int. J. Nanomed.* **2023**, *18*, 1335–1345. [CrossRef] [PubMed]
32. Peng, P.; Liu, C.; Li, Z.; Xue, Z.; Mao, P.; Hu, J.; Xu, F.; Yao, C.; You, M. Emerging ELISA Derived Technologies for in Vitro Diagnostics. *TrAC Trends Anal. Chem.* **2022**, *152*, 116605. [CrossRef]
33. Hu, T.; Wu, Z.; Sang, W.; Ding, B.; Chen, K.; Li, X.; Shen, Y.; Ni, Z. A Sensitive Electrochemical Platform Integrated with a 3D Graphene Aerogel for Point-of-Care Testing for Tumor Markers. *J. Mater. Chem. B* **2022**, *10*, 6928–6938. [CrossRef]
34. Zhu, J.; Feng, Y.; Chai, H.; Liang, F.; Cheng, Z.; Wang, W. Performance-Enhanced Clogging-Free Viscous Sheath Constriction Impedance Flow Cytometry. *Lab Chip* **2023**, *23*, 2531–2539. [CrossRef] [PubMed]
35. Luan, X.; Liu, P.; Huang, D.; Zhao, H.; Li, Y.; Sun, S.; Zhang, W.; Zhang, L.; Li, M.; Zhi, T.; et al. piRT-IFC: Physics-Informed Real-Time Impedance Flow Cytometry for the Characterization of Cellular Intrinsic Electrical Properties. *Microsyst. Nanoeng.* **2023**, *9*, 77. [CrossRef]
36. Feng, Y.; Zhu, J.; Chai, H.; He, W.; Huang, L.; Wang, W. Impedance-Based Multimodal Electrical-Mechanical Intrinsic Flow Cytometry. *Small* **2023**, *19*, 2303416. [CrossRef]
37. Tang, D.; Jiang, L.; Xiang, N.; Ni, Z. Discrimination of Tumor Cell Type Based on Cytometric Detection of Dielectric Properties. *Talanta* **2022**, *246*, 123524. [CrossRef]
38. Zhu, S.; Zhu, Z.; Ni, C.; Zhou, Z.; Chen, Y.; Tang, D.; Guo, K.; Yang, S.; Liu, K.; Ni, Z.; et al. Liquid Biopsy Instrument for Ultra-Fast and Label-Free Detection of Circulating Tumor Cells. *Research* **2024**, *7*, 0431. [CrossRef] [PubMed]
39. Xie, X.; Zhao, J.; Liu, T.; Li, L.; Qin, Y.; Song, X.; Ge, Y.; Zhao, J. Label-Free and Real-Time Impedance Sensor Integrated Liver Chip for Toxicity Assessment: Mechanism and Application. *Sens. Actuators B Chem.* **2023**, *393*, 134282. [CrossRef]
40. Zhu, J.; Pan, S.; Chai, H.; Zhao, P.; Feng, Y.; Cheng, Z.; Zhang, S.; Wang, W. Microfluidic Impedance Cytometry Enabled One-Step Sample Preparation for Efficient Single-Cell Mass Spectrometry. *Small* **2024**, *20*, 2310700. [CrossRef]
41. Caselli, F.; Reale, R.; De Ninno, A.; Spencer, D.; Morgan, H.; Bisegna, P. Deciphering Impedance Cytometry Signals with Neural Networks. *Lab Chip* **2022**, *22*, 1714–1722. [CrossRef] [PubMed]
42. Zhbanov, A.; Lee, Y.S.; Son, M.; Kim, B.J.; Yang, S. Improved Hematology Analysis Based on Microfluidic Impedance Spectroscopy: Erythrocyte Orientation and Anisotropic Dielectric Properties of Flowing Blood. *Anal. Chem.* **2025**, *97*, 4871–4880. [CrossRef] [PubMed]
43. Huang, L.; Jasim, I.; Alkorjia, O.; Agca, C.; Oksman, A.; Agca, Y.; Goldberg, D.E.; Benson, J.D.; Almasri, M. An Impedance Based Microfluidic Sensor for Evaluation of Individual Red Blood Cell Solute Permeability. *Anal. Chim. Acta* **2023**, *1267*, 341226. [CrossRef] [PubMed]
44. Oshabaheebwa, S.; Delianides, C.A.; Patwardhan, A.A.; Evans, E.N.; Sekyonda, Z.; Bode, A.; Apio, F.M.; Mutuluuza, C.K.; Sheehan, V.A.; Suster, M.A.; et al. A Miniaturized Wash-Free Microfluidic Assay for Electrical Impedance-Based Assessment of Red Blood Cell-Mediated Microvascular Occlusion. *Biosens. Bioelectron.* **2024**, *258*, 116352. [CrossRef]
45. Fu, H.; Qin, Z.; Li, X.; Pan, Y.; Xu, H.; Pan, P.; Song, P.; Liu, X. Paper-Based All-in-One Origami Nanobiosensor for Point-of-Care Detection of Cardiac Protein Markers in Whole Blood. *ACS Sens.* **2023**, *8*, 3574–3584. [CrossRef]
46. Sayad, A.; Uddin, S.M.; Yao, S.; Wilson, H.; Chan, J.; Zhao, H.; Donnan, G.; Davis, S.; Skafidas, E.; Yan, B.; et al. A Magnetoimpedance Biosensor Microfluidic Platform for Detection of Glial Fibrillary Acidic Protein in Blood for Acute Stroke Classification. *Biosens. Bioelectron.* **2022**, *211*, 114410. [CrossRef]
47. Petchakup, C.; Chen, Y.Y.C.; Tay, H.M.; Ong, H.B.; Hon, P.Y.; De, P.P.; Yeo, T.W.; Li, K.H.H.; Vasoo, S.; Hou, H.W. Rapid Screening of Urinary Tract Infection Using Microfluidic Inertial-Impedance Cytometry. *ACS Sens.* **2023**, *8*, 3136–3145. [CrossRef]
48. Na, J.-T.; Chun-Dong, X.; Wang, Y.-X.; Li, Y.-J.; Wang, Y.; Liu, B.; Qin, K.-R. Fabricating a Multi-Component Microfluidic System for Exercise-Induced Endothelial Cell Mechanobiology Guided by Hemodynamic Similarity. *Talanta* **2023**, *253*, 123933. [CrossRef]
49. Huang, C.-H.; Lei, K.F. Quantitative Study of Tumor Angiogenesis in Three-Dimensional Matrigel Barrier Using Electric Impedance Measurement Technique. *Sens. Actuators B Chem.* **2022**, *370*, 132458. [CrossRef]

50. Liang, C.-C.; Chen, P.-Y.; Liu, N.-C.; Lee, I.-C. Comparison between Dynamic versus Static Models and Real-Time Monitoring of Neuronal Dysfunction in an Amyloid-β Induced Neuronal Toxic Model on a Chip Platform. *Lab Chip* **2024**, *24*, 1887–1902. [CrossRef]
51. Ceccarelli, M.C.; Lefevre, M.C.; Marino, A.; Pignatelli, F.; Krukiewicz, K.; Battaglini, M.; Ciofani, G. Real-Time Monitoring of a 3D Blood–Brain Barrier Model Maturation and Integrity with a Sensorized Microfluidic Device. *Lab Chip* **2024**, *24*, 5085–5100. [CrossRef] [PubMed]
52. Dhwaj, A.; Roy, N.; Jaiswar, A.; Prabhakar, A.; Verma, D. 3D-Printed Impedance Micropump for Continuous Perfusion of the Sample and Nutrient Medium Integrated with a Liver-On-Chip Prototype. *ACS Omega* **2022**, *7*, 40900–40910. [CrossRef] [PubMed]
53. Liang, F.; Huang, X.; Huang, B.; He, Y.; Luo, H.; Shi, J.; Wang, L.; Aimé, C.; Peng, J.; Chen, Y. A Microfluidic Tool for Real-Time Impedance Monitoring of in Vitro Renal Tubular Epithelial Cell Barrier. *Sens. Actuators B Chem.* **2023**, *392*, 134077. [CrossRef]
54. Fernandes, J.; Karra, N.; Bowring, J.; Reale, R.; James, J.; Blume, C.; Pell, T.J.; Rowan, W.C.; Davies, D.E.; Swindle, E.J.; et al. Real-Time Monitoring of Epithelial Barrier Function by Impedance Spectroscopy in a Microfluidic Platform. *Lab Chip* **2022**, *22*, 2041–2054. [CrossRef] [PubMed]
55. Khalid, M.A.U.; Kim, K.H.; Salih, A.R.C.; Hyun, K.; Park, S.H.; Kang, B.; Soomro, A.M.; Ali, M.; Jun, Y.; Huh, D.; et al. High Performance Inkjet Printed Embedded Electrochemical Sensors for Monitoring Hypoxia in a Gut Bilayer Microfluidic Chip. *Lab Chip* **2022**, *22*, 1764–1778. [CrossRef]
56. Chen, X.; Shen, M.; Liu, S.; Wu, C.; Sun, L.; Song, Z.; Shi, J.; Yuan, Y.; Zhao, Y. Microfluidic Impedance Cytometry with Flat-End Cylindrical Electrodes for Accurate and Fast Analysis of Marine Microalgae. *Lab Chip* **2024**, *24*, 2058–2068. [CrossRef]
57. Chen, J.; Zhong, J.; Lei, H.; Ai, Y. Label-Free Multidimensional Bacterial Characterization with an Ultrawide Detectable Concentration Range by Microfluidic Impedance Cytometry. *Lab Chip* **2023**, *23*, 5029–5038. [CrossRef]
58. Muhsin, S.A.; Al-Amidie, M.; Shen, Z.; Mlaji, Z.; Liu, J.; Abdullah, A.; El-Dweik, M.; Zhang, S.; Almasri, M. A Microfluidic Biosensor for Rapid Simultaneous Detection of Waterborne Pathogens. *Biosens. Bioelectron.* **2022**, *203*, 113993. [CrossRef]
59. Chen, J.; Zhong, J.; Chang, Y.; Zhou, Y.; Koo, S.H.; Tan, T.Y.; Lei, H.; Ai, Y. Rapid and Accurate Antimicrobial Susceptibility Testing Using Label-Free Electrical Impedance-Based Microfluidic Platform. *Small* **2024**, *20*, 2303352. [CrossRef]
60. Tang, T.; Liu, X.; Yuan, Y.; Zhang, T.; Kiya, R.; Yang, Y.; Yamazaki, Y.; Kamikubo, H.; Tanaka, Y.; Li, M.; et al. Parallel Impedance Cytometry for Real-Time Screening of Bacterial Single Cells from Nano- to Microscale. *ACS Sens.* **2022**, *7*, 3700–3709. [CrossRef]
61. Zhang, S.; Han, Z.; Qi, H.; Liu, S.; Liu, B.; Sun, C.; Feng, Z.; Sun, M.; Duan, X. Convolutional Neural Network-Driven Impedance Flow Cytometry for Accurate Bacterial Differentiation. *Anal. Chem.* **2024**, *96*, 4419–4429. [CrossRef] [PubMed]
62. Piedimonte, P.; Sola, L.; Cretich, M.; Gori, A.; Chiari, M.; Marchisio, E.; Borga, P.; Bertacco, R.; Melloni, A.; Ferrari, G.; et al. Differential Impedance Sensing Platform for High Selectivity Antibody Detection down to Few Counts: A Case Study on Dengue Virus. *Biosens. Bioelectron.* **2022**, *202*, 113996. [CrossRef] [PubMed]
63. Almalaysha, M.; Singh, A.; Muhsin, S.A.; Carlson, A.V.; Trout, K.E.; Morey, A.; Zhang, S.; Channaiah, L.H.; Almasri, M. A Highly Sensitive Microfluidic Biosensor for Rapid and Accurate Detection of Salmonella in Raw Chicken Products. *Sens. Actuators Rep.* **2025**, *9*, 100257. [CrossRef]
64. Coatrini-Soares, A.; Coatrini-Soares, J.; Neto, M.P.; de Mello, S.S.; Pinto, D.D.S.C.; Carvalho, W.A.; Gilmore, M.S.; Piazzetta, M.H.O.; Gobbi, A.L.; Brandão, H.d.M.; et al. Microfluidic E-Tongue to Diagnose Bovine Mastitis with Milk Samples Using Machine Learning with Decision Tree Models. *Chem. Eng. J.* **2023**, *451*, 138523. [CrossRef]
65. Dowdell, K.S.; Olsen, K.; Paz, E.F.M.; Sun, A.; Keown, J.; Lahr, R.; Steglitz, B.; Busch, A.; LiPuma, J.J.; Olson, T.; et al. Investigating the Suitability of Online Flow Cytometry for Monitoring Full-Scale Drinking Water Ozone System Disinfection Effectiveness. *Water Res.* **2024**, *257*, 121702. [CrossRef]
66. Çam, S.B.; Çiftci, E.; Gürbüz, N.; Altun, B.; Korkusuz, P. Allogeneic Bone Marrow Mesenchymal Stem Cell-Derived Exosomes Alleviate Human Hypoxic AKI-on-a-Chip within a Tight Treatment Window. *Stem Cell Res. Ther.* **2024**, *15*, 105. [CrossRef]
67. Wang, W.; Jin, Y.; Huang, Y.; Zhao, Z.; Li, M.; Mao, H.; Li, R.; Xiong, J. A Self-Driven Multifunctional Microfluidic Sweat Analysis System for Efficient Sweat Collection and Real-Time Monitoring. *Sens. Actuators B Chem.* **2024**, *414*, 135920. [CrossRef]
68. Zahed, A.; Kim, D.K.; Jeong, S.H.; Reza, S.; Sharifuzzaman; Pradhan, G.B.; Song, H.; Asaduzzaman; Park, J.Y. Microfluidic-Integrated Multimodal Wearable Hybrid Patch for Wireless and Continuous Physiological Monitoring. *ACS Sens.* **2023**, *8*, 2960–2974. [CrossRef]
69. Butement, J.T.; Wang, X.; Siracusa, F.; Miller, E.; Pabortsava, K.; Mowlem, M.; Spencer, D.; Morgan, H. Discrimination of Microplastics and Phytoplankton Using Impedance Cytometry. *ACS Sens.* **2024**, *9*, 5206–5213. [CrossRef]
70. da Silva, T.A.; Juncá, M.A.C.; Braunger, M.L.; Riul, A., Jr.; Barbin, D.F. Application of a Microfluidic Electronic Tongue Based on Impedance Spectroscopy for Coconut Water Analysis. *Food Res. Int.* **2024**, *187*, 114353. [CrossRef]
71. Xue, Y.; Hassan, Q.; Noroozifar, M.; Sullan, R.M.A.; Kerman, K. Microfluidic Flow Injection Analysis System for the Electrochemical Detection of Dopamine Using Diazonium-Grafted Copper Nanoparticles on Multi-Walled Carbon Nanotube-Modified Surfaces. *Talanta* **2024**, *266*, 125030. [CrossRef] [PubMed]

72. Park, S.; Kim, S.; Lee, S.; Tsukruk, V.V.; Park, S.; Lim, H. Advanced Microfluidic-Based Wearable Electrochemical Sensors for Continuous Biochemical Monitoring. *Adv. Electron. Mater.* **2025**, *85*, 2500010. [CrossRef]
73. Zhou, J.; Dong, J.; Hou, H.; Huang, L.; Li, J. High-Throughput Microfluidic Systems Accelerated by Artificial Intelligence for Biomedical Applications. *Lab Chip* **2024**, *24*, 1307–1326. [CrossRef] [PubMed]
74. Liu, J.; Du, H.; Huang, L.; Xie, W.; Liu, K.; Zhang, X.; Chen, S.; Zhang, Y.; Li, D.; Pan, H. AI-Powered Microfluidics: Shaping the Future of Phenotypic Drug Discovery. *ACS Appl. Mater. Interfaces* **2024**, *16*, 38832–38851. [CrossRef]
75. Nelson, M.D.; Goenner, B.L.; Gale, B.K. Utilizing ChatGPT to Assist CAD Design for Microfluidic Devices. *Lab Chip* **2023**, *23*, 3778–3784. [CrossRef]
76. Page, M.J.; McKenzie, J.E.; Bossuyt, P.M.; Boutron, I.; Hoffmann, T.C.; Mulrow, C.D.; Shamseer, L.; Tetzlaff, J.M.; Akl, E.A.; Brennan, S.E.; et al. The PRISMA 2020 statement: An updated guideline for reporting systematic reviews. *BMJ* **2021**, *372*, n71. [CrossRef]

MDPI AG
Grosspeteranlage 5
4052 Basel
Switzerland
Tel.: +41 61 683 77 34

Micromachines Editorial Office
E-mail: micromachines@mdpi.com
www.mdpi.com/journal/micromachines

www.ingramcontent.com/pod-product-compliance
Lightning Source LLC
Chambersburg PA
CBHW040839250926
51850CB00007BA/1355
9783725862030